MEMS Mirrors

Special Issue Editor
Huikai Xie

MDPI • Basel • Beijing • Wuhan • Barcelona • Belgrade

MDPI

Special Issue Editor
Huikai Xie
University of Florida
USA

Editorial Office
MDPI
St. Alban-Anlage 66
Basel, Switzerland

This edition is a reprint of the Special Issue published online in the open access journal *Micromachines* (ISSN 2072-666X) in 2017–2018 (available at: http://www.mdpi.com/journal/micromachines/special_ issues/mems_mirrors).

For citation purposes, cite each article independently as indicated on the article page online and as indicated below:

Lastname, F.M.; Lastname, F.M. Article title. *Journal Name* **Year**, *Article number*, page range.

First Editon 2018

Cover image courtesy of Huikai Xie.

ISBN 978-3-03842-867-1 (Pbk)
ISBN 978-3-03842-868-8 (PDF)

Table of Contents

About the Special Issue Editor

Huikai Xie received his B.S., M.S., and Ph.D. degrees all in electrical and computer engineering from Beijing Institute of Technology, Tufts University, and Carnegie Mellon University, respectively. He joined the University of Florida in 2002. He also worked at Tsinghua University (1992–1996), Bosch Corp. (2001–2002), and US Air Force Research Lab (2007,2008 and 2009 summer). His research interests include MEMS/NEMS, integrated sensors, microactuators, integrated power passives, CNT-CMOS integration, optical MEMS, LiDAR, micro-spectrometers, optical bioimaging, and endomicroscopy. He has published over 280 technical papers, and holds more than 30 US patents. He is an associate editor of several international journals including Micromachines and IEEE Sensors Letters. He is a fellow of IEEE and SPIE.

micromachines

MDPI

Editorial

Editorial for the Special Issue on MEMS Mirrors

Huikai Xie

Department of Electrical and Computer Engineering, University of Florida, Gainesville, FL 32611, USA;
hkx@ufl.edu

Received: 22 February 2018; Accepted: 23 February 2018; Published: 27 February 2018

MEMS mirrors can steer, modulate, and switch light, as well as control the wavefront for focusing or phase modulation. MEMS mirrors have found enormous commercial success in projectors, displays, and fiberoptic communications. Micro-spectrometers based on MEMS mirrors are starting to appear in the consumer market as well. There are also many breakthroughs in applying MEMS mirrors for endoscopic imaging. There is a new wave of opportunities for MEMS mirrors coming up, such as the micro-LiDAR for autonomous driving and robotics, optical cross-connect (OXC) for cloud data centers, and optical scanners for virtual reality/augmented reality. Of course, there are numerous challenges that researchers and engineers must overcome to fully utilize the potential of MEMS mirrors. For example, modeling and control are inherently complex due to the multiphysics, multi-DOF, and nonlinear nature of the micro-actuators for MEMS mirrors. Reliability is always a huge hurdle for commercialization and the tradeoffs among the speed, aperture, and scan range are often overwhelming.

There are 15 papers published in this special issue, covering MEMS mirrors based on all of the commonly used actuation mechanisms including electrostatic [1–5], electromagnetic [6–8], piezoelectric [9], and electrothermal [10–14]. Half of these papers explore various aspects of MEMS mirrors, such as working in harsh environments [1], spring hardening compensation [3], optimal packaging conditions for resonance operation [5], input saturation control [8], extremely large scan angle [9], overshoot suppression [10], electrothermal actuator modeling [13], and design optimization [14]. There is also a paper reporting a passive micromirror [15]. The other half of the papers are focused on a range of applications of MEMS mirrors including grayscale photolithography [2], biomedical imaging [4,11,12], laser display [6], and LiDAR [7].

In particular, Zamkotsian et al. tested a deformable mirror (DM) array of single crystalline silicon hexagonal tip-tilt-piston mirrors in a cryo-vacuum chamber designed for reaching 10^{-6} mbar and 160 K. The DM array survived, which provides a viable solution for DM arrays in harsh environments such as in space [1]. Izawa et al. applied the spring softening effect of electrostatic comb drives to compensate for the spring hardening effect of torsion-bar springs [3]. Zhao et al. identified an optimal range of vacuum for operating electrostatic comb-drive micromirrors [5]. Tan et al. proposed a control design framework based on composite nonlinear feedback (CNF) and the integral sliding mode (ISM) technique to improve MEMS micromirrors' performance under input saturation [8]. Gu-Stoppel et al. presented a lead zirconate titanate (PZT)-actuated micromirror that achieves an extremely large scan angle of up to 106° and a high frequency of 45 kHz simultaneously [9]. M. Li et al. effectively eliminated the overshoot and oscillation of electrothermally-actuated micromirrors simply by setting the product of the thermal response time and the fundamental resonance frequency to be greater than $Q/2\pi$ [10]. Torres et al. reported the modeling of a MEMS mirror structure with four actuators driven by the phase-change of VO_2 thin film [13]. Saleem et al. presented the parametric design optimization of an electrothermally actuated micromirror for the deflection angle, input power, and micromirror temperature rise from ambient conditions [14]. Sabry et al. fabricated a silicon micromachined three-dimensional curved mirror for optical fiber light collimation [15].

On the applications side, Deng et al. successfully demonstrated maskless grayscale photolithography using Texas Instruments' digital micromirror devices (DMDs) [2]. H. Li et al. employed a novel lever-based compliant mechanism to enable large vertical displacements of a reflective mirror for the axial scanning of a multi-photon fluorescence imaging microscopy system [4]. Li et al. fabricated a two-axis electromagnetic micromirror with Ni electroplated on the mirror plate to eliminate Joule heating and applied the micromirror to a laser display system, effectively reducing laser speckles [6]. Ye et al. fabricated a Ti-alloy-based electromagnetic micromirror with a very large aperture of 12 mm and a rapid scanning frequency of 1.24 kHz using electrical discharging and explored its potential application for micro-LiDAR [7]. Lara-Castro et al. proposed a vertical scanning electrothermal bimorph micromirror design by employing polysilicon as the heater material [11]. Tanguy et al. demonstrated a two-axis micromirror with a pair of electrothermal actuators and a set of passive torsion bars and applied it to an ultra-compact Mirau interferometer-based optical coherence tomography (OCT) imaging probe [12].

I would like to take this opportunity to thank all the authors for submitting their papers to this special issue. I also want to thank all the reviewers for dedicating their time and helping to improve the quality of the submitted papers.

Conflicts of Interest: The author declares no conflict of interest.

References

1. Zamkotsian, F.; Lanzoni, P.; Barette, R.; Helmbrecht, M.; Marchis, F.; Teichman, A. Operation of a MOEMS Deformable Mirror in Cryo: Challenges and Results. *Micromachines* **2017**, *8*, 233. [CrossRef]
2. Deng, Q.; Yang, Y.; Gao, H.; Zhou, Y.; He, Y.; Hu, S. Fabrication of Micro-Optics Elements with Arbitrary Surface Profiles Based on One-Step Maskless Grayscale Lithography. *Micromachines* **2017**, *8*, 314. [CrossRef]
3. Izawa, T.; Sasaki, T.; Hane, K. Scanning Micro-Mirror with an Electrostatic Spring for Compensation of Hard-Spring Nonlinearity. *Micromachines* **2017**, *8*, 240. [CrossRef]
4. Li, H.; Duan, X.; Li, G.; Oldham, K.R.; Wang, T.D. An Electrostatic MEMS Translational Scanner with Large Out-of-Plane Stroke for Remote Axial-Scanning in Multi-Photon Microscopy. *Micromachines* **2017**, *8*, 159. [CrossRef]
5. Zhao, R.; Qiao, D.; Song, X.; You, Q. The Exploration for an Appropriate Vacuum Level for Performance Enhancement of a Comb-Drive Microscanner. *Micromachines* **2017**, *8*, 126. [CrossRef]
6. Li, F.; Zhou, P.; Wang, T.; He, J.; Yu, H.; Shen, W. A Large-Size MEMS Scanning Mirror for Speckle Reduction Application. *Micromachines* **2017**, *8*, 140. [CrossRef]
7. Ye, L.; Zhang, G.; You, Z. Large-Aperture kHz Operating Frequency Ti-alloy Based Optical Micro Scanning Mirror for LiDAR Application. *Micromachines* **2017**, *8*, 120. [CrossRef]
8. Tan, J.; Sun, W.; Yeow, J.T.W. An Enhanced Robust Control Algorithm Based on CNF and ISM for the MEMS Micromirror against Input Saturation and Disturbance. *Micromachines* **2017**, *8*, 326. [CrossRef]
9. Gu-Stoppel, S.; Giese, T.; Quenzer, H.-J.; Hofmann, U.; Benecke, W. PZT-Actuated and -Sensed Resonant Micromirrors with Large Scan Angles Applying Mechanical Leverage Amplification for Biaxial Scanning. *Micromachines* **2017**, *8*, 215. [CrossRef]
10. Li, M.; Chen, Q.; Liu, Y.; Ding, Y.; Xie, H. Modelling and Experimental Verification of Step Response Overshoot Removal in Electrothermally-Actuated MEMS Mirrors. *Micromachines* **2017**, *8*, 289. [CrossRef]
11. Lara-Castro, M.; Herrera-Amaya, A.; Escarola-Rosas, M.A.; Vázquez-Toledo, M.; López-Huerta, F.; Aguilera-Cortés, L.A.; Herrera-May, A.L. Design and Modeling of Polysilicon Electrothermal Actuators for a MEMS Mirror with Low Power Consumption. *Micromachines* **2017**, *8*, 203. [CrossRef]
12. Tanguy, Q.A.A.; Bargiel, S.; Xie, H.; Passilly, N.; Barthès, M.; Gaiffe, O.; Rutkowski, J.; Lutz, P.; Gorecki, C. Design and Fabrication of a 2-Axis Electrothermal MEMS Micro-Scanner for Optical Coherence Tomography. *Micromachines* **2017**, *8*, 146. [CrossRef]
13. Torres, D.; Zhang, J.; Dooley, S.; Tan, X.; Sepúlveda, N. Modeling of MEMS Mirrors Actuated by Phase-Change Mechanism. *Micromachines* **2017**, *8*, 138. [CrossRef]

14. Saleem, M.M.; Farooq, U.; Izhar, U.; Khan, U.S. Multi-Response Optimization of Electrothermal Micromirror Using Desirability Function-Based Response Surface Methodology. *Micromachines* **2017**, *8*, 107. [CrossRef]

15. Sabry, Y.M.; Khalil, D.; Saadany, B.; Bourouina, T. In-Plane Optical Beam Collimation Using a Three-Dimensional Curved MEMS Mirror. *Micromachines* **2017**, *8*, 134. [CrossRef]

micromachines

MDPI

Article

Operation of a MOEMS Deformable Mirror in Cryo: Challenges and Results

Frederic Zamkotsian [1,*], Patrick Lanzoni [1], Rudy Barette [1], Michael Helmbrecht [2], Franck Marchis [3] and Alex Teichman [2]

[1]   Aix Marseille Univ, CNRS, LAM, Laboratoire d'Astrophysique de Marseille, 38 rue Frederic Joliot Curie, 13388 Marseille CEDEX 13, France; patrick.lanzoni@lam.fr (P.L.); rudy.barette@lam.fr (R.B.)

[2]   Iris AO, 2930 Shattuck Avenue #304, Berkeley, CA 94705, USA; info@irisao.com or michael.helmbrecht@irisao.com (M.H.); alex.teichman@irisao.com (A.T.)

[3]   Carl Sagan Center, SETI Institute, 189 Bernardo Ave, Mountain View, CA 94043, USA; fmarchis@seti.org

*   Correspondence: frederic.zamkotsian@lam.fr; Tel.: +33-4-9504-4151

Received: 24 May 2017; Accepted: 14 July 2017; Published: 27 July 2017

Abstract: Micro-opto-electro-mechanical systems (MOEMS) Deformable Mirrors (DM) are key components for next generation optical instruments implementing innovative adaptive optics systems, both in existing telescopes and in the future ELTs. Characterizing these components well is critical for next generation instruments. This is done by interferometry, including surface quality measurement in static and dynamical modes, at ambient and in vacuum/cryo. We use a compact cryo-vacuum chamber designed for reaching 10–6 mbar and 160 K in front of our custom Michelson interferometer, which is able to measure performance of the DM at actuator/segment level and at the entire mirror level, with a lateral resolution of 2 μm and a sub-nanometer z-resolution. We tested the PTT 111 DM from Iris AO: an array of single crystalline silicon hexagonal mirrors with a pitch of 606 μm, able to move in tip, tilt, and piston (stroke 5–7 μm, tilt ±5 mrad). The device could be operated successfully from ambient to 160 K. An additional, mainly focus-like, 500 nm deformation of the entire mirror is measured at 160 K; we were able to recover the best flat in cryo by correcting the focus and local tip-tilts on all segments, reaching 12 nm rms. Finally, the goal of these studies is to test DMs in cryo and vacuum conditions as well as to improve their architecture for stable operation in harsh environments.

Keywords: MEMS mirror arrays; MOEMS; cryogenic testing; adaptive optics; wavefront correction

1. Introduction

Wavefront correction is a key issue in a wide range of applications, from physics to biology or astronomy. Collimating and focusing with high accuracy a very large number of photons in high energy lasers, correcting the wavefront through diffuse or inhomogeneous media for sharp retinal and tissue imaging, or correcting, in closed loop, the atmospheric turbulence for revealing the faintest or most remote objects in the Universe, requires high performance wavefront correction systems.

In astronomy, high performance adaptive optical (AO) systems are being studied around the world for next generation instrumentation of 10 m-class telescopes as well as for future extremely large optical telescopes. Adaptive optics systems are based on a combination of three elements: the wavefront sensor for measuring the shape of the wavefront arriving in the telescope; the deformable mirror for correcting the wavefront; and, finally, the real time computer for closing the loop of the system at a frequency ranging from 0.5 to 3 kHz, in order to follow the evolution of the atmospherical perturbations (Figure 1).

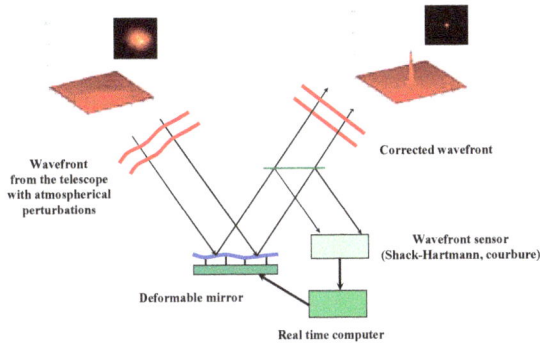

Figure 1. Schematic of a wavefront correction system.

Four main types of AO systems have been built or are under development: Single-Conjugate Adaptive Optics (SCAO), Multi-Conjugate Adaptive Optics (MCAO), Multi-Object Adaptive Optics (MOAO), and Extreme Adaptive Optics (ExAO). These AO systems are associated with different types of WaveFront Sensors (WFS), combined with natural guide stars or laser guide stars, and different architectures of Deformable Mirrors (DM). Numerous science cases will use these AO systems: SCAO, the "classical" AO system, will provide accurate narrow field imagery and spectroscopy; MCAO, wide field imagery and spectroscopy; MOAO, distributed partial correction AO, and high dynamic range AO for the detection and the study of circumstellar disks and extra-solar planets. Corrected fields will vary from few arcsec to several arcmin.

These systems require a large variety of deformable mirrors with very challenging parameters. For a 8 m telescope, the number of actuators varies from a few 10 up to 5000; these numbers increase impressively for a 40 m telescope, ranging from a few 100 to over 50,000 (the inter-actuator spacing from less than 200 µm to 1 mm, and the deformable mirror size from 10 mm to a few 100 mm). Conventional technology cannot provide this wide range of deformable mirrors. The development of new technologies based on micro-opto-electro-mechanical systems (MOEMS) is promising for future deformable mirrors. The major advantages of the micro-deformable mirrors (MDM) are their compactness, scalability, and specific task customization using elementary building blocks. This technology permits the development of a complete generation of new mirrors. However this technology also has some limitations. For example, pupil diameter is an overall parameter and for a 40 m primary telescope, the internal pupil diameter cannot be reduced below 0.5 m. According to the maximal size of the wafers (8 inches), a deformable mirror based on MOEMS technology cannot be built into one piece. New AO architectures have been proposed to avoid this limitation [1].

This new family of deformable mirrors will have also to fulfill some additional requirements: their ability to work in vacuum and at cryogenic temperatures. In order to study the early Universe, astronomers need to image and characterize astronomical objects (galaxies, quasars, large scale structures, etc.) in the infra-red part of the electromagnetic spectrum. At these wavelengths, the background noise of the instruments themselves must be reduced drastically by cooling them down at cryogenic temperatures, the longer the wavelength, the lower the temperature. Another challenge is to introduce MOEMS devices in future space instruments for reaching unprecedented performance. For all these applications, designing, testing, and operating MOEMS deformable mirrors in harsh environments (vacuums, cryogenic temperatures) is a critical issue.

For several years Laboratoire d'Astrophysique de Marseille (LAM) has been involved in the conception of new MOEMS devices as well as in the characterization of these components for the future instrumentation of ground-based and space telescopes. These studies include programmable slits for application in multi-object spectroscopy (JWST, European networks, EUCLID, BATMAN), deformable mirrors for adaptive optics, and programmable gratings for spectral tailoring.

We are particularly engaged in a European development of micromirror arrays (MMA) called MIRA for generating reflective slit masks in future Multi-object spectroscopy (MOS) instruments; this technique is a powerful tool for space and ground-based telescopes for the study of the formation and evolution of galaxies. MMA with 100×200 μm^2 single-crystal silicon micromirrors were successfully designed, fabricated, and tested. Arrays are composed of 2048 micromirrors (32×64) with a peak-to-valley deformation less than 10 nm, and a tilt angle of 24° for an actuation voltage of 130 V. The micromirrors were actuated successfully before, during, and after cryogenic cooling, down to 162 K. The micromirror surface deformation was measured at cryo and is below 30 nm peak-to-valley [2,3]. In order to fill large focal planes (mosaicing of several chips), we are currently developing large micromirror arrays integrated with their electronics.

LAM is also leading the conception and realization of new MOEMS-based instruments, including the development of the Digital-Micromirror-Device (DMD)-based MOS instrument, to be mounted on the Telescopio Nazionale Galileo (TNG) by mid-2018 and called BATMAN [4].

In this paper, we present the specific set-up for the interferometric characterization of a segmented deformable mirror from Iris AO, in vacuum and at cryogenic temperatures. The results on the mirror surface operated from ambient down to 160 K are shown for the first time and analyzed.

2. Deformable Mirrors

Three main Micro-Deformable Mirrors (MDM) architectures are under study in different laboratories and companies. First, the bulk micro-machined continuous-membrane deformable mirror, studied by Delft University and OKO Company (Rijswijk, The Netherlands), is a combination of bulk silicon micromachining with standard electronics technology [5]. This mirror is formed by a thin flexible conducting membrane, coated with a reflective material, and stretched over an electrostatic electrode structure. This mirror shows very good mirror quality, but the mean deformed surface is a concave surface, and the number of actuators cannot be scalable to hundreds of electrodes. Second, the segmented, micro-electro-mechanical deformable mirror realized by Iris AO [6] consists of a set of segmented piston-tip-tilt moving surfaces, fabricated in dense array. Third, the surface micro-machined continuous-membrane deformable mirror made by Boston Micromachines Corporation (BMC) is based on a single compliant optical membrane supported by multiple attachments to an underlying array of surface-normal electrostatic actuators [7]. This device has been demonstrated recently in several AO systems, including the Gemini Planet Imager (GPI) instrument on Gemini telescope. The third concept shows limited strokes for large driving voltages, and the mirror surface quality may need further improvement for Extreme AO. All of these devices are based on silicon or polysilicon materials.

Two candidates, BMC deformable mirrors and Iris AO deformable mirrors, are foreseen for characterization in vacuum and at cryogenic temperatures, but none has yet been fully operated at cryogenic temperatures. Their architectures are presented in this paragraph.

2.1. BMC Deformable Mirror

BMC produces advanced MEMS deformable mirrors. The concept is based on an array of electrostatic actuators linked one by one to a continuous top mirror surface with etch holes (Figure 2). Their main parameters are approaching the requirements values, i.e., large number of actuators (up to 4096, see Figure 2), large stroke (up to 5.5 μm for lower actuator count devices), and good surface quality.

BMC is currently conducting several actions for developing DMs devoted to space applications. In 2011, a classical DM has been tested with a Sounding Rocket in the US, in the Planetary Imaging Concept Testbed Using a Rocket Experiment (PICTURE). PICTURE-B has been launched in November 2015. This experiment measures directly optical light scattered by the debris disk around Epsilon-Eri star. In parallel, several studies have been engaged to modify the DM architecture and make it compatible with space environment [8].

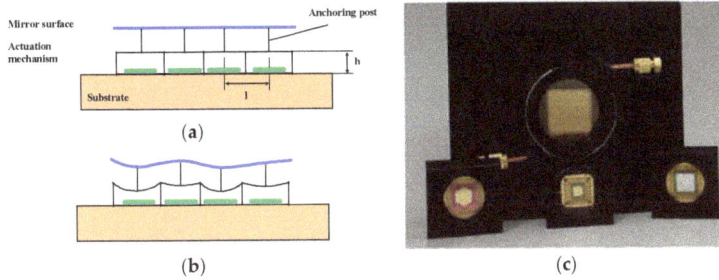

Figure 2. Continuous membrane Micro-Deformable Mirrors (MDM) from Boston Micromachines Corporation (BMC): schematic view of the (**a**) MDM non-actuated, (**b**) MDM actuated; (**c**) devices picture (courtesy from BMC).

2.2. Iris AO Deformable Mirror

Iris AO has been manufacturing piston-tilt-tilt (PTT) MEMS DMs with high-optical-quality mirror segments since 2002. University research that demonstrated the basic concept was first reported in 2001 [9] and early demonstrations of products were reported some years later [10]. Current products span apertures of 3.5 mm–7.7 mm with 111–489 actuators. A 939 actuator DM will be available in late 2017 and development is underway on a 3000 actuator PTT DM.

Figure 3 shows photographs of two different segmented DMs manufactured by Iris AO with 111 (PTT111) and 489 (PTT489) actuators. The DM segments are independent and are controlled by three actuators to give three degrees of freedom. Thus, the PTT111 DM has a total of 37 independent PTT segments and the PTT489 DM has 163. The PTT111 DM shown here is coated with a protected-silver coating. The PTT489 DM is shown with a gold coating. The coatings are typical coatings used for macroscale optics. They are much thicker than typical coatings used on MEMS and thin-membrane mirrors, so they are very robust. This is possible because the segment design and fabrication technology is very tolerant to various coatings. Even high-reflectance dielectric coatings for wavelengths ranging from 255 nm–1600 nm are available as options.

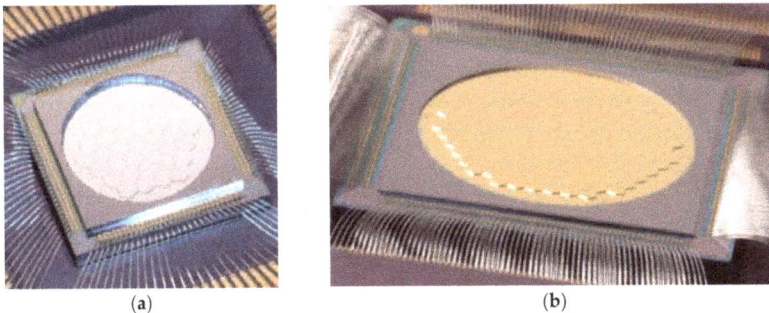

Figure 3. (**a**) Photograph of protected-silver coated PTT111 Deformable Mirror (DM) mounted in a ceramic package. The 111 actuator DM is comprised of 37 PTT segments tiled in a hex-packed array with an inscribed aperture of 3.5 mm; (**b**) Photograph of gold-coated PTT489 DM. The 489 actuator DM is comprised of 163 PTT segments with an inscribed aperture of 7.7 mm. The 700 µm vertex-to-vertex segment footprint for the PTT111 and PTT489 DM are identical.

Figure 4 shows an exploded-view schematic diagram of one of the mirror segments. During manufacturing, the rigid mirror segments are bonded to actuator arrays at the chip level prior to the microstructure release step. After removing the sacrificial layer, the segments elevate above the underlying electrodes because of engineered residual stresses in the bimorph and actuator platform materials. The proprietary material stack is chosen to minimize elevation due to changes in temperature. The scaling in the vertical direction is highly exaggerated in Figure 4. In reality, the platform elevation is up to 40 µm whereas the segments are 350 µm on a side (700 µm across). The three underlying actuator electrodes enable piston-tip-tilt (PTT) positioning. Factory calibration and an open-loop controller conveniently convert from a desired position to actuator voltages [11].

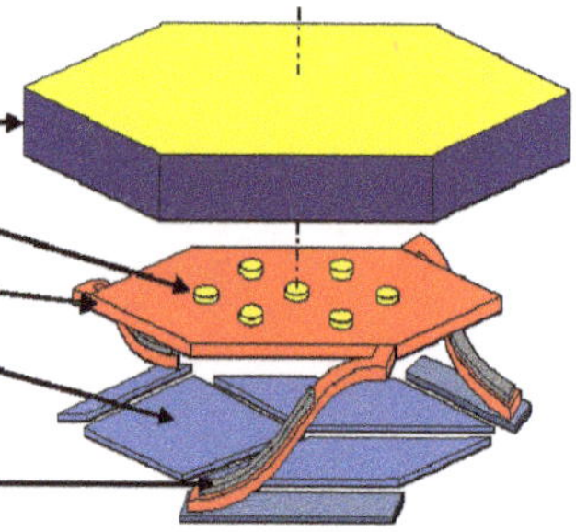

Figure 4. Exploded-view schematic diagram of a 700 µm circumscribed diameter (606.2 µm flat-to-flat pitch) mirror segment. Scaling is highly exaggerated in the vertical direction. Actuator platform heights are in the range of 20–40 µm for these segments, which enables 8 µm of stroke.

The DM is manufactured using typical MEMS and integrated circuit materials such as polycrystalline silicon (polysilicon), silicon dioxides, silicon nitrides, and a proprietary bimorph material with similar coefficient of thermal expansion (CTE) to that of polysilicon. After the DM is fabricated using highly stable MEMS materials, it is mounted onto a ceramic pin-grid array (PGA) package using an epoxy. The DM is sealed in nitrogen by epoxying a cover window over the DM.

3. Cryogenic Interferometric Test Set-Up

Over the last few years, LAM has developed expertise in the characterization of micro-optical components, especially for small-scale deformation characterization on their surface from room temperature down to cryogenic temperatures. The aim is to be able to use these devices in all type of spectrographs, from ground-based to space telescopes as well as from Visible to IR instruments. The background of this test set-up development was in the framework of the NASA study of NIRSpec for the JWST.

3.1. Interferometric Bench

The measurement of the shape and the deformation parameters of the MOEMS devices is made thanks to an interferometric characterization bench, fully developed in-house (hardware and software). Characterizations are done in static or dynamic modes, including measurements of optical surface quality at different scales, actuator stroke, maximum mirror deformation, and cut-off frequency; the latest parameter has not yet been measured in this study.

This bench is a high-resolution and low-coherence Twyman-Green interferometer. It is a modular set-up around three main parts: illumination, interferometric cavity, and imaging system. *The illumination* is provided by a halogen lamp with an interference filter (typical example: $\lambda_0 = 650$ nm, $\Delta\lambda = 10$ nm). Fixing the temporal light coherence eliminates all parasitic fringe patterns induced by classical high coherence sources such as lasers. *The interferometric cavity* is clear of any optical components in order to keep the highest optical quality by avoiding any differential aberration introduced by these additional elements. *The imaging system* offers two magnification configurations by a simple lens change: (1) a high in-plane resolution or (2) a large field of view authorizing either a very sharp (around 2 μm) analysis of the micro-mirror structure inside a small field (typically 1 mm), or the whole device study with a larger size (up to 40 mm).

Out-of-plane measurements are performed with phase-shifting interferometry showing very high resolution (standard deviation < 1 nm). The bench is mounted on a damped optical table, and surrounded by a Plexiglas enclosure. Features such as optical quality or electro-mechanical behavior are extracted from these high precision three-dimensional component maps. Range is increased without loosing accuracy by using two-wavelength phase-shifting interferometry authorizing large steps measurements. Dynamic analysis like vibration mode and cut-off frequency could be also measured with time-averaged interferometry [12].

3.2. Cryogenic Experiment

In front of our interferometric setup has been installed a cryo chamber for reaching temperatures as low as 100 K when not loaded, using a cryogenic generator, and vacuum pressure in the range of 10^{-6} mbar (Figure 5a). The chamber is equipped with an internal screen radiatively insulating the sample from the chamber. Temperature sensors and local heaters are used for controlling the environment, and connected to a custom built control electronics and control-command software. The component is monitored continuously thanks to a glass window at the entrance of the chamber. As we are using low coherence interferometry, the fringe contrast can be kept at maximum by balancing the optical path in the reference arm of the interferometer with a compensation plate identical to the chamber window. In our set-up, the chamber window and the compensation plate have been manufactured at the same time in the same glass material [13].

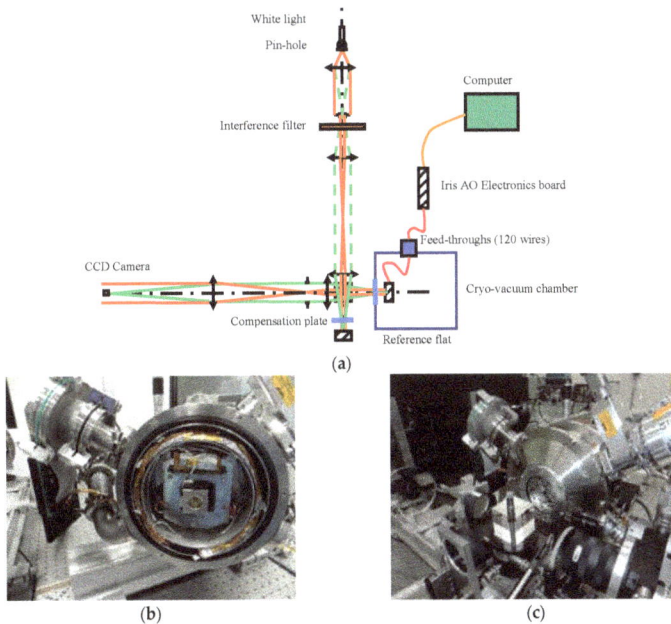

Figure 5. (a) Schematic description of our interferometric measurement set-up with the cryo-vacuum chamber; (b) PTT111 device mounted in the cryogenic chamber for characterization in a space environment; (c) Front window is closed and the cryogenic chamber is installed in front of our interferometric setup. The segmented deformable mirror could be successfully actuated before, during, and after cryogenic cooling at 160 K.

3.3. PTT111 Mounting

The PTT111 device is packaged in PGA chip carrier. The PGA is inserted in a ZIF-holder integrated on a PCB board. Large metallic surfaces on the PCB facilitate cooling down the system; eliminating the solder-stop layer eases outgassing of the PCB base material during evacuation of the chamber. The PCB itself is mounted via a fix-point-plane-plane attachment system to a solid aluminum block, the latter being interconnected to the cryo-generator (Figure 5b). Thick copper wires between the PCB and the aluminum block further enhance thermal transport between the sample chip and the cryostat. Two ribbon cables allow interconnecting up to 120 electrical wires through two 100-pin and 20-pin feed-throughs. Outside of the cryo chamber, the wires are connected to the Iris AO control electronics. Temperature sensors are connected to the aluminum block and to the window frame on top of the device.

The chamber is then closed by a flange and placed in front of the interferometer (Figure 5c). Along the reference path, two compensation plates are placed for compensating the chamber window (large plate) and the device window (small plate). By this way, we keep a high contrast for the interferometric fringes.

4. PTT111 Surface Characterization

These first experiments are done using an engineering grade device where segments #23 and #24 are inactive because of defects in the DM that occurred during manufacturing. The manufacturer calls these locked segments or lockouts as they are not controllable. Because the segments are independent, the defects only affect the locked segments and not neighboring one. The segment thickness is 25 µm and the coating is protected silver. The maximum array stroke after flattening is 3.01 µm, and the maximum tilt angle is 5 mrad. Figure 6 is a picture of the device made on our bench (without the interferometric fringes). The two lockouts segments are at the upper right.

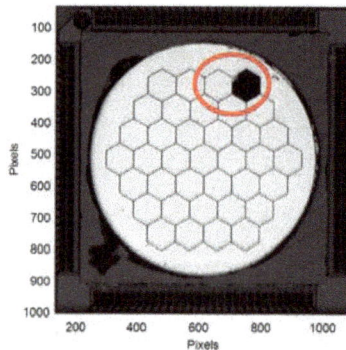

Figure 6. PTT111 device tested in our experiment (engineering grade device). Segments #23 and #24 (in the red circle) are lockouts.

The device is driven by the *Graphical User Interface* (GUI) provided by Iris AO for the integration and pre-characterization phases. The interferometric measurements are done with the LAM-developed software, in Matlab, and linked with the Matlab driver provided by Iris AO. The GUI is showing a view of the mirror with numbered segments, and the global Zernike coefficient as well as local (at segment level) piston/tip/tilt positions could be tuned for each segment.

A calibration step has been done by Iris AO for measuring each actuator response (3 actuators/segment). Then, a "best flat" condition is calculated in order to minimize the residual wavefront error on the surface, and applied. Figure 7 is a screen shot made at the beginning of the experiment when the best flat condition is applied; at the left hand side is the interferometric image of the mirror with the "best flat" condition; at the right bottom is the view of the Iris AO GUI for driving the device.

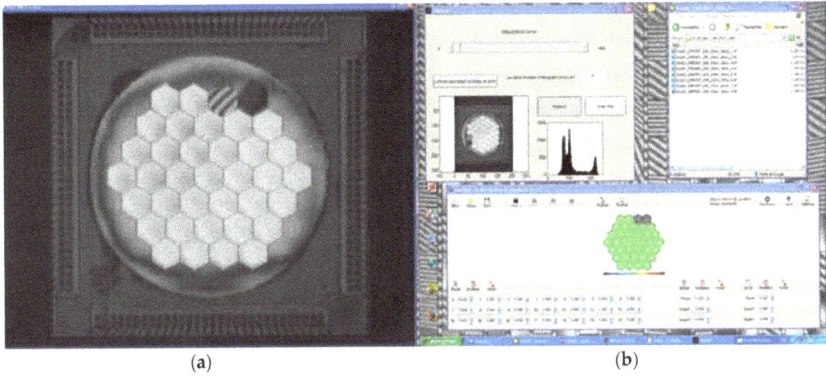

(a)   (b)

Figure 7. Screen shot during the experiment; (**a**) The interferometric image of the mirror when the "best flat" condition is applied; (**b**) The Iris AO GUI for driving the device.

We could then apply a series of commands to the mirror. In Figure 8, different mirror configurations are presented. From top left to bottom right, we can see:

- a pure piston (150 nm) on the central M1 segment,
- a global astigmatism on the mirror, using the global Zernike coefficient set at 0.125,
- a series of three identical tilts on all segments in X direction, with 0.25 mrad, 1.5 mrad, and 4.9 mrad respectively.

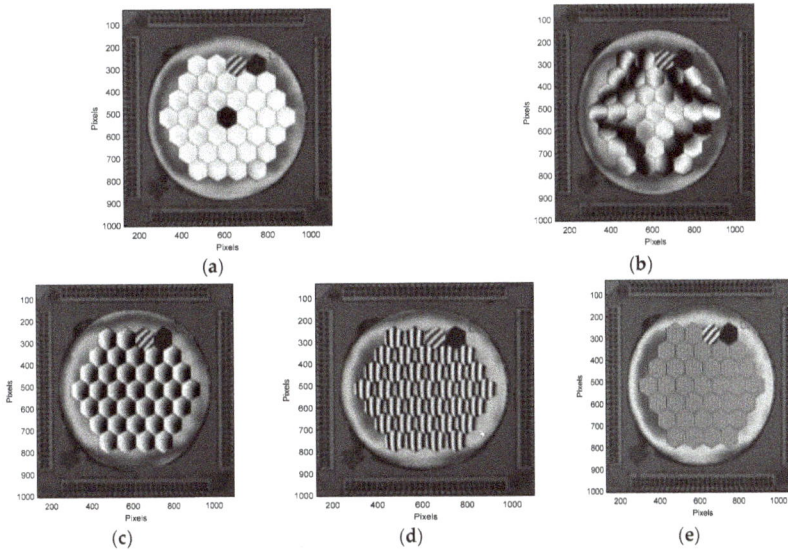

Figure 8. (**a**) 150 nm piston on segment#1; (**b**) astigmatism (0.125 on Zernike coefficient); (**c–e**) Increasing tilt values along the X direction are applied to all segments (0.25, 1.5 and 4.9 mrad).

11

4.1. Best Flat at Ambient

From the interferometric measurement, we obtain the surface deformation of the deformable mirror; in the whole paper, deformation values are always given in nanometers, and maps are displayed with a ±250 nm range for better comparison between related Figures 9, 11, and 13a–15a.

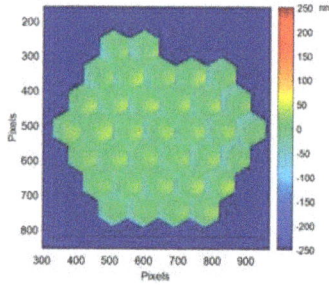

Figure 9. Best flat surface deformation at ambient (17 nm RMS, 123 nm PtV) when the best flat condition calibrated by Iris AO is applied; the gravity effect has not been removed.

The best flat residual over the whole mirror is very good with 17 nm RMS, 123 nm PtV (Figure 9). Note that this best flat provided by Iris AO is not corrected from the gravity effect occurring in our experiment, as the PTT111 is mounted in vertical position.

This result shows the high quality of the mirror architecture and of the fabrication process. This flatness is a combination of a very good reproducibility of the actuator platform position after his elevation thanks to the bimorph flexures (Figure 4), and the choice of thick single-crystalline Silicon for the segment material. This position is very stable; long term measurement has been done at ambient on position stability [6] and reproducibility, but this has not been done yet in cryo.

At segment level, the residual wavefront error is with a slight convex shape, observed on most of the segments (see Section 4.6).

We measure the mirror surface shape with high accuracy and sort out tilt and piston values for wach segment. They are displayed on segment maps in Figure 10. Over the maps, the tilts have a value of 32 µrad RMS (140 µrad PtV) while the pistons expands on 6.2 nm RMS (28 nm PtV). The residual tilts are mainly due to the gravity effect as the device is mounted in vertical position in our measurement set-up.

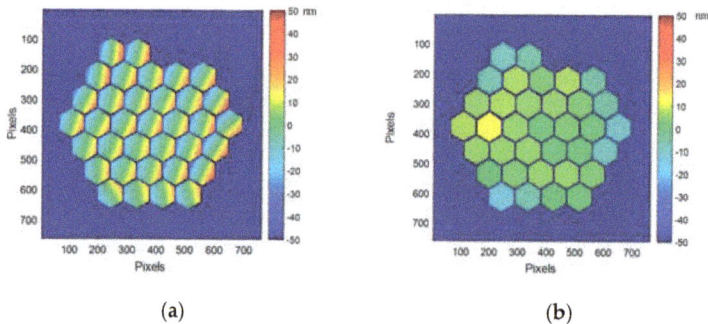

(a)   (b)

Figure 10. Residual tip-tilts (a) and pistons (b) at ambient, when the best flat condition from Iris AO is applied; the gravity effect has not been removed.

4.2. Best Flat at 160 K, First Run

The device is then cooled down slowly from ambient temperature (293 K) down to 160 K, with the device constantly operating in its best flat condition. The PTT111 device is operating properly at all temperatures between 293 K and 160 K, and in vacuum.

Every 10 K an interferometric measurement is done in order to follow the differential deformation of the mirror at whole mirror level as well as at segment level. Several patterns are applied and measured in order to see the ability of the device to behave as at room temperature; the applied "patterns" are best flat, pure pistons on some segments, and different tilts on the segments. Due to the vibrations induced by the cryo pump on the sample, we have to stop it during the measurement, leading to a limited increase of the temperature during the measurement duration. Phase shifting interferometry parameters have been adjusted in order to minimize the measurement time to a few minutes.

In Figure 11, the best flat surface deformation at cryo (160 K) is given with the original best flat condition as calibrated at ambient. A global convex deformation is observed reaching a deformation of 86 nm RMS, 501 nm PtV. Some additional deformations (mainly tilts) are observed on some segments at the upper left side (segments #26, 27, and 28).

The global convex shape in cryo is due to the packaging "shrinking" in cryo. The Coefficient of Thermal Expansion (CTE) mismatch between die/package materials induces a global effect on the mirror when cooled down at 160 K.

The mirror is operating perfectly in cryo, and a "new" best flat condition will be developed and described in the next paragraphs.

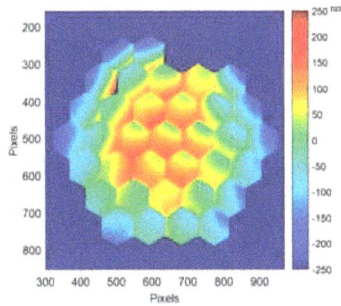

Figure 11. Best flat mirror deformation at cryo (160 K), first run, with the original best flat calibrated at ambient (86 nm RMS, 501 nm PtV).

We measure and sort out for each segment tilt and piston values. They are displayed on segment maps and shown in graphs, in Figure 12. The tilts have a value of 200 µrad RMS (950 µrad PtV) while the pistons expands on 74 nm RMS (239 nm PtV). Tilts and pistons are not behaving the same way. Pistons are behaving within the three concentric rings of PTT111 37 segments, with a common motion for the central area (segments #1 to 7), the middle ring (segments #8 to 19), and the outer ring (segments #20 to 37); in Figure 12, residual pistons at ambient (black signs) are 6.2 nm RMS (28 nm PtV), and 74 nm RMS (239 nm PTV) in cryo (red signs); this effect is clearly related to the shrinkage of the overall device. As for the tilts, they don't show a clear pattern; they are scattered away from the original positions at ambient. In Figure 12, residual tilts at ambient (black signs) are 32 µrad RMS (140 µrad PtV), and 200 µrad RMS (950 µrad PTV) in cryo (red signs); this differential evolution is possibly due to the different modification of the complex structure underneath each segment: the actuator platform, the bonding pads, the mirror segment, the coating, and the underlying 3 legs bimorph structure.

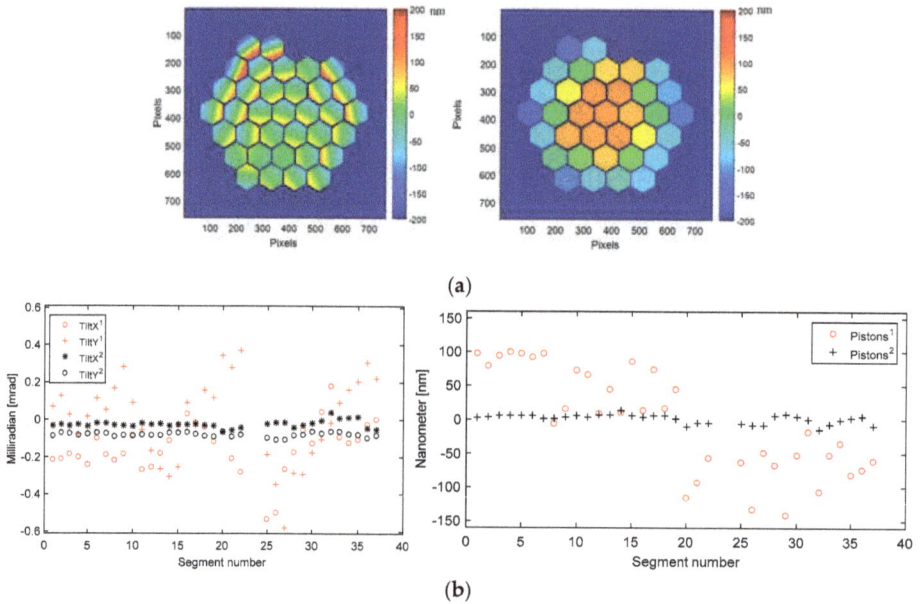

Figure 12. (**a**) Segment maps of the residual tip-tilts and pistons at cryo (160 K), first run; (**b**) Values for each of the 37 segments of PTT111 (in red tilt X and tilt Y at cryo, in black, for reference, values at ambient). Residual tilts at ambient (black signs) are 32 μrad RMS (140 μrad PtV), and 200 μrad RMS (950 μrad PtV) in cryo (red signs). Residual pistons at ambient (black signs) are 6.2 nm RMS (28 nm PtV), and 74 nm RMS (239 nm PtV) in cryo (red signs).

These effects will be corrected in a second step (see Section 4.4).

4.3. Best Flat at Ambient (Back 1)

When the device is warmed up back to ambient temperature we measure its surface shape in the original best flat mode again, and we get a slightly different result, as revealed in Figure 13.

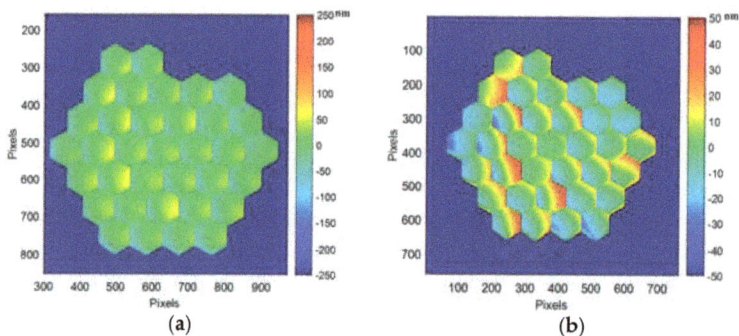

Figure 13. Mirror surface deformation at ambient (293 K) when best flat Iris AO condition is applied, after one cooling cycle: (**a**) Surface deformation map (22 nm RMS, 123 nm PtV); (**b**) Difference between the two surface deformation maps before and after one cooling cycle.

The degradation of the surface deformation is going up from 17 nm to 22 nm RMS (123 nm to 152 nm PtV), the differences being mainly due to the additional tilts on each segment by 0.15 mrad PtV over the device, while the pistons stay within the same range as before the cooling cycle. This is probably due to the stress relaxation process in the complex structure supporting each segment, leading to a slightly different evolution from segment to segment. As for the segment mirror surface, there is no difference (below 1 nm), thanks to the high quality of the segment made of single-crystalline silicon material. This effect could be described as an annealing process.

In order to improve the best flat quality, we decided to develop an improved best flat procedure by measuring, with high accuracy, the tip-tilt and piston residuals and combining them with the original best flat values calibrated by Iris AO. We then obtain the *improved best flat* shown in Figure 14. The mirror surface deformation is then as low as 10 nm RMS, 79 nm PtV.

In the graphs of Figure 14, residual tilts with improved best flat (red signs) are 6.5 µrad RMS (36 µrad PtV), while they were 32 µrad RMS (140 µrad PTV) with the original best flat (black signs). Residual pistons with improved best flat (red signs) are 1.6 nm RMS (7 nm PtV), while they were 6.2 nm RMS (28 nm PtV) with the original best flat (black signs).

Improved best flat condition will be used in the following experiments.

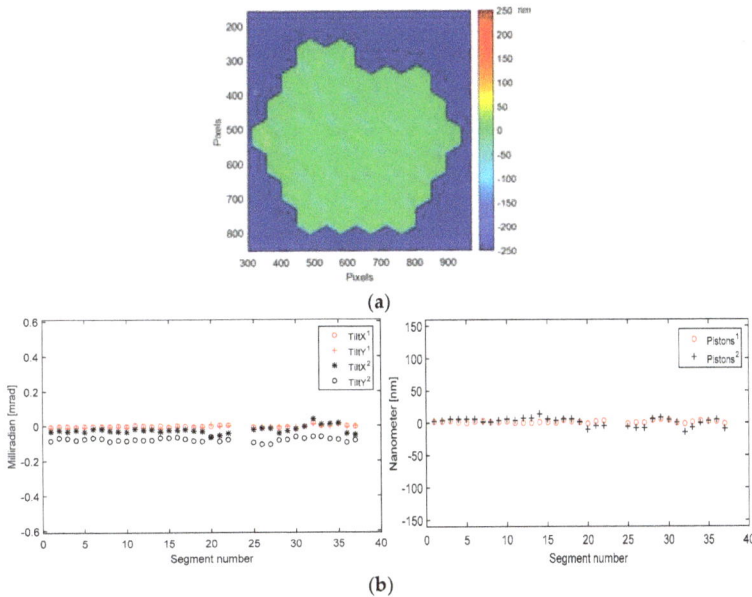

Figure 14. Mirror surface deformation at ambient (293 K) when *improved best flat* condition is applied, after one cycle of cooling: (**a**) Surface deformation map (10 nm RMS, 79 nm PtV); (**b**) Residual tilts and pistons values for each of the 37 segments of PTT111. Residual tilts with improved best flat (red signs) are 6.5 µrad RMS (36 µrad PtV), while they were 32 µrad RMS (140 µrad PTV) with the original best flat (black signs). Residual pistons with improved best flat (red signs) are 1.6 nm RMS (7 nm PtV), while they were 6.2 nm RMS (28 nm PtV) with the original best flat (black signs).

4.4. Best Flat at 160 K, Second Run

In order to demonstrate full operation of PTT111 at cryogenic temperature, we decide to cool down the device and optimize in-situ all actuators for generating a cryo best flat.

Our strategy is a weighted addition of the consecutive measurement residual errors and, using Iris AO electronics, we are loading these calculated values actuator by actuator, departing from the original values provided by Iris AO, and applyling them to the device. Our *cryo best flat* condition is a combination of [best flat calibrated by Iris-AO at ambient, improved best flat at ambient, improved best flat in cryo (first run), improved best flat in cryo (second run)]. Then, in a single measurement step and applying this best flat condition, we got, at 160 K, a mirror surface deformation as low as 12 nm RMS, 113 nm PtV (Figure 15).

Our *cryo best flat* at 160 K is then very close to our *improved best flat* at ambient (293 K), showing our ability to operate properly PTT111 device in cryo. The deformation difference is 2 nm RMS, 34 nm PtV between 160 K and 293 K. This additional deformation is due mainly to the mirror segment deformation, as revealed in the following Section 4.6.

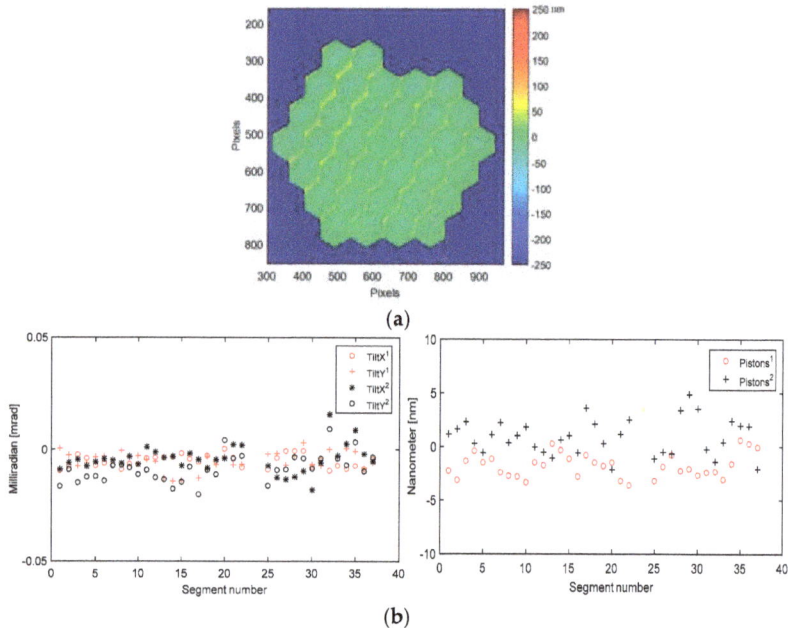

(a)

(b)

Figure 15. Mirror surface deformation at cryo (160 K), second run, when *cryo best flat* condition is applied: (**a**) Surface deformation map (12 nm RMS, 113 nm PtV); (**b**) Residual tilts and pistons values for each of the 37 segments of PTT111. Residual tilts with *cryo best flat* (red signs) are 3.5 μrad RMS (17 μrad PtV), while they were 6.5 μrad RMS (36 μrad PtV) with the *improved best flat* at ambient (black signs). Residual pistons with *cryo best flat* (red signs) are 1.2 nm rms (4.3 nm PtV), while they were 1.6 nm RMS (7 nm PtV) with the *improved best flat* at ambient (black signs).

In the graphs of Figure 15, residual tilts with *cryo best flat* (red signs) are 3.5 μrad RMS (17 μrad PtV), while they were 6.5 μrad RMS (36 μrad PtV) with the *improved best flat* at ambient (black signs). Residual pistons with *cryo best flat* (red signs) are 1.2 nm rms (4.3 nm PtV), while they were 1.6 nm RMS (7 nm PtV) with the *improved best flat* at ambient (black signs).

A second loop of best flat optimisation is useless as the remaining mirror surface deformation is only due to the contributions of individual segment deformations; this is clearly visible in the surface deformation map of Figure 15.

The additional deformation of PTT111 at cryo is 501 nm PtV (Figure 11). The *cryo best flat* is compensating this deformation, minimizing the whole mirror deformation, down to 12 nm RMS (123 nm PtV). The maximum stroke of this device being 3.01 μm at ambient, then the operational stroke is reduced to 2.5 μm at cryo (16.7% stroke reduction).

4.5. Best Flat at Ambient (Back 2)

When the device is warmed up back to ambient temperature we measure its surface shape in the improved best flat mode again, and we get nearly exactly the same surface, as shown in Figure 14. The device seems to now be stabilized with no evolution after a complete cycle at cryogenic temperature.

A thermal cycling test might be useful to confirm this result.

4.6. Analysis at Segment Level

Thanks to our set-up spatial resolution, we have several thousand measurement points per segment. It is then possible to measure, at segment level, the deformation induced by the strong temperature change from ambient to cryo. Figure 16a,b are identical to Figures 14a and 15a, with the mirror segment marked, and an expended scale for revealing the local effects on each segment. The surface deformation by segments at ambient (values 2 in blue) has a mean value of 7.2 nm with a standard deviation of 1.5 nm, and at 160 K (values 1 in red), a mean value of 8.5 nm with a standard deviation of 1.6 nm.

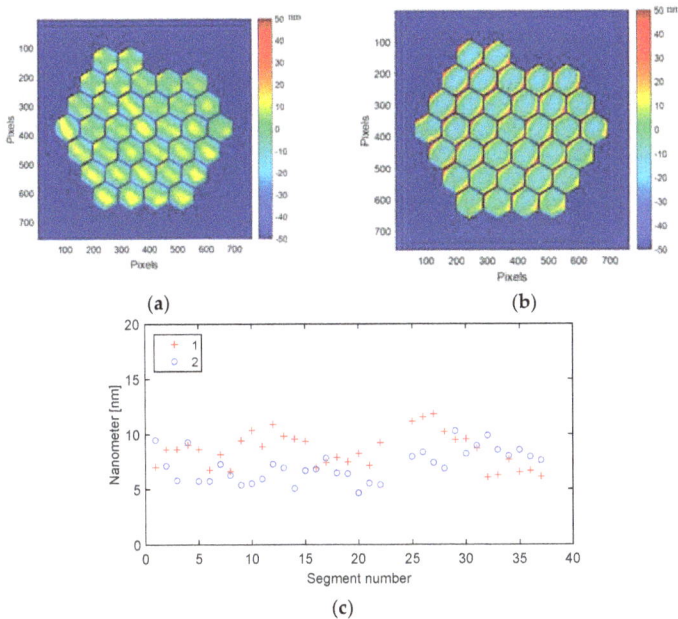

Figure 16. DM segments surface deformation: (**a**) At ambient (293 K), 5 nm RMS, 24 nm PtV; (**b**) At 160 K 8 nm RMS, 47 nm PtV; (**c**) Surface deformation by segments at ambient (values 2 in blue) has a mean value of 7.2 nm with a standard deviation of 1.5 nm, and at 160 K (values 1 in red), a mean value of 8.5 nm with a standard deviation of 1.6 nm.

By selecting a typical segment, a closer analysis of the segment evolution at cryogenic temperature could be done, especially on its shape. In Figure 17, on segment #21, we can clearly see that the convex cylindrical shape at ambient is changing to an astigmatic concave shape at cryo. At ambient (293 K), the segment surface deformation is 5 nm RMS (24 nm PtV), while at 160 K the deformation is still low, at 8 nm RMS (47 nm PtV).

A very interesting feature is observed when looking at the deformation difference between ambient and 160 K (Figure 17c): it reveals a pure concave axisymetrical change of 4.9 nm RMS (71 nm PtV). This is due to the CTE mismatch between the single-crystalline silicon and the silver-protected coating deposited on top of the segment. All segments are behaving in the same way as shown in Figure 18.

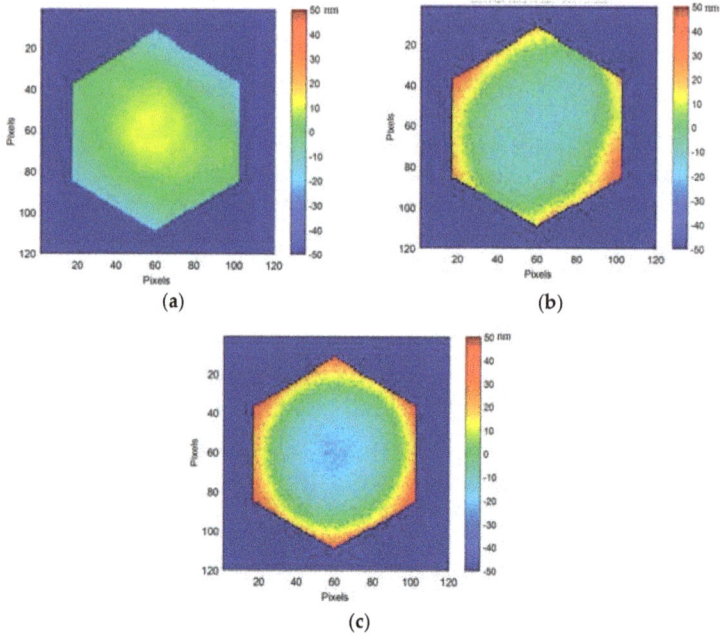

Figure 17. DM segment #21 surface deformation: (**a**) At ambient (293 K), 5 nm RMS, 24 nm PtV; (**b**) At 160 K 8 nm RMS, 47 nm PtV; (**c**) Deformation difference between ambient and 160 K, 4.9 nm RMS, 71 nm PtV.

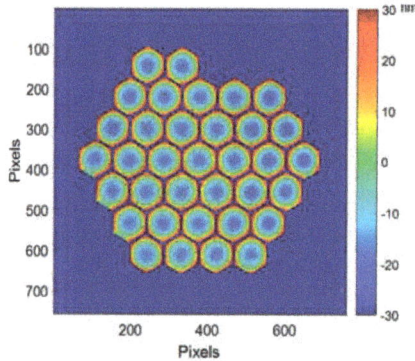

Figure 18. Segment deformation induced by the temperature change between ambient and cryo (160 K); map of all segments of PTT111.

The mean deformation at ambient is in the range of 25 nm, while it rises to 50 nm in cryo. This deformation difference is still within the requirement of almost all foreseen wavefront correction systems. This deformation difference at segment level is the major contribution to the whole mirror surface deformation, as described in Section 4.4.

5. Conclusions

Innovative wavefront correction systems in existing telescopes on the ground and in space, as well as in the future ELTs, need efficient MOEMS Deformable Mirrors (DM) able to perform at room temperature as well as in cryogenic and vacuum environments.

Using a specific interferometric bench coupled with a cryo-vacuum chamber, a PTT 111 DM from Iris AO has been successfully tested from ambient temperature to 160 K. The device is properly operating in cryo, revealing an additional, mainly focus-like, 500 nm deformation at 160 K; we were able to recover the best flat in cryo by correcting the focus and local tip-tilts on all segments, reaching 12 nm RMS on the entire surface.

Tests on DMs with different mirror thicknesses (25 µm and 50 µm) and different coatings (silver and gold) are currently under way.

Finally, the goal of these studies is to test DMs in cryo and vacuum conditions as well as to improve their architecture for stable operation in a harsh environment.

Author Contributions: F.Z. and M.H. supervised the work; M.H., F.M. and A.T. provided the devices and their driving electronics; P.L., F.Z. and R.B. designed and realized the experimental set-up; P.L. and F.Z. performed the experiments; F.Z. and P.L. carried out calculations and data treatment for the results interpretation; F.Z., P.L. and M.H. analyzed the results; F.Z. wrote the paper, and M.H. and P.L. contributed for some paragraphs.

Conflicts of Interest: The authors declare no conflict of interest.

References

1. Zamkotsian, F.; Dohlen, K. Prospects for MOEMS-based adaptive optical systems on extremely large telescopes. In Proceedings of the Conference Beyond Conventional Adaptive Optics, Venice, Italy, 7–10 May 2001.

2. Waldis, S.; Zamkotsian, F.; Clerc, P.-A.; Noell, W.; Zickar, M.; De Rooij, N. Arrays of high tilt-angle micromirrors for multiobject spectroscopy. *IEEE J. Sel. Top. Quantum Electron.* **2007**, *13*, 168–176. [CrossRef]

3. Canonica, M.; Zamkotsian, F.; Lanzoni, P.; Noell, W.; de Rooij, N. The two-dimensional array of 2048 tilting micromirrors for astronomical spectroscopy. *J. Micromech. Microeng.* **2013**, *23*, 055009. [CrossRef]

4. Zamkotsian, F.; Ramarijaona, H.; Moschetti, M.; Lanzoni, P.; Riva, M.; Tchoubaklian, N.; Jaquet, M.; Spano, P.; Bon, W.; Alata, R.; et al. Building BATMAN: A new generation spectro-imager on TNG telescope. In Proceedings of the SPIE Conference on Astronomical Instrumentation 2016 (Proc. SPIE 9908), Edinburgh, UK, 26 June–1 July 2016.

5. Vdovin, G.; Middelhoek, S.; Sarro, P.M. Technology and applications of micromachined silicon adaptive mirrors. *Opt. Eng.* **1997**, *36*, 1382–1390. [CrossRef]

6. Helmbrecht, M.A.; He, M.; Kempf, C.J.; Marchis, F. Long-term stability and temperature variability of Iris AO segmented MEMS deformable mirrors. In Proceedings of the SPIE 9909 Conference on Adaptive Optics Systems V, Edinburgh, UK, 26 June 2016.

7. Cornelissen, S.; Bifano, T.G. Advances in MEMS deformable mirror development for astronomical adaptive optics. In Proceedings of the SPIE 8253 Conference on MOEMS, San Francisco, CA, USA, 21–26 January 2012.

8. Bierden, P. MEMS deformable mirrors for space imaging. In Proceedings of the ESA Workshop on Innovative Technologies for Space Optics, Noordwijk, The Netherlands, 23–26 November 2015.

9. Helmbrecht, M.A.; Srinivasan, U.; Rembe, C.; Howe, R.T.; Muller, R.S. Micromirrors for adaptive-optics arrays, transducers 2001. In Proceedings of the 11th International Conference on Solid State Sensors and Actuators, Munich, Germany, 10–14 June 2001; pp. 1290–1293.

10. Helmbrecht, M.A.; Juneau, T.; Hart, M.; Doble, N. Performance of a high-stroke, segmented MEMS deformable-mirror technology. In Proceedings of the SPIE 6113 Conference on MEMS/MOEMS Components and Their Applications III, San Jose, CA, USA, 21 January 2006.

11. Helmbrecht, M.A.; Juneau, T. Piston-tip-tilt positioning of a segmented MEMS deformable mirror. In Proceedings of the SPIE 6467 Conference on MEMS Adaptive Optics, San Jose, CA, USA, 9 February 2007.

Micromachines **2017**, *8*, 233

12. Liotard, A.; Zamkotsian, F. Static and dynamic micro-deformable mirror characterization by phase-shifting and time-averaged interferometry. In Proceedings of the SPIE 5494 Conference on Astronomical Telescopes and Instrumentation, Glasgow, UK, 21–25 June 2004.
13. Zamkotsian, F.; Grassi, E.; Waldis, S.; Barette, R.; Lanzoni, P.; Fabron, C.; Noell, W.; de Rooij, N. Interferometric characterization of MOEMS devices in cryogenic environment for astronomical instrumentation. In Proceedings of the SPIE 6884 Conference on Reliability, Packaging, Testing, and Characterization of MEMS/MOEMS VII, San Jose, CA, USA, 7 February 2008.

micromachines

MDPI

Article

Fabrication of Micro-Optics Elements with Arbitrary Surface Profiles Based on One-Step Maskless Grayscale Lithography

Qinyuan Deng [1,2] , Yong Yang [1,*], Hongtao Gao [1], Yi Zhou [1,2], Yu He [1] and Song Hu [1]

[1] State Key Laboratory of Optical Technologies for Microfabrication, Institute of Optics and Electronics, Chinese Academy of Sciences, Chengdu 610209, China; dqy_storm@163.com (Q.D.); gaohongtao@ioe.ac.cn (H.G.); alanzhouyi@163.com (Y.Z.); heyu@ioe.ac.cn (Y.H.); husong@ioe.ac.cn (S.H.)
[2] University of Chinese Academy of Sciences, Beijing 100049, China
*   Correspondence: yangyong@ioe.ac.cn; Tel.: +86-028-8510-0167

Received: 1 September 2017; Accepted: 17 October 2017; Published: 23 October 2017

Abstract: A maskless lithography method to realize the rapid and cost-effective fabrication of micro-optics elements with arbitrary surface profiles is reported. A digital micro-mirror device (DMD) is applied to flexibly modulate that the exposure dose according to the surface profile of the structure to be fabricated. Due to the fact that not only the relationship between the grayscale levels of the DMD and the exposure dose on the surface of the photoresist, but also the dependence of the exposure depth on the exposure dose, deviate from a linear relationship arising from the DMD and photoresist, respectively, and cannot be systemically eliminated, complicated fabrication art and large fabrication error will results. A method of compensating the two nonlinear effects is proposed that can be used to accurately design the digital grayscale mask and ensure a precise control of the surface profile of the structure to be fabricated. To testify to the reliability of this approach, several typical array elements with a spherical surface, aspherical surface, and conic surface have been fabricated and tested. The root-mean-square (RMS) between the test and design value of the surface height is about 0.1 μm. The proposed method of compensating the nonlinear effect in maskless lithography can be directly used to control the grayscale levels of the DMD for fabricating the structure with an arbitrary surface profile.

Keywords: maskless lithography; micro-optics elements; arbitrary surface; exposure dose; nonlinear effect

1. Introduction

During past decades, much research effort has been devoted to micro-optical elements with arbitrary surface profiles, which can usually achieve extraordinary properties far more than macro components, and has important applications in optical communication, sensors, special illumination, and other fields [1–5]. However, the limited fabrication methods for such micro-optical elements have restricted its development. Electron-beam lithography [6,7] and focused-ion beam [8] can realize a high-resolution fabrication of the structure with complicated surface profiles in principle, but require a long-term and expensive device. The direct laser writing technique [9–11] is a promising and economic method for the fabrication of microstructures, but the scanning mode will limit the improvement of work efficiency. The thermal reflow method [12–14] cooperated with conventional binary lithography is usually applied for micro lens array's generation efficiently, but this method is difficult to control the surface profile precisely. Grayscale lithography [15–17] using a gray-tone mask is an effective method to obtain various exposure dose distributions on the photoresist by modulating the intensity of ultraviolet (UV) light. However, the physical gray-tone mask is usually fabricated by a direct writing lithography method and each grayscale mask can only be applied for a fixed structure. This is inconvenient for the flexible research of various structures and may lead to unnecessary cost.

Recently, maskless photolithography has been proposed for microstructure fabrication [18–22]. A digital micro-mirror device (DMD) has been adopted to replace the traditional physical mask. A DMD is composed of an array of micro-mirrors, each of which can be independently controlled by a computer, so a digital image acting as mask can be dynamically displayed in real-time. The great advantages of maskless photolithography are that no expensive physical mask is necessary. The digital image can be a flexible design according to the profile of the structure to be fabricated by computer. Due to its capability of low cost and high flexible, maskless lithography has received significant attention in the microfabrication field.

To date, the main approaches of fabrication art for achieving micro-optic elements based on the maskless photolithography are as follows: The first is to transfer the CAD data of the surface function of the structure to be fabricated into a serial slice along the direction of high, with each slice being a binary image. These binary images are generated by the DMD under computer control in real-time, and then are delivered by the imaging system to the surface of the photoresist where a superimposed exposure dose proportional to the profile function is obtained. Using this approach, Totsu et al. have fabricated the positive photoresist patterns of spherical and aspherical micro lens arrays with the diameter of each lens being 100 μm [23]. Zhong et al. also adopted the technique for the fabrication of continuous relief micro-optic elements [24]. Although these works can achieve the micro-optic elements at low cost with time savings, a common point of these methods is the tedious preprocessing of slicing for each design and multiple exposures are needed.

In this paper, we present a new fabrication art approach on the basis of maskless photolithography. By generating a grayscale map under an appropriate exposure time, a one-step exposure control for arbitrary surface profiles can be achieved, so both the slicing process and multiple exposures are avoided. The grayscale level is generated by the multiple reflection technique which adopts the means of pulse width modulation of the DMD. Although similar grayscale lithography based on DMD has been adopted by Wang et al. for the fabrication of diffractive optics [25,26], they do not describe the detailed experiment processing for the nonlinear effect existing in the fabrication procedure. In the digital grayscale lithography, a serious problem caused by DMD is that the exposure dose will not linearly change with respect to grayscale levels under a constant exposure time. Accurate general theoretical formulae for describing such nonlinear relationships are lacking. Additionally, the relationship between the exposure depth and exposure dose is also nonlinear due to the property of the photoresist. To compensate the two nonlinear effects and then to generate the appropriate grayscale map, we adopt the approach of calibrating the relationship between the exposure depth and grayscale levels under an appropriate exposure time that requires no specific knowledge about the nonlinear effects of the DMD and photoresist. During this procedure, a reasonable grayscale level range which maintains a smooth-slow increment and stable intensity distribution needs to be considered to ensure a well-controlled surface profile and suitable surface roughness. After generating the grayscale mask on the basis of the adjustment curve, a one-step exposure fabrication art has been built to obtain desired exposure depth. This method can ensure a precise control of the surface profile of the micro-optics element to be fabricated, and the process of slicing and multiple exposures are not needed.

To verify the reliability of this method, several typical array elements with a spherical surface, aspheric surface, and conic surface have been fabricated and tested, whose diameter is 200 μm and height is 6 μm. The photoresist molds were reversely replicated in polydimethysiloxane (PDMS) elastomer, which was widely used in micro-optics elements because of its excellent optical properties.

2. Experiment Setup and Methods

2.1. Maskless Photolithography System

Figure 1a shows the maskless photolithography experimental setup in our laboratory, which consists of a uniform illumination system at 365 nm wavelength, a high-speed optical projection system for dynamic UV-light patterning, and a three-axis computer-controlled stage for X-Y location and focus control. The light from a mercury lamp is filtered to obtain the UV light at a wavelength of 365 nm, which is introduced into the collimating lens device to provide uniform illumination. The DMD chip (Wintech DLP 4100 0.7″ XGA, Wintech Digital Systems Technology Corp., Carlsbad, CA, USA) from Texas Instruments plays the role of a mask that reflect the uniform incident UV light pixel-by-pixel to generate image frames. This image will be transferred by the projection objective to a photoresist-coated substrate. The DMD consists of a 1024×768 micromirror array with a cell size of 13.68 µm, the demagnification of the projection objective is 6.84. Thus, the theoretical resolution of the UV-light pattern projected on the photoresist surface is 2 µm. The XYZ stage has a travel range of 100 mm $\times$ 100 mm in the X-Y plane and 10 mm in the vertical direction with a resolution of 50 nm, which enables us to achieve a large exposure area at the substrate and a precise control of focus.

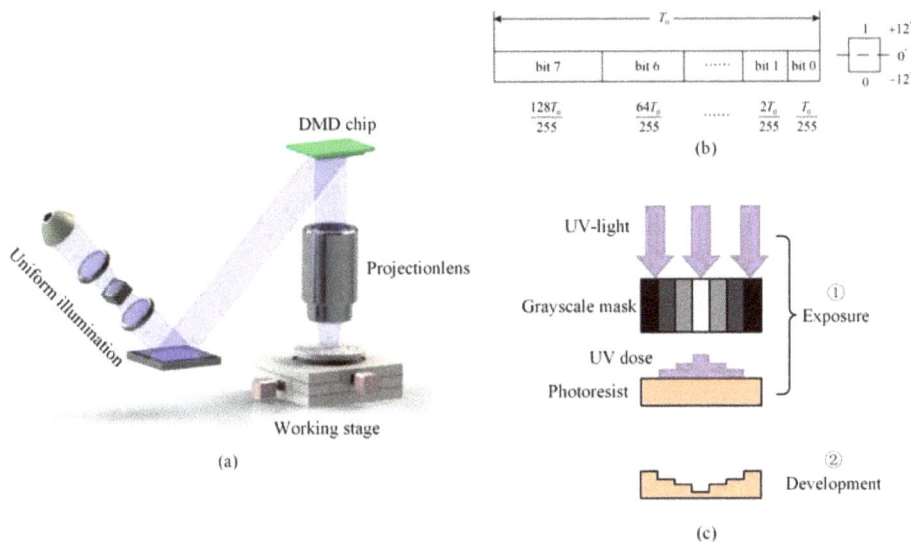

Figure 1. (a) Schematic view of the digital micro-mirror device (DMD)-based maskless photolithography system; (b) pulse width modulation based on 8-bit planes; and (c) the fabrication procedure based on grayscale mask exposure.

Figure 1b illustrates the grayscale control based on the pulse width modulation of 8-bit planes, which enable 256 grayscale levels. The single-frame time T_0 of a grayscale image is divided into eight different time intervals controlled by an 8-bit binary sequence. Each micro-mirror unit of the DMD can be individually controlled by a computer in the direction of $\pm 12°$ ("1" or "0") to determine whether its working state is ON or OFF in each bit plane. By rapidly (typically 20 µs) changing the rotation direction of the micromirror based on the pulse width modulation technique, we obtain a finely-tuned grayscale level which is in proportion to the time duty cycle of the ON states.

Figure 1c shows the fabrication procedure based on grayscale mask exposure. This procedure consists of two steps: exposure and development. In the first step, the high grayscale level will result in a greater exposure dose distribution on the photoresist surface due to its larger duty cycle time of the ON states. Then, in the second step, the photoresist pattern whose exposure depth is in proportion to the exposure dose will be obtained after development. By using the one-step maskless grayscale lithography, one can flexibly control the grayscale level to modulate the exposure dose required for the fabrication of micro-optic elements with an arbitrary surface.

2.2. Nonlinear Effects

Essentially, the aim of generating the grayscale is to control the exposure depth of the photoresist. If the grayscale is of a linear dependence relationship on the exposure depth, then only one scale factor is needed to be calibrated. However, due to the two reasons discussed below, an adjustment curve instead of one scale factor must be determined.

Usually, the exposure dose is defined as the product of the intensity of the incidence light $I(x, y)$ and exposure time T:

$$E(x,y) = I(x,y) \times T \tag{1}$$

In the maskless lithography system, the exposure dose $E(x, y)$ is proportional to the intensity $I(x, y)$ under a pre-setting exposure time. In general, the dependence of the intensity on the grayscale level deviates from the linear relationship due to the peculiarity of the DMD that cannot be systemically eliminated. To estimate the relationship between intensity and grayscale, the intensity data tested by a UV radiation illuminometer (UIT-250, Ushio America, Inc., Cypress, CA, USA) is presented in Table 1, which just gives the data for grayscales over 30 because the intensity below 30 is basically zero. Figure 2a shows an intuitive presentation about this nonlinear relationship. We note that the intensity of a lower grayscale increases slowly at a stable state. In contrast, the intensity of a higher grayscale increases rapidly at an unstable energy level, which is disadvantageous for the control of the surface roughness. There are few specific expressions to describe this nonlinear relationship, but it does exist in the DMD [27]. We assume that this phenomenon may be caused by the nonlinear control of pulse width modulation in Figure 1b. The high level bit planes have larger duty cycle times and contribute a significant intensity, but the low bit plane just has a small duty cycle time and imparts a small intensity. Thus, the final presentation is that the intensity varies exponentially with the grayscale and is unstable at high grayscale levels.

Table 1. The tested data between intensity and grayscale level.

Grayscale	Intensity (mW/cm²)	Standard Deviation (mW/cm²)	Grayscale	Intensity (mW/cm²)	Standard Deviation (mW/cm²)
30	0	0	150	6.51	0
40	0.13	0	160	7.85	0.02
50	0.26	0	170	9.95	0.02
60	0.43	0	180	11.5	0.02
70	0.7	0	190	15.73	0.02
80	1.03	0	200	18.91	0.02
90	1.5	0	210	26.15	0.015
100	2	0	220	32.46	0.02
110	2.78	0.005	230	36.22	0.04
120	3.63	0.005	240	42.6	0.025
130	4.47	0.01	250	42.63	0.015
140	5.37	0.005	-	-	-

In addition, to test the dependence of the exposure depth on the exposure dose, we performed an exposure test on the photoresist (AZ-9260, Clariant Corporation, Muttenz, Switzerland), which is spin-coated on the substrate at 2000 rpm followed by prebaking at 100 °C for 10 min to obtain a photoresist layer of about 10.5 μm. A grayscale grating of 200 level (18.91 mW/cm²) with a 400 μm period that consists of 200 pixels in the horizontal direction is applied to perform this experiment for a convenient measurement. The exposure time changes from 1 s to 14 s which enables a constant exposure dose increment. After development in the developer (AZ 400K, Clariant Corporation, 1:2, 40 s) we extracted the exposure depth by the stylus profiler, as presented in Table 2. Due to the properties of the photoresist, the relationship between the exposure depth and the exposure dose is nonlinear, also. Some studies [28,29] have reported that the exposure depth is a logarithmic function of the dose and can be determined by the contrast curve, which is defined as the linear slope of the contrast curve as follows:

$$\gamma = \frac{1}{\ln E_{cl} - \ln E_{th}} = \frac{h(x,y)/H}{\ln E(x,y) - \ln E_{th}}, \text{ where } 0 < h(x,y) < H \text{ and } E_{th} < E(x,y) < E_{cl} \tag{2}$$

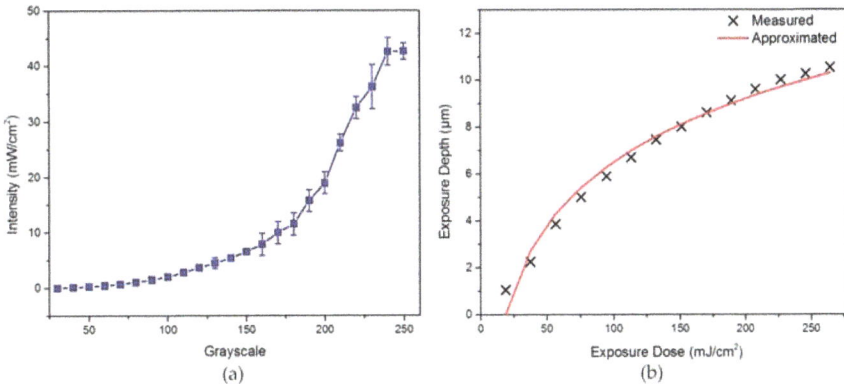

Figure 2. (**a**) Relation between intensity of ultraviolet (UV) light at a wavelength of 365 nm and grayscale levels; and (**b**) the relation between the exposure depth and exposure dose.

Table 2. The tested exposure depth data under different exposure doses.

Exposure Dose (mJ/cm^2)	Exposure Depth (µm)	Standard Deviation (µm)	Exposure Dose (mJ/cm^2)	Exposure Depth (µm)	Standard Deviation (µm)
18.9	1.05	0.025	151.2	8	0.25
37.8	2.25	0.2	170.1	8.6	0.23
56.7	3.85	0.1	189	9.12	0.19
75.6	5	0.15	207.9	9.6	0.18
94.5	5.9	0.15	226.8	10	0.15
113.4	6.7	0.2	245.7	10.26	0.18
132.3	7.45	0.25	264.6	10.53	0.19

In Equation (2), the parameter E_{th} is the threshold dose to initiate a photoresist reaction, E_{cl} is the clearing dose required for removing the photoresist layer H completely, and $E(x, y)$ is the required dose for target exposure depth $h(x, y)$, which can be obtained by direct inversion of Equation (2) as follows:

$$h(x,y) = H \times \gamma \times \ln\left(\frac{E(x,y)}{E_{th}}\right) \tag{3}$$

We have calculated the theoretical curve of the exposure depth with respect to the dose. Figure 2b shows the sampling data and theoretical curve about the exposure depth. We note that these two curves are basically approximated. From the above, we know that both the relationship between intensity and grayscale level of the DMD and the relationship between the exposure dose and the exposure depth are all nonlinear.

2.3. Grayscale Mask Design

To compensate for the two types of nonlinear effects shown in Figure 2, and then generate an accurate grayscale map, a valid method is to find a reliable calibration curve which can provide a precise relationship between the exposure depth and the grayscale value. According to this calibration curve, the grayscale compensation for arbitrary surface designs is achievable. There are two points that deserve consideration in the calibration processing: the calibration curve should maintain smooth-slow growth and a stable intensity level because the rapid increment of the exposure depth and an unstable intensity level will lead to inaccurate control of the surface profile and a terrible surface roughness. Thus, an exposure test was carried out for the grayscale range of 30 to 150 which is a reasonable increment and a stable intensity level according to the curve shown in Figure 2a. The photoresist layer of about 8 µm was prepared at 2500 rpm, which could be flexibly adjusted according to the required thickness. Here we adopt the grayscale maps of a grating to perform this adjustment, as shown in Figure 3a. The period of the grating is set as 400 μm, which consists of 200 pixels in the lateral direction. The exposure time was set as 20 s, which was estimated by the curve shown in Figure 2b to ensure a sufficient exposure dose. The test data measured by the stylus profiler was given in Table 3 and plotted in Figure 3b. This adjustment curve between the exposure depth and grayscale level provides a reference which adequately considers the nonlinear effect between the exposure dose and grayscale level, and the nonlinear effect between the exposure depth and exposure dose, simultaneously.

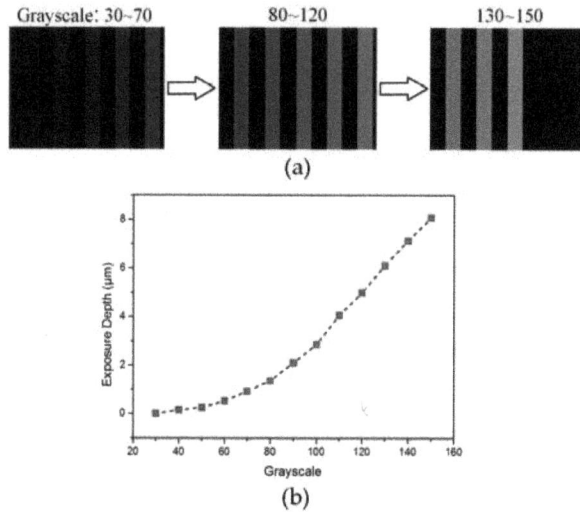

Figure 3. (a) Grayscale maps (30 to 150 grayscale levels) of a grating with a period of 400 μm (200 pixels in the horizontal direction) for the calibration; and (b) calibration curve between the exposure depth and grayscale level.

Table 3. The exposure depth measurement result under a 20 s exposure time.

Grayscale	Exposure Depth (μm)	Grayscale	Exposure Depth (μm)
30	0	100	2.86
40	0.14	110	4.06
50	0.25	120	4.98
60	0.52	130	6.1
70	0.91	140	7.12
80	1.36	150	8.08
90	2.1	-	-

Since only discrete information about the relationship between the grayscale and the exposure depth can be obtained, in a practical application, to generate a precise grayscale map for a structure to be fabricated, a numerical fitting processing will be adopted. We assume that the exposure depth between two adjacent sampling points is a linear correlation due to the narrow sampling step, thus a linear interpolation method is applied to extracted suitable grayscale level. Although this processing may introduce some errors, it imposes little influence on the generation of the grayscale map due to the average resolution being less than 0.1 μm, which is calculated according to the linear interpolation in a single sample interval.

As an example, we design a micro-lens array (MLA) with a spherical surface (Figure 4a), whose aperture B is 200 μm and height h is 6 μm. This spherical surface can be expressed as:

$$z(x,y) = \sqrt{r^2 - x^2 - y^2} - r + h, \text{where } 0 \le z \le h \text{ and } 0 \le \sqrt{x^2 + y^2} \le B \qquad (4)$$

In the above equation, the parameter r is the radius of this spherical surface and can be numerically calculated according to the aperture and the height. According to Figure 3b, we generate the grayscale map of a single lens which consists of 100×100 pixels. From the cross-section of the grayscale map presented in Figure 4a, we note that the variation of grayscale value is stair-stepping. The small steps may impart some influence on the surface roughness, but does not break the outline due to the continuous surface. Finally, the grayscale mask corresponding to this spherical MLA is generated to perform the exposure, as shown in Figure 4b.

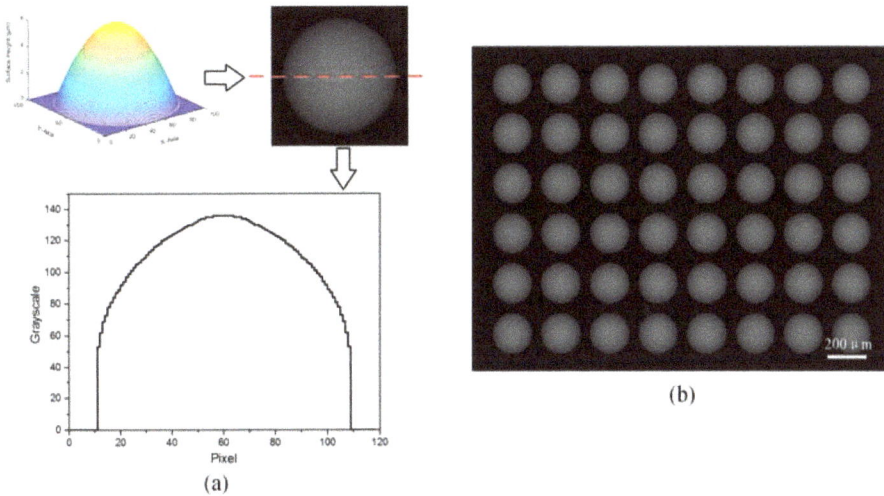

Figure 4. (a) Designed spherical micro-lens array (MLA) model and cross-section of the grayscale map of a single lens; and (b) grayscale mask with a 6×8 lens array.

3. Results and Discussion

To verify the availability of our method, experiment was carried out for grayscale map shown in Figure 4b. For the photolithography, we used positive photoresist (AZ-9260, Clariant Corporation). The photoresist was spin coated on a glass substrate at 2500 rpm followed by a prebaking at 100 °C for 10 min. Then a photoresist film about 8 μm thickness was obtained. Next, the grayscale map was exposed on the surface of photoresist. The total time for the exposure of the grayscale map is 20 s. After development in the alkaline developer (AZ 400K, Clariant Corporation, 1:2) for approximately 40 s, the lens-shaped profile of the photoresist, which corresponding to the reversed profile of the designed model in Figure 4b was obtained. Figure 5a,b shows the microscope and scanning electron microscope (SEM) images of this concave spherical MLA in photoresist.

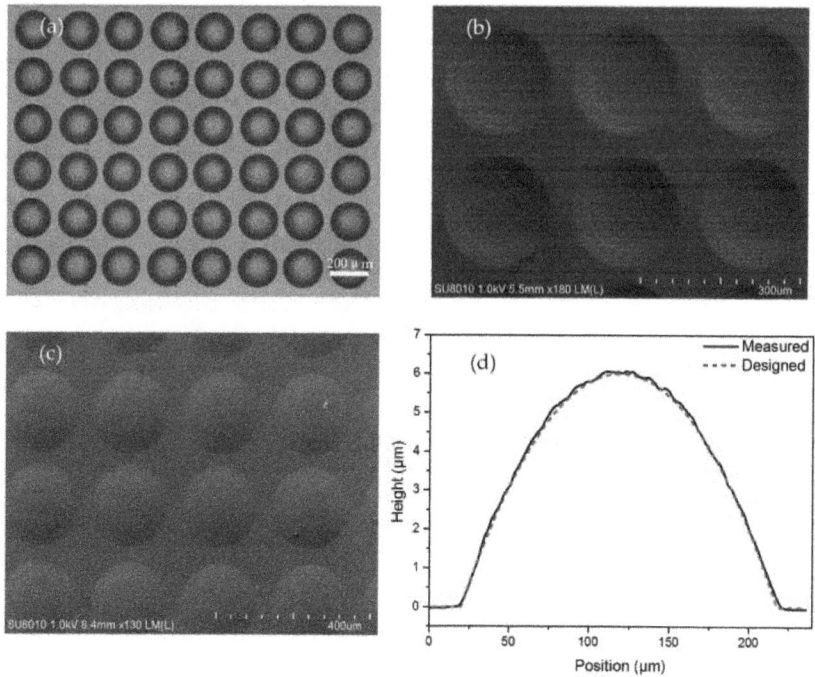

Figure 5. Microscope image (**a**) and scanning electron microscope (SEM) image (**b**) of a concave spherical MLA in photoresist; (**c**) SEM image of a convex spherical MLA in polydimethysiloxane (PDMS); and (**d**) the measured and designed cross-sections of the convex spherical MLA.

To obtain the available spherical MLA, the photoresist mold in Figure 5b was reversely transferred in the PDMS elastomer. Here the PMDS Sylgard 184 (from Dow Corning, Midland, MI, USA) was prepared by mixing the PDMS with diluter at a proportion of 10:1 in weight, after which it was kept under vacuum for dehydration for 20 min. Then the PDMS solution was applied to the mold and kept at 100 °C for 15 min. Figure 5c shows the SEM image of a convex spherical MLA on the top of PDMS. To evaluate the surface profile of the PDMS MLA, a cross-section of the convex lens was profiled by the stylus profiler, as shown in Figure 5d. It is obvious that the measurement profile (black solid line) agrees well with the designed profile (red dash line), and the largest difference between them was 0.21 μm (3.5% of the total height). To analyze the deviation between the practical curve and the theoretical curve, we calculated the root-mean-square (RMS) value as follows:

$$R = \left\{ \frac{1}{N} \sum_{i=1}^{N} \left[h(i) - h_{\text{design}}(i) \right]^2 \right\}^{1/2} \tag{5}$$

The calculated RMS deviation was 0.08 μm, which was enough for the MLA application in visible light.

To estimate the optical performance of the convex MLA in PDMS, both focusing and imaging experiments were carried out, as shown in Figures 6 and 7. In the focusing experiment, a laser beam at a wavelength of 532 nm was introduced to illuminate the whole PDMS MLA. A 6 × 8 light spot array with uniform intensity was captured by a charge-coupled device (CCD) placed in the MLA's focal plane. Figure 6b,c presents the images of focused light spots and intensity distribution, respectively. Figure 6d shows an image of the normalized intensity of a single typical spot. The full width at half maximum (FWHM) for this particular spot is about 30 μm. The focal length of the MLA was estimated to be about 1.7 mm. Then we tested the imaging ability of the MLA on a microscopy setup. As exhibited in Figure 7, a mask with the letter "M" was placed between the white light source and convex MLA, and the image array was observed by using a CCD camera mounted with an objective lens. The uniform light spots and clear "M" letters indicate that the fabricated convex spherical MLA in PDMS has excellent optical properties, which has important applications in array illumination and micro-imaging.

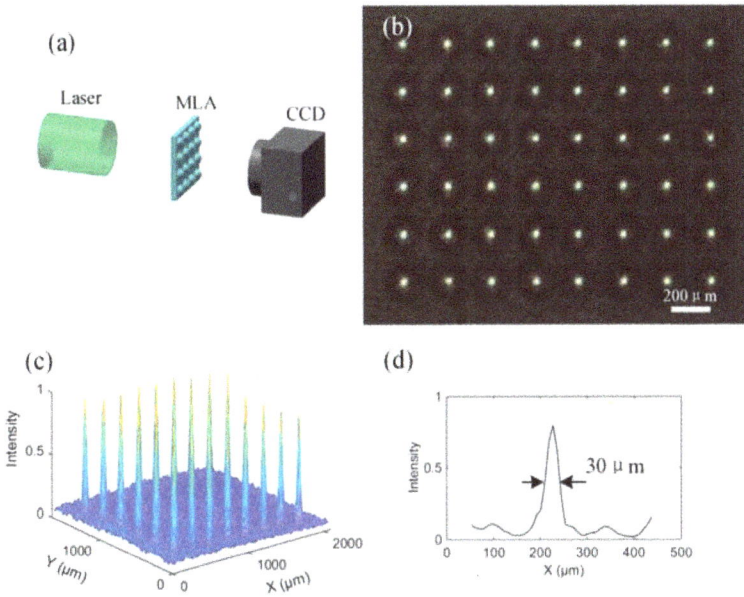

Figure 6. (**a**) Test of the focusing performance of convex spherical MLA in PDMS; (**b**) the image of focused light spots; (**c**) the normalized intensity distribution of focused light spots; and (**d**) an image of the normalized intensity of a single typical single spot.

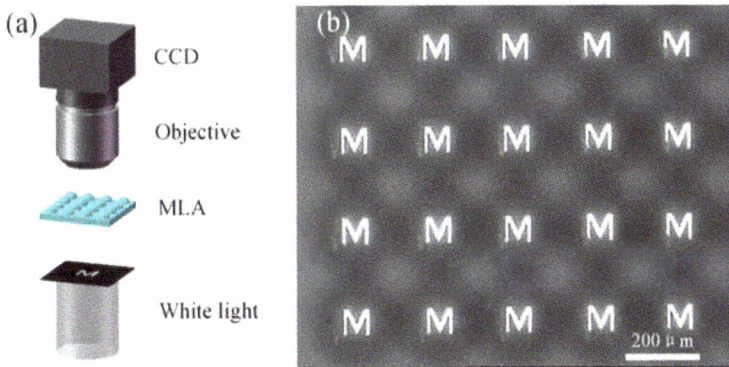

Figure 7. (**a**) Test of the imaging performance of convex spherical MLA in PDMS; and (**b**) the arrayed images of the letter "M" observed by charge-coupled device (CCD).

To further demonstrate that this one-step maskless grayscale lithography is capable with the fabrication of micro-optics with arbitrary surface, we also designed MLAs with an aspherical surface and a conical surface. The expression of the aspheric surface is:

$$z(x,y) = h \times \exp\left(-\alpha(x^2+y^2)/2\right), \text{where } 0 \leq z \leq h, \text{ and } 0 \leq \sqrt{x^2+y^2} \leq B \tag{6}$$

and the expression of the conical surface is:

$$z(x,y) = h \times \left(1 - \frac{\sqrt{x^2+y^2}}{B}\right), \text{where } 0 \leq z \leq h, \text{ and } 0 \leq \sqrt{x^2+y^2} \leq B \tag{7}$$

In the above equations, the parameters h, α, and B are set as 6 µm, 0.0009, and 200 µm, respectively. Figure 8 shows the fabricated MLAs with an aspherical surface and a conical surface in PDMS. The cross-section of these two kinds of microlenses presented a good agreement with the designs. The RMS deviation of the aspherical and conical microlens were calculated as 0.1 µm and 0.11 µm, respectively, which further indicated that this method enables a precise control of the customized surface profile and is prospective in the fabrication of micro-optics.

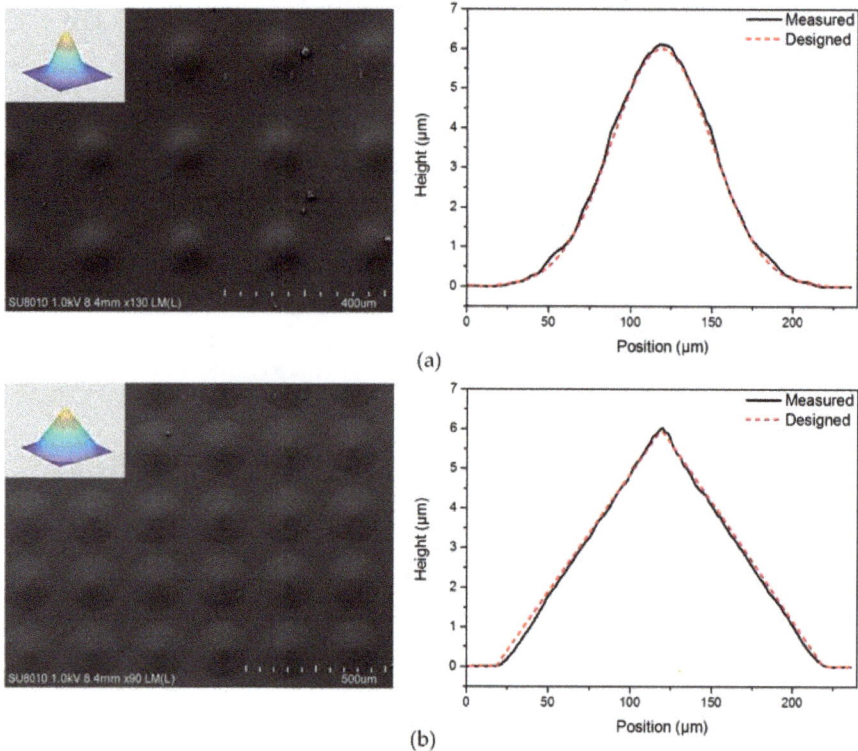

(a)

(b)

Figure 8. (**a**) SEM image of the aspherical MLA in PDMS and the cross-sections of measurement and design; and (**b**) SEM image of the conic MLA in PDMS and the cross-sections of measurement and design.

4. Conclusions

This work reports a one-step maskless grayscale lithography method based on DMD for the rapid and cost-effective fabrication of micro-optical elements with arbitrary surface. The reference curve between the exposure depth and grayscale level effectively compensates the nonlinear effect between intensity and the grayscale level of the DMD, as well as the nonlinear relationship between the exposure depth and dose, and can be used to generate appropriate grayscale map and ensure a precise control of surface profile of the structure to be fabricated. Using this method, we successfully fabricated several typical MLA with a spherical surface, aspherical surface, and conic surface with an aperture of 200 μm and a height of 6 μm in the photoresist. Then these photoresist models were reversely transferred in PMDS for optical testing. The cross-section measurement results agree well with the designed profile. The root mean square (RMS) between the test and design value of the surface height is about 0.1 μm. Additionally, we tested the focusing and imaging ability of the replicated convex spherical MLA in PDMS. Both the uniform focus light spots and clear image of the letter "M" indicated that the generated MLA in PDMS could achieve excellent optical properties. These results demonstrate that the proposed method of compensating the nonlinear effect in maskless lithography can be directly used to control the grayscale levels of the DMD for fabricating the structure with an arbitrary surface profile.

Acknowledgments: This research was financed by the Chinese National Natural Science Foundation (grant No. 61604154) and the West Light Foundation of The Chinese Academy of Sciences YA15k001.

Author Contributions: Qinyuan Deng designed and carried out experiments and wrote the paper; Yong Yang and Yu He helped to perform SEM measurements and optical test; and Hongtao Gao, Yi Zhou, and Song Hu revised the paper.

Conflicts of Interest: The authors declare no conflict of interest.

References

1. Lima Monteiro, D.W.D. *CMOS-Based Integrated Wavefront Sensor*; Delft University Press: Delft, The Netherlands, 2002.
2. Sun, Y.; Forrest, S.R. Enhanced light out-coupling of organic light-emitting devices using embedded low-index grids. *Nat. Photon.* **2008**, *2*, 483–487. [CrossRef]
3. Seabra, A.C.; Araes, F.G.; Romero, M.A.; Neto, L.G.; Nabet, B. Increasing the optical coupling efficiency of planar photodetectors: Electron beam writing of an integrated microlens array on top of an MSM device. In Proceedings of the SPIE 4089, Optics in Computing 2000, Quebec City, QC, Canada, 24 May 2000; pp. 890–894.
4. Chen, Y.; Elshobaki, M.; Ye, Z.; Park, J.M.; Noack, M.A.; Ho, K.M.; Chaudhary, S. Microlens array induced light absorption enhancement in polymer solar cells. *Phys. Chem. Chem. Phys. PCCP* **2013**, *15*, 4297–4302. [CrossRef] [PubMed]
5. Cho, M.; Daneshpanah, M.; Moon, I.; Javidi, B. Three-dimensional optical sensing and visualization using integral imaging. *Proc. IEEE* **2006**, *99*, 556–575.
6. Balslev, S.; Rasmussen, T.; Shi, P.; Kristensen, A. Single mode solid state distributed feedback dye laser fabricated by gray scale electron beam lithography on a dye doped SU-8 resist. *J. Micromech. Microeng.* **2005**, *15*, 2456. [CrossRef]
7. Graells, S.; Aćimović, S.; Volpe, G.; Quidant, R. Direct growth of optical antennas using e-beam-induced gold deposition. *Plasmonics* **2010**, *5*, 135–139. [CrossRef]
8. Henry, M.D.; Shearn, M.J.; Chhim, B.; Scherer, A. Ga(+) beam lithography for nanoscale silicon reactive ion etching. *Nanotechnology* **2010**, *21*, 245303. [CrossRef] [PubMed]
9. Kohoutek, T.; Hughes, M.A.; Orava, J.; Mastumoto, M.; Misumi, T.; Kawashima, H.; Suzuki, T.; Ohishi, Y. Direct laser writing of relief diffraction gratings into a bulk chalcogenide glass. *J. Opt. Soc. Am. B* **2012**, *29*, 2779–2786. [CrossRef]

10. Malinauskas, M.; Žukauskas, A.; Purlys, V.; Gaidukevičiu, A.; Balevičius, Z.; Piskarskas, A.; Fotakis, C.; Pissadakis, S.; Gray, D.; Gadonas, R. 3D microoptical elements formed in a photostructurable germanium silicate by direct laser writing. *Opt. Lasers Eng.* **2012**, *50*, 1785–1788. [CrossRef]

11. Yang, Q.; Tong, S.; Chen, F.; Deng, Z.; Bian, H.; Du, G.; Yong, J.; Hou, X. Lens-on-lens microstructures. *Opt. Lett.* **2015**, *40*, 5359–5362. [CrossRef] [PubMed]

12. Chang, C.Y.; Yang, S.Y.; Huang, L.S.; Chang, J.H. Fabrication of plastic microlens array using gas-assisted micro-hot-embossing with a silicon mold. *Infrared Phys. Technol.* **2006**, *48*, 163–173. [CrossRef]

13. Yang, H.; Chao, C.K.; Wei, M.K.; Lin, C.P. High fill-factor microlens array mold insert fabrication using a thermal reflow process. *J. Micromech. Microeng.* **2004**, *14*, 1197. [CrossRef]

14. He, M.; Yuan, X.C.; Ngo, N.Q.; Bu, J.; Kudryashov, V. Simple reflow technique for fabrication of a microlens array in solgel glass. *Opt. Lett.* **2003**, *28*, 731–733. [CrossRef] [PubMed]

15. Levy, U.; Desiatov, B.; Goykhman, I.; Nachmias, T.; Ohayon, A.; Meltzer, S.E. Design, fabrication, and characterization of circular dammann gratings based on grayscale lithography. *Opt. Lett.* **2010**, *35*, 880–882. [CrossRef] [PubMed]

16. Zhang, J.; Guo, C.; Wang, Y.; Miao, J.; Tian, Y.; Liu, Q. Micro-optical elements fabricated by metal-transparent-metallic-oxides grayscale photomasks. *Appl. Opt.* **2012**, *51*, 6606–6611. [CrossRef] [PubMed]

17. Mori, R.; Hanai, K.; Matsumoto, Y. Three dimensional micro fabrication of photoresist and resin materials by using gray-scale lithography and molding. *IEEJ Trans. Sens. Micromach.* **2006**, *124*, 359–363. [CrossRef]

18. Kessels, M.V.; Nassour, C.; Grosso, P.; Heggarty, K. Direct write of optical diffractive elements and planar waveguides with a digital micromirror device based UV photoplotter. *Opt. Commun.* **2010**, *283*, 3089–3094. [CrossRef]

19. Iwasaki, W.; Takeshita, T.; Peng, Y.; Ogino, H.; Shibata, H.; Kudo, Y.; Maeda, R.; Sawada, R. Maskless lithographic fine patterning on deeply etched or slanted surfaces, and grayscale lithography, using newly developed digital mirror device lithography equipment. *Jpn. J. Appl. Phys.* **2012**, *51*, 06FB05. [CrossRef]

20. Song, S.H.; Kim, K.; Choi, S.E.; Han, S.; Lee, H.S.; Kwon, S.; Park, W. Fine-tuned grayscale optofluidic maskless lithography for three-dimensional freeform shape microstructure fabrication. *Opt. Lett.* **2014**, *39*, 5162–5165. [CrossRef] [PubMed]

21. Kumaresan, Y.; Rammohan, A.; Dwivedi, P.K.; Sharma, A. Large area ir microlens arrays of chalcogenide glass photoresists by grayscale maskless lithography. *ACS Appl. Mater. Interfaces* **2013**, *5*, 7094–7100. [CrossRef] [PubMed]

22. Aristizabal, S.L.; Cirino, G.A.; Montagnoli, A.N.; Sobrinho, A.A.; Rubert, J.B.; Mansano, R.D. Microlens array fabricated by a low-cost grayscale lithography maskless system. *Opt. Eng.* **2013**, *52*, 125101. [CrossRef]

23. Totsu, K.; Fujishiro, K.; Tanaka, S.; Esashi, M. Fabrication of three-dimensional microstructure using maskless gray-scale lithography. *Sens. Actuators A Phys.* **2006**, *130*, 387–392. [CrossRef]

24. Zhong, K.; Gao, Y.; Li, F.; Luo, N.; Zhang, W. Fabrication of continuous relief micro-optic elements using real-time maskless lithography technique based on dmd. *Opt. Laser Technol.* **2014**, *56*, 367–371. [CrossRef]

25. Wang, P.; Dominguez-Caballero, J.A.; Friedman, D.J.; Menon, R. A new class of multi-bandgap high-efficiency photovoltaics enabled by broadband diffractive optics. *Progress Photovolt. Res. Appl.* **2015**, *23*, 1073–1079. [CrossRef]

26. Wang, P.; Mohammad, N.; Menon, R. Chromatic-aberration-corrected diffractive lenses for ultra-broadband focusing. *Sci. Rep.* **2016**, *6*, 21545. [CrossRef] [PubMed]

27. Rammohan, A.; Dwivedi, P.K.; Martinez-Duarte, R.; Katepalli, H.; Madou, M.J.; Sharma, A. One-step maskless grayscale lithography for the fabrication of 3-dimensional structures in SU-8. *Sens. Actuators B Chem.* **2010**, *153*, 125–134. [CrossRef]

28. Smith, B.W.; Suzuki, K. *Microlithography: Science and Technology*; CRC Press, Taylor & Francis Group: Boca Raton, FL, USA, 2007.

29. Hur, J.G. Maskless fabrication of three-dimensional microstructures with high isotropic resolution: Practical and theoretical considerations. *Appl. Opt.* **2011**, *50*, 2383–2390. [CrossRef] [PubMed]

micromachines

MDPI

Article

Scanning Micro-Mirror with an Electrostatic Spring for Compensation of Hard-Spring Nonlinearity

Takashi Izawa, Takashi Sasaki and Kazuhiro Hane *

Department of Finemechanics, Tohoku University, Aramaki-aza Aoba 6-6-01, Aoba-ku, Sendai 980-8579, Japan; takashi.izawa@aisin.co.jp (T.I.); t_sasaki@hane.mech.tohoku.ac.jp (T.S.)
* Correspondence: hane2@hane.mech.tohoku.ac.jp; Tel.: +81-22-795-6962

Received: 1 July 2017; Accepted: 1 August 2017; Published: 4 August 2017

Abstract: A scanning micro-mirror operated at the mechanical resonant frequency often suffer nonlinearity of the torsion-bar spring. The torsion-bar spring becomes harder than the linear spring with the increase of the rotation angle (hard-spring effect). The hard-spring effect of the torsion-bar spring generates several problems, such as hysteresis, frequency shift, and instability by oscillation jump. In this paper, a scanning micro-mirror with an electrostatic-comb spring is studied for compensation of the hard-spring effect of the torsion-bar spring. The hard-spring effect of the torsion-bar spring is compensated with the equivalent soft-spring effect of the electrostatic-comb spring. The oscillation curve becomes symmetric at the resonant frequency although the resonant frequency increases. Theoretical analysis is given for roughly explaining the compensation. A 0.5 mm square scanning micro-mirror having two kinds of combs, i.e., an actuator comb and a compensation comb, is fabricated from a silicon-on-insulator wafer for testing the compensation of the hard-spring in a vacuum and in atmospheric air. The bending of the oscillation curve is compensated by applying a DC voltage to the electrostatic-comb spring in vacuum and atmosphere. The compensation is attributed by theoretical approach to the soft-spring effect of the electrostatic-comb spring.

Keywords: scanning micro mirror; nonlinear spring; resonant vibration; microelectromechanical systems

1. Introduction

A scanning micro-mirror is one of the key devices of micro-electro-mechanical systems. A scanning micro-mirror is a fundamental component of laser projection displays, which simply consists of a two-dimensional scanning micro-mirror and a collimated laser beam. A scanning micro-mirror is also promising for laser reflectometry. The reflected laser light from scanned object is detected for visualizing the object. Distance measurement by laser scanning by utilizing a scanning micro-mirror is much awaited key technology of automatic operation of automotive car [1]. The requirements and progresses on scanning micro-mirror are extensively reviewed [2], where several operational characteristics of micro-mirrors are described in detail. For the above purposes, wider and faster scanning is needed. In the case of a scanning micro-mirror operated at wide angle and high frequency, the mechanical resonance of torsional oscillation is often used to increase the rotational angle and to minimize the necessary force and energy for oscillational scanning. The oscillation curve is the oscillational amplitude plotted as a function of the frequency of the applied AC voltage. The oscillation curve is usually symmetric with respect to the peak resonant frequency when the scanning angle is small. By increasing the oscillation amplitude, i.e., rotation angle, the torsional spring becomes harder than the linear spring as a function of rotation, which is called the hard-spring effect [3,4]. The hard-spring effect generates a shift and a bending of the oscillation curve towards higher frequencies, which often causes instability of operation, such as hysteresis phenomenon of the oscillation curve. On the other hand, it is noteworthy from the bending of the oscillation curve that the electrostatic vertical comb for rotation shows a soft-spring property when the vertical comb does not have the height offset between the fixed and movable combs [5,6].

Resonant frequency tuning of mechanical resonators are often needed to compensate the frequency shift caused by the fabrication error and environmental changes, such as temperature. Several methods for tuning the resonant frequency were studied on the basis of electrostatic force [7–13] and thermal expansion [14]. In case of the electrostatic method, the effective springs by electrostatic force were incorporated in the laterally oscillating resonators [7–12]. The capacitance of the electrostatic springs was modulated as a function of displacement to generate the linear springs. The nonlinearity of a laterally-oscillating resonator was also tuned by applying a DC voltage [15]. On the other hand, the torsional resonant frequency of scanning micro-mirror was also tuned by applying a DC voltage to the electrostatic vertical combs [13]. The electrostatic vertical combs also function as an additional spring to the mechanical torsion-bar spring. However, there are few reports on the tuning of the nonlinearity of the torsion-bar spring. The thermal expansion of a Y-shaped torsional spring adjusted the tension of the spring for the parametric operation [16]. In the case of the electrostatic method, there is no report for the compensation of the hard-spring effect of the torsion-bar spring.

In this report, a method for compensating a hard-spring effect of the torsion-bar spring of the scanning micro-mirror is proposed by using the electrostatic vertical combs without a height offset between the movable and fixed combs. The hard-spring effect of the torsion-bar spring is compensated by applying a DC voltage to the electrostatic combs. An analytical model is given on the basis of variable capacitance of vertical comb electrodes for a rough explanation of the compensation of the hard-spring effect. A 0.5 mm square scanning micro-mirror is designed and fabricated from a silicon-on-insulator wafer. The oscillation curve is measured by varying the DC voltage applied to the electrostatic combs around the resonant frequency. The hard-spring effect is compensated by the applied DC voltage.

2. Principle

Figure 1 shows a part of the rotational spring of the proposed scanning micro-mirror. The mirror is supported by two silicon torsion-bars, one of which is shown in Figure 1. In addition to the silicon torsion-bar, the electrostatic vertical combs are also fabricated between the mirror plate and the silicon torsion-bar as shown in Figure 1. The torsion-bar works as a rotational spring for the rotation of the mirror plate. The rotational angle θ of the torsion-bar is proportional to the applied torque and the proportionality constant (i.e., the spring constant of the torsion bar) is given by k_{m0}. If a nonlinearity of the torsion-bar exists at a large rotation angle, the spring constant is expressed by [17]:

$$k_m = k_{m0}(1 + \alpha \theta^2) \tag{1}$$

Here, α is the nonlinearity coefficient. When α is positive, the torsion bar is a hard-spring, the oscillation amplitude, measured as a function of frequency around a rotational resonant frequency (oscillation curve), bends toward the higher frequency. When α is negative, it becomes a soft-spring. In the conventional torsion-bar, having a rectangular cross-section, the hard-spring effect is generated due to axial tension and is often observed at a large oscillation amplitude.

Here we consider the electrostatic combs as a spring, where a DC voltage is applied to the combs. In order to obtain the symmetry of spring at rotation angle $\pm\theta$, there is no offset for the rotational axis from the central plane of the fixed combs. The cross-sectional structure of the electrostatic combs is shown in Figure 2. Therefore, the torque generated by applying a DC voltage is always a resorting torque as a function of rotation angle.

The energy stored in the comb electrodes is the electrostatic energy given by:

$$U_e = \frac{CV^2}{2} = \frac{\varepsilon_0 V^2 N}{2g} A(\theta) \tag{2}$$

where C is the capacitance of the combs and V is the applied DC voltage. Using the overlapped area $A(\theta)$ of facing comb fingers shown in Figure 2, the capacitance C is expressed by $C = N\varepsilon_0 A(\theta)/g$, where ε_0 is the permittivity of vacuum, N is number of comb finger gaps, and g is the gap between the facing comb fingers. Then, the restoring torque is given by differentiating the stored energy U_e as shown in Equation (3):

$$T_e = -\frac{\partial U_e}{\partial \theta} = -\frac{1}{2} N \varepsilon_0 \frac{V^2}{g} \frac{\partial A}{\partial \theta} \tag{3}$$

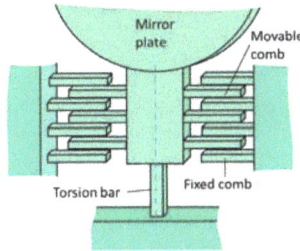

Figure 1. Schematic diagram of torsion-bar with electrostatic spring consisting of comb electrodes.

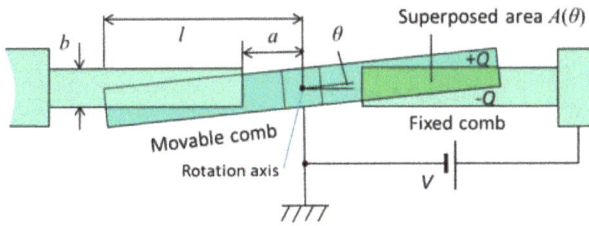

Figure 2. Schematic diagram of the cross-section of the electrostatic spring consisting of comb electrodes.

The electrostatic spring also shows a nonlinear effect. The simplest expression of the spring constant k_e of electrostatic spring contains the lowest nonlinear coefficient β by neglecting the higher order terms as:

$$k_e = k_{e0}(1 - \beta \theta^2) \tag{4}$$

If the nonlinear term $k_{e0}\beta$ of Equation (4) can be equal to the nonlinear term $k_{m0}\alpha$ of Equation (1), the nonlinear effect of total spring is apparently compensated and the oscillation curve becomes symmetric at the resonant frequency neglecting the higher order components. When the spring constant of the electrostatic spring is given by Equation (4), then the torque T_e of the electrostatic spring can be calculated by multiplying the spring constant by rotation angle as follows:

$$T_e = k_{e0}(1 - \beta \theta^2)\theta \tag{5}$$

where the torque is expressed as a cubic equation of θ.

Using the dimension parameters of the comb fingers as shown in Figure 2, the superposed area $A(\theta)$ of the comb fingers is expressed approximately as:

$$A(\theta) = \begin{cases} b(l-a) - \frac{1}{2}(l^2 - a^2)\theta & (0 < \theta < \frac{b}{l}) \\ \frac{(b-a\theta)^2}{2\theta} & (\frac{b}{l} < \theta < \frac{b}{a}) \end{cases} \tag{6}$$

The area $A(\theta)$ is an even function of θ and decreases with the increase in θ. Figure 3 shows the area $A(\theta)$ using the dimensions of the designed scanning micro-mirror described later in this paper. The symbols l, a, b, and g are the distance between the rotation axis and the end of movable comb finger, the distance from the rotation axis to the end of fixed comb finger, thickness of comb fingers, and gap between the fingers of movable and fixed combs. Torque T_e is obtained by differentiating the stored energy of comb capacitors, and thus proportional to the differential of area as shown in Equation (3). Figure 4 shows the calculated values $-\partial A/\partial\theta$ as a function of θ using the designed dimensions of the fabricated scanning micro-mirror.

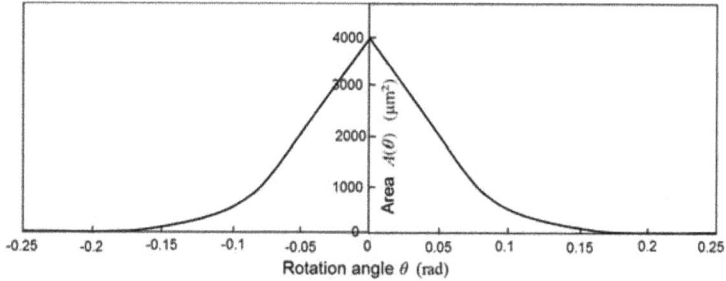

Figure 3. Superposed area of comb fingers as a function of rotation angle.

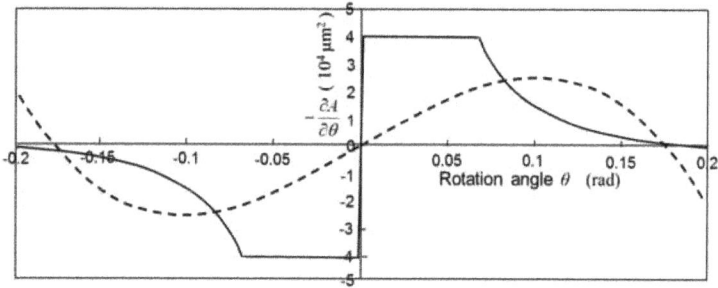

Figure 4. Derivation of the superposed area of comb fingers as a function of rotation angle (solid curve) and the curve by the least-square method (dotted curve).

In order to obtain a rough estimation, Equation (5) may approximate the calculated values of the torque proportional to $-\partial A/\partial\theta$. The least-square method is applied to obtain the approximate equation. In this method, the equation:

$$Z = \int_0^{\theta_m} \left(\frac{-1}{2} N \varepsilon_0 \frac{V^2}{g} \frac{\partial A}{\partial \theta} - k_{e0}(1 - \beta\theta^2)\theta \right)^2 d\theta \tag{7}$$

is minimized to obtain the coefficients k_{e0} and β. The maximum angle of mechanical rotation is given by θ_m. For a simple case, the maximum angle is assumed to be the angle where the overlapped area $A(\theta)$ becomes zero. When the value of b is much smaller than a and l, the angle θ_m can be expressed as $\theta_m = \tan^{-1}(b/a) \approx b/a$. In addition, the angle where the shape of the overlapped area changes from the triangle to the quadrangle with the increase of rotation angle can also approximate to $\tan^{-1}(b/l) \approx b/l$. The torque approximate to a third order equation is given by:

$$T_e = \frac{1}{2} N \varepsilon_0 \frac{V^2}{g} \cdot \frac{15a^3}{32b} \left\{ \left(20\ln\frac{l}{a} - 7\left(1 - \frac{a^2}{l^2}\right) \right) - \frac{7a^2}{3b^2} \left(12\ln\frac{l}{a} - 5\left(1 - \frac{a^2}{l^2}\right) \right)\theta^2 \right\}\theta \tag{8}$$

Therefore, the coefficients k_{e0} and $k_{e0}\beta$ are given, respectively, as:

$$k_{e0} = \frac{1}{2} N \varepsilon_0 \frac{V^2}{g} \cdot \frac{15a^3}{32b} \left(20\ln\frac{l}{a} - 7\left(1 - \frac{a^2}{l^2}\right) \right) \tag{9}$$

$$k_{e0}\beta = \frac{1}{2}N\varepsilon_0 \frac{V^2}{g} \cdot \frac{35a^5}{32b^3}\left(12\ln\frac{l}{a} - 5\left(1 - \frac{a^2}{l^2}\right)\right) \tag{10}$$

The torque approximate to Equation (8) is shown by the dotted curve in Figure 4, which shows a large nonlinearity due to the strong angle dependence of the capacitance.

When the scanner is driven periodically by an external torque T_0, the motion equation is expressed as follows:

$$I_\theta \frac{d^2\theta}{dt^2} + \gamma_\theta \frac{d\theta}{dt} + k_{m0}(1 + \alpha\theta^2)\theta + k_{e0}(1 - \beta\theta^2)\theta = T_0\sin(\omega t + \phi) \tag{11}$$

Here, I_θ and γ_θ are the rotational inertia and the dumping coefficient of scanning micro-mirror, and ω is the angular frequency of the external torque. Since the nonlinearity of electrostatic combs are considered to be soft-springs, if the hard-spring effect of the torsion bars is compensated by the soft-spring effect of the electrostatic comb, then the hard-spring nonlinearity of the scanning micro-mirror can be suppressed by the compensated condition. The compensated condition is given by equalizing the nonlinear coefficients:

$$k_{m0}\alpha = k_{e0}\beta = \frac{1}{2}N\varepsilon_0 \frac{V^2}{g} \cdot \frac{35a^5}{32b^3}\left(12\ln\frac{l}{a} - 5\left(1 - \frac{a^2}{l^2}\right)\right) \tag{12}$$

In addition, the equivalent spring constant of the system is increased by the addition of the electrostatic spring as expressed by the following approximate equation:

$$k_{m0} + k_{e0} = k_{m0} + \frac{1}{2}N\varepsilon_0 \frac{V^2}{g} \cdot \frac{15a^3}{32b}\left(20\ln\frac{l}{a} - 7\left(1 - \frac{a^2}{l^2}\right)\right) \tag{13}$$

On the other hand, the spring constant of the torsion-bar having a square cross-section is given by:

$$k_{m0} = \frac{Ewb^3}{3(1 - v)L}\left\{1 - \frac{192w}{\pi^5 b}\tanh\left(\frac{\pi b}{2w}\right)\right\} \tag{14}$$

where E is the Young's modulus of silicon and v is the Poisson ratio. The symbols L, w, and b represent the length, width, and thickness of torsion-bars, respectively. When the mirror rotates, the torsion of the bar generates a tension in the bar. Due to the tension, the spring constant of the torsion-bar increases (i.e., hard-spring effect), and the increase of the spring constant is expressed by [18]:

$$\Delta k_m = k_{m0}\alpha\theta^2 = \frac{E}{16L^3}\left(\frac{w^5 b}{10} + \frac{w^3 b^3}{9} + \frac{wb^5}{10}\right)\theta^2 \tag{15}$$

Therefore, considering the hard-spring effect, the spring constant of the torsion-bar is given by:

$$k_m = k_{m0}(1 + \alpha\theta^2) = \frac{Ewb^3}{3(1 - v)L}\left\{1 - \frac{192w}{\pi^5 b}\tanh\left(\frac{\pi b}{2w}\right)\right\} + \frac{E}{16L^3}\left(\frac{w^5 b}{10} + \frac{w^3 b^3}{9} + \frac{wb^5}{10}\right)\theta^2 \tag{16}$$

Using the design parameters, the values of the analytical equations are obtained. It is assumed that $L = 200$ μm, $w = 20$ μm, and $b = 20$ μm. Since $E = 1.6 \times 10^{11}$ Pa and $v = 0.3$, the spring constant of the torsion bar is given by $k_m = 6.7 \times 10^{-6} + 2.5 \times 10^{-8}\cdot\theta^2$ (Nm) with the rotational angle θ in units of radians.

Without applying a voltage to the electrostatic comb, the resonant frequency of scanner is given by the ratio of spring constant and inertia:

$$f_{m0} = \frac{1}{2\pi}\sqrt{\frac{k_{m0}}{I_\theta}} \tag{17}$$

Therefore, the resonant frequency of the scanning micro-mirror with the application of a voltage to the electrostatic combs may be approximately obtained by introducing the total spring constant into Equation (17):

$$f_{me} = \frac{1}{2\pi}\sqrt{\frac{k_m + k_e}{I_\theta}} = \frac{1}{2\pi}\sqrt{\frac{k_{m0} + k_{e0} + (k_{m0}\alpha - k_{e0}\beta)\theta^2}{I_\theta}} = \frac{1}{2\pi}\sqrt{\frac{k_{m0} + k_{e0}}{I_\theta}} \times \sqrt{1 + \frac{k_{m0}\alpha - k_{e0}\beta}{k_{m0} + k_{e0}}\theta^2}$$

$$\approx \frac{1}{2\pi}\sqrt{\frac{k_{m0} + k_{e0}}{I_\theta}}\left(1 + \frac{1}{2} \times \frac{k_{m0}\alpha - k_{e0}\beta}{k_{m0} + k_{e0}}\theta^2\right) \tag{18}$$

The frequency shift Δf_{me0} by operating the electrostatic combs at a small angle is expressed by using the linear part of the spring constants of the torsion-bars and the electrostatic combs under the condition of $k_{e0} \ll k_{m0}$ as:

$$f_{me0} = f_{m0} + \Delta f_{me0} = \frac{1}{2\pi}\sqrt{\frac{k_{m0} + k_{e0}}{I_\theta}} = \frac{1}{2\pi}\sqrt{\frac{k_{m0}}{I_\theta}\left(1 + \frac{k_{e0}}{k_{m0}}\right)} \approx f_{m0}\left(1 + \frac{1}{2} \cdot \frac{k_{e0}}{k_{m0}}\right) \tag{19}$$

Therefore:

$$\frac{\Delta f_{me0}}{f_{m0}} \approx \frac{1}{2} \cdot \frac{k_{e0}}{k_{m0}} \tag{20}$$

Moreover, when the rotation angle is large, the nonlinear coefficients influence the oscillation frequency. The excess increase in resonant frequency from the resonant frequency at the small angle is expressed by:

$$f_{me} = f_{me0} + \Delta f_{me}(\theta) \tag{21}$$

The increased rate proportional to the square of θ can be defined as a nonlinear rate R_S as follows:

$$R_S = \frac{\Delta f_{me}}{f_{me0}} \approx \frac{1}{2} \cdot \frac{k_{m0}\alpha - k_{e0}\beta}{k_{m0}}\theta^2 \tag{22}$$

The nonlinear rate R_S corresponds to the bending of oscillation curve.

Under our designed conditions, the rotational inertia is given by $I_\theta = 4.2 \times 10^{-16}$ Nm and, thus, the calculated resonant frequency f_{m0} is approximately 29 kHz. From the designed parameters of the electrostatic combs, the values are given as, $l = 300$ μm, $a = 100$ μm, $b = 20$ μm, and $g = 5$ μm. The number of comb finger gaps is obtained from the four combs on the both sides and they are totally $N = 80$. Using the permittivity $\varepsilon_0 = 8.85 \times 10^{-12}$ F/m, the spring constant of electrostatic combs is obtained analytically as $k_e = (2.6 \times 10^{-11} - 8.5 \times 10^{-10} \cdot \theta^2)V^2$. When the designed values are applied to Equations (20) and (22), we obtain:

$$\frac{\Delta f_{me0}}{f_{me0}} \approx \frac{1}{2}\left(\frac{k_{e0}}{k_{m0}}\right) = \frac{1}{2} \cdot \frac{2.63 \times 10^{-11}V^2}{6.96 \times 10^{-6}} \tag{23}$$

$$R_S = \frac{1}{2} \cdot \frac{2.49 \times 10^{-8} - 8.47 \times 10^{-10}V^2}{6.96 \times 10^{-6}}\theta^2 \tag{24}$$

From Equation (24), the comb voltage necessary for compensating the nonlinearity of the torsion-bar is estimated roughly to be 5.4 V from the condition of $R_S = 0$.

3. Design and Fabrication

Based on the principle described in Section 2, a one-dimensional scanning micro-mirror with the electrostatic compensation combs is designed and fabricated. Figure 5a shows the oblique schematic diagram of the scanning micro-mirror, which consists of a mirror plate, two torsion-bars, two pairs of actuator combs, and four pairs of the compensation combs. The mirror plate is 500 μm square and 20 μm in thickness, which is equal to the thickness of the top silicon layer of SOI wafer. The torsion-bars are 250 μm in length (symbol L in Section 2) and 20 μm in width and thickness (symbols w and b). Figure 5b shows the top view of the scanning micro-mirror. In order to rotate the micro-mirror, the vertical comb-drive actuators are installed at the edges of mirror plate as shown in Figure 5a. The movable fingers of the actuator combs are 200 μm in length, 5 μm in width, and 20 μm thick. The fixed fingers of the actuator combs are 200 μm in length, 8 μm in width, and 200 μm in thickness, which is same as thickness of the silicon substrate of SOI wafer. Therefore, the movable combs and the fixed combs are different in height, the former is on the top silicon layer of SOI wafer and the latter is on the silicon substrate. The height difference is 21 μm, which is equal to the addition of the thicknesses of the top silicon layer and the buried oxide layer of SOI wafer. The overlap length of the actuator comb fingers is 190 μm, and the gap between the movable and fixed fingers is 5 μm. The distance between the bottom of the movable fingers and the top of the fixed fingers of the actuator comb is 5 μm. The number of the finger gaps is 44 for each side of the micro mirror.

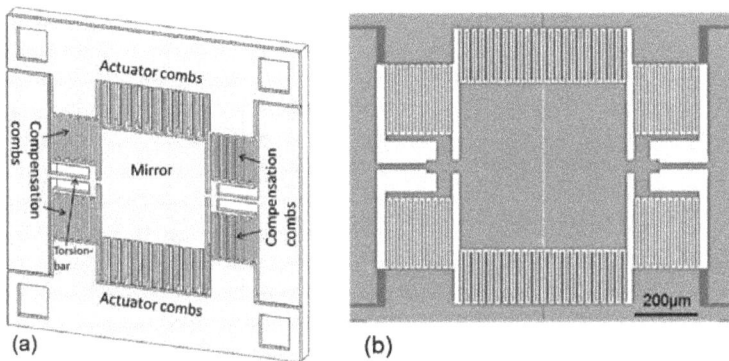

Figure 5. Schematic diagrams of scanner: (**a**) oblique view; and (**b**) top view.

On the other hand, the compensation combs are located around the torsion-bars as shown in Figure 5. The movable and fixed fingers of the compensation combs are 205 μm in length, 5 μm in width, and 20 μm (symbol b) in thickness. The gap (g) between the movable and fixed comb fingers is 5 μm. The distance (l) from the rotation axis to the end of movable fingers is 300 μm, and the distance (a) from the rotation axis to the end of fixed fingers is 100 μm. The width of the compensation comb fingers is 5 μm. The number of the finger gaps of each compensation comb is 40.

The fabrication steps are shown in Figure 6. The SOI wafer used for the fabrication consists of a 20 μm thick top silicon layer, 1 μm thick buried oxide layer, and 200 μm thick silicon substrate. The top silicon layer is coated by a resist polymer (OFPR800-200cp, Tokyo Ohka. Kogyo Company, Ltd., Kawasaki, Japan) (a and b), and patterned by deep reactive ion etching (c). After removing the resist polymer (d), the back side of the wafer is coated by the resist polymer and patterned (e). The silicon substrate is etched from the backside by the deep reactive etching (f). After removing the resist polymer (g), the buried oxide layer is partially etched by a buffered hydrofluoric acid solution. Finally, the device is dried after replacing the acid solution with water, ethanol, and isopropyl alcohol to prevent comb fingers from sticking.

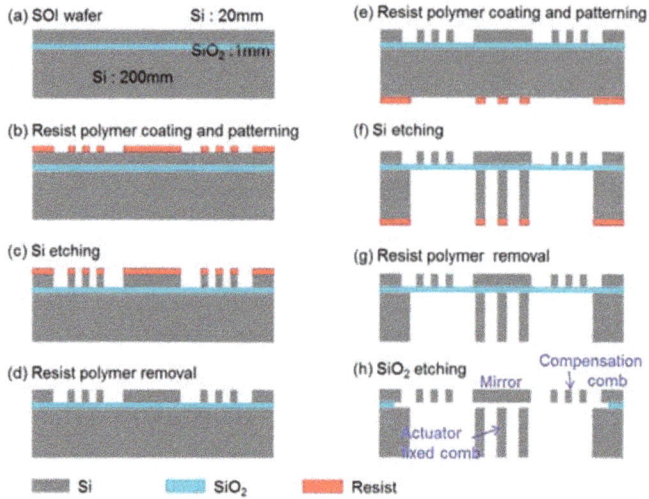

Figure 6. Schematic diagram of the fabrication processes.

4. Fabrication Results and Operation Characteristics in Vacuum

Figure 7a shows the whole view of the fabricated scanning micro-mirror. The compensation combs are fabricated on the same plane of the top silicon layer as shown in Figure 7b. Figure 7c shows the magnified view of the actuator combs, where the height difference between the movable comb (top silicon layer) and the fixed comb (silicon substrate) is seen from the defocused image of the lower comb. The mirror plate can be rotated around the silicon torsion-bars by the initial toque generated by the actuator combs having the height difference.

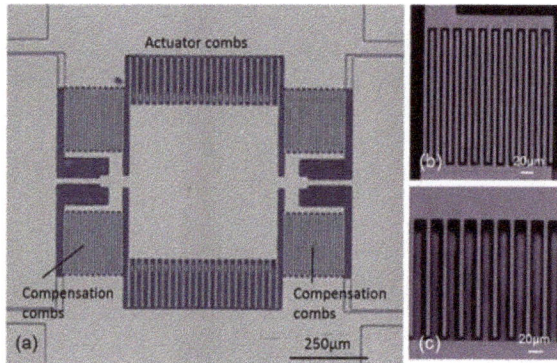

Figure 7. Optical micrograph of the fabricated scanning micro-mirror: (**a**) whole view; (**b**) compensation combs; and (**c**) actuator combs.

The mechanical scan angle was measured from the deflection angle (optical scan angle) of a laser beam impinging on the mirror plate. The deflection angle was obtained from the length of the laser scan line on screen and the distance between the mirror plate and the screen. The scanning micro-mirror was placed in a vacuum chamber and the air dumping effect was nearly removed at a pressure of 30 Pa. Figure 8a shows the mechanical scan angle (a half of the optical scan angle) measured as a function of the frequency of the applied voltage. The voltage ($E(t)$) applied to the actuator combs was an AC voltage of 10 V amplitude with 10 V DC voltage ($E(t) = 10\sin(2\pi ft) + 10$ V, t: time, f: frequency). Without applying DC voltage V for the compensation combs, the peak mechanical rotation angle θ is approximately 9.7 degrees at the frequency of 25.628 kHz. The rotation angle increases gradually with increase in the frequency of $E(t)$ before reaching the peak amplitude, and rapidly decreases with the increase in the frequency after the peak amplitude. Since the decrease from the peak amplitude is steeper than that for the increase to the peak amplitude, the oscillation amplitude curve is not symmetrical with respect to the peak frequency (f_P), and the peak frequency is shifted to higher frequency from the symmetrical position. This is mainly caused by a hard-spring effect of the torsion-bars. The quality factor of the oscillation can be roughly obtained from the full width at half maximum which is about 3000, although the oscillation curve is affected by the hard-spring effect.

In order to evaluate the nonlinearity of oscillation, the nonlinear rate R_S can be approximately obtained from the measured oscillation curve as shown in Figure 8b. The nonlinear rate R_S is nearly equal to the normalized frequency difference (($f_P - f_C$)/f_C) between the peak frequency (f_P) and the center frequency (f_C). The center frequency is obtained from the averaged frequency of the two frequencies (f_1 and f_2) where a horizontal line at a low angle level crosses the oscillation curve as shown in Figure 8b. Although the actual center frequency is the resonant frequency (f_{em0}) corresponding to the peak frequency of the oscillation curve at a very small oscillation angle, we roughly estimate f_{em0} from f_C assuming the symmetry of oscillation curve at the low angle level. The horizontal line at the low angle level is obtained around 10% of the peak angle.

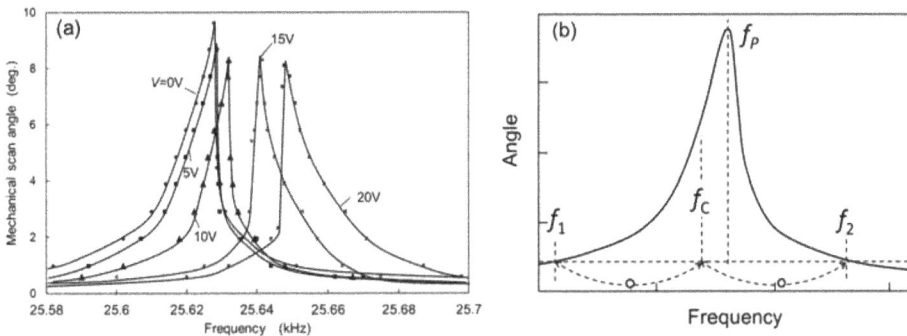

Figure 8. (**a**) Mechanical scan angle measured as a function of the frequency with the voltage V applied to the compensation combs as a parameter; and (**b**) the frequencies for obtaining the value of R_S.

The peak frequency shifts to higher frequency by increasing the voltage V applied to the compensation combs. At the same time, the shape of the oscillation curve changes from the bend toward higher frequency to the bend toward lower frequency gradually with the increase of V from 0 to 20 V as shown in Figure 8a. The change of oscillation curve is caused by the change in the nonlinearity of the total spring by compensating the hard-spring effect of the torsion-bars with the soft-spring effect of the electrostatic spring. The peak frequency shift is caused by the increase in the linear part of spring constant by adding the electrostatic spring to the torsion-bar spring.

Figure 9a shows the resonant frequency shift normalized by the frequency at zero voltage ($V = 0$) as a function of the voltage applied to the compensation combs. The resonant frequency was roughly obtained from the center frequency (f_C) as described above. The measured frequency shift varies nearly quadratically as a function of the applied voltage to the compensation combs. The measured frequency shift is roughly explained by the calculation using Equation (23), which is quadratically dependent on the applied voltage.

Figure 9b shows the measured nonlinear rate R_S with applied voltage, where R_S defined as $R_S = (f_P - f_C)/f_C$ (the frequencies are shown in the inset of Figure 8). The measured R_S decreases with the increase in the voltage applied to compensation combs, and becomes zero at the voltage around 11.5 V. Therefore, by adjusting the applied voltage to the compensation combs, the nonlinearity of the torsion-bar spring can be compensated. The nonlinear rate calculated by Equation (24) is also shown in Figure 9b. The quadratic dependence of the measured R_S is explained by the calculated R_S, although the magnitude of the calculation is small compared to the measured value. The difference between the measurement and the calculation may be caused by the rough analytical calculation of the nonlinear rate of electrostatic combs. Since the actuator combs also generated electrostatic force, this may also cause additional nonlinear effect. Different actuation by piezoelectric or electromagnetic force can further clarify the difference between calculation and measurement.

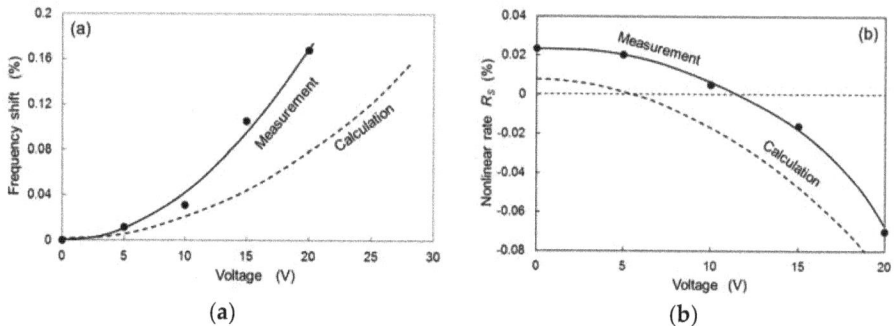

Figure 9. (a) Resonant frequency shift and (b) nonlinearity rate as a function of the voltage applied to compensation combs.

5. Operation Characteristics in Atmospheric Air

The scanning micro-mirror was also operated in the atmospheric air and the operational characteristics were investigated. Compared to the operation in vacuum, a much higher voltage (~6 times) was needed for the operation of the actuator comb at the same oscillation angle. Large driving energy is required to overcome the air friction [4]. It was reasonable to consider that the actuator combs also generated some nonlinear effect since the torque was nonlinearly dependent on the rotation angle. The nonlinear effect of the actuator combs was also included in the measurement in vacuum, although the influence was smaller. The high voltage for the actuator combs seemed to cause a greater nonlinear effect, which might be proportional to the square of the AC voltage applied to the actuator combs. However, the analytical treatment of the nonlinearity of the actuator combs was more complex, since the torque was dependent not only on the rotation angle, but also on the alternating voltage. Here, although the analytical explanation cannot be provided, we describe only the operational characteristics and the compensation for the bending of oscillation curve by the application of a DC voltage to the compensation combs.

Figure 10 shows the mechanical scan angles measured as a function of the frequency of the applied voltage to the actuator combs around the resonant frequency with the DC voltage applied to the compensation combs as a parameter. The voltage ($E(t)$) applied to the actuator combs to obtain nearly the same oscillation angle as in vacuum was $E(t) = 60\sin(2\pi ft) + 60$ V. The quality factor of the oscillation is obtained at about 53, which is smaller by a factor of 57 than that in vacuum. By increasing the voltage to compensation combs, the peak frequency shifts to higher frequency. The oscillation curve bends toward the higher frequency at the compensation voltage V ranging from 0 V to 60 V, and the bending is shifted towards the lower frequency at V from 100 V to 120 V as shown in Figure 10.

Figure 11 shows the nonlinear rate R_S measured as a function of the DC voltage V applied to the compensation combs. The value of R_S is approximately 1% without applying V, which is larger than that of measured value in vacuum. Even if the value of R_S is large, the nonlinearity can be compensated by increasing the DC voltage as shown in Figure 11. The nonlinear rate is nearly compensated at V of 87 V. Therefore, the proposed method is effective for compensating the nonlinear hard-spring effect caused not only by the hard-spring effect of the torsion-bar, but also by other effects.

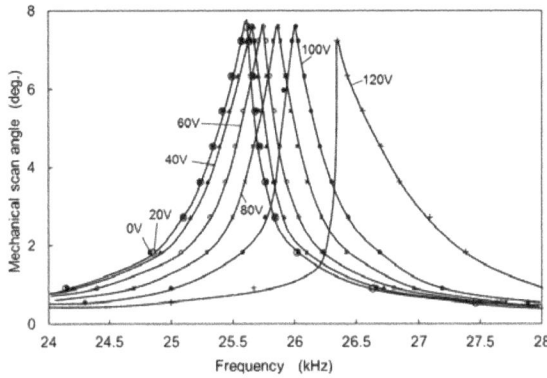

Figure 10. Mechanical scan angle measured as a function of the frequency with the voltage V applied to the compensation combs as a parameter.

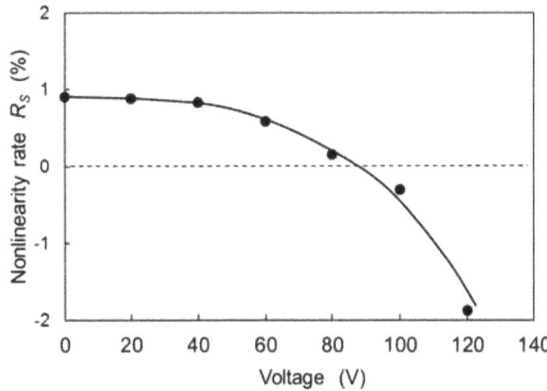

Figure 11. Nonlinear rate as a function of the voltage applied to the compensation combs.

The proposed method can be applied to improve the control of scanning micro-mirror. When the scanning micro-mirror has such a large bend which causes amplitude jump in the oscillation curve [4], the operation frequency is usually adjusted to the frequency near the amplitude jump for maximizing the oscillation amplitude. However, the condition of the amplitude jump is likely to shift due to operation conditions, such as temperature. If the amplitude jump in the oscillation curve is removed by decreasing the hard-spring effect, the micro-mirror can be controlled more stably by maintaining the peak amplitude with a simple feedback loop.

6. Conclusions

The hard-spring nonlinearity of the torsion-bars of the scanning micro-mirror was compensated by a soft-spring effect of the electrostatic combs operated at DC voltage. The analytical model based on the capacitive electrostatic force was derived for explanation of the proposed method. The oscillation curve of the scanning micro-mirror suffering from the bend toward higher frequency was corrected to be nearly symmetric and further bent toward a lower frequency with the increase in the applied voltage to the compensation combs. In addition, the resonant frequency shifted by increasing the total spring constant. A 0.5 mm square scanner was fabricated from a 20 μm thick top silicon layer of a SOI wafer. The fabricated micro-mirror oscillated in vacuum at the resonant frequency around 26 kHz with a mechanical rotation angle of ±8 degrees. The nonlinear rate of 0.02% was compensated at the compensation voltage of 12 V in vacuum, which was roughly explained by the analytical calculation. Moreover, a 1% nonlinear rate at the atmospheric pressure was compensated at the voltage of 87 V. Therefore, the proposed method using the soft-spring effect of electrostatic combs was effective for the compensation of the hard-spring nonlinearity.

Acknowledgments: This work was supported by KAKENHI (17H01267) and Center of Innovation (COI). The fabrication was carried out in Micro/Nano Center of Tohoku University.

Author Contributions: T.I. designed, fabricated, performed the experiments, and analyzed experimentally and theoretically, T.S. advised in fabrication and experiment, K.H. analyzed theoretically and supervised the research.

Conflicts of Interest: The authors declare no conflict of interest.

References

1. Hofmann, U.; Aikio, M.; Janes, J.; Senger, F.; Stenchly, V.; Hagge, J.; Quenzer, H.-J.; Weiss, M.; von Wantoch, T.; Mallas, C.; et al. Resonant biaxial 7-mm MEMS mirror for omnidirectional Scanning. *J. Micro/Nanolithogr. MEMS MOEMS* **2014**, *13*, 011103. [CrossRef]
2. Holmström, S.T.S.; Baran, U.; Urey, H. MEMS laser scanners: A review. *J. MEMS* **2014**, *23*, 259–275. [CrossRef]
3. Jeong, J.-W.; Kim, S.; Solgaard, O. Split-frame gimbaled two-dimensional MEMS scanner for miniature dual-axis confocal microendoscopes fabricated by front-side processing. *J. Microelectromech. Syst.* **2012**, *7*, 308–315. [CrossRef]
4. Chu, H.M.; Tokuda, T.; Kimata, M.; Hane, K. Compact low-voltage operation micromirror based on high-vacuum seal technology using metal can. *J. Microelectromech. Syst.* **2010**, *19*, 927–935. [CrossRef]
5. Ataman, C.; Urey, H. Modeling and characterization of comb-actuated resonant microscanners. *J. Micromech. Microeng.* **2006**, *16*, 9–16. [CrossRef]
6. Ishikawa, N.; Ikeda, K.; Okazaki, K.; Sawada, R. Scanning angle control of the optical micro scanner actuated by electrostatic force. In Proceedings of the 2014 IEEE International Conference on Optical MEMS and Nanophotonics (OMN), Glasgow, UK, 17–21 August 2014; pp. 121–122.
7. Lee, K.B.; Cho, Y.-H. A triangular electrostatic comb array for micromechanical resonant frequency tuning. *Sens. Actuator A Phys.* **1998**, *70*, 112–117. [CrossRef]
8. Scheibner, D.; Mehner, J.; Reuter, D.; Kotarsky, U.; Gessner, T.; Dotzel, W. Characterization and self-test of electrostatically tunable resonators for frequency selective vibration measurements. *Sens. Actuator A Phys.* **2004**, *111*, 93–99. [CrossRef]
9. Lee, K.-B.; Lin, L.; Cho, Y.-H. A closed-form approach for frequency tunable comb resonators with curved finger contour. *Sens. Actuator A Phys.* **2008**, *141*, 523–529. [CrossRef]
10. Adams, S.G.; Bertsch, F.M.; Shaw, K.A.; MacDonald, N.C. Independent tuning of linear and nonlinear stiffness coefficient. *J. Microelectromech. Syst.* **1998**, *7*, 172–180. [CrossRef]
11. Jensen, B.D.; Mutlu, S.; Miller, S.; Kuribayashi, K.; Allen, J.J. Shaped comb fingers for tailored electromechanical restoring force. *J. Microelectromech. Syst.* **2003**, *12*, 373–383. [CrossRef]
12. Morgan, B.; Ghodssi, R. Vertically-shaped tunable MEMS resonators. *J. Microelectromech. Syst.* **2008**, *17*, 85–92. [CrossRef]
13. Eun, Y.; Kim, J.; Lin, L. Resonant-frequency tuning of angular vertical comb-drive microscanner. *Micro Nano Syst. Lett.* **2014**, *2*, 4. [CrossRef]
14. Remtema, T.; Lin, L. Active frequency tuning for micro resonators by localized thermal stressing effects. *Sens. Actuator A Phys.* **2001**, *91*, 326–332. [CrossRef]
15. Yao, J.J.; MacDonald, N.M. A micromachined, single-crystal silicon, tunable resonator. *J. Micromech. Microeng.* **1996**, *6*, 257–264. [CrossRef]
16. Kim, J.; Kawai, Y.; Inomata, N.; Ono, T. Parametrically driven resonant micro-mirror scanner with tunable spring. In Proceedings of the 26th IEEE International Conference on Micro Electro Mechanical Systems (MEMS), Taipei, Taiwan, 20–24 January 2013; pp. 580–583.

17. Kundu, S.K.; Ogawa, S.; Kumagai, S.; Fujishima, M.; Hane, K.; Sasaki, M. Nonlinear spring effect of tense thin-film torsion bar combined with electrostatic driving. *Sens. Actuator A Phys.* **2013**, *195*, 83–89. [CrossRef]
18. Senturia, S.D. *Microsystem Design*; Springer: Dordrecht, The Netherlands, 2001.

micromachines

MDPI

Article

An Electrostatic MEMS Translational Scanner with Large Out-of-Plane Stroke for Remote Axial-Scanning in Multi-Photon Microscopy

Haijun Li [1], Xiyu Duan [2], Gaoming Li [1], Kenn R. Oldham [3] and Thomas D. Wang [1,2,3,*]

[1] Department of Internal Medicine, University of Michigan, Ann Arbor, MI 48109, USA; haijunl@umich.edu (H.L.); gaomingl@umich.edu (G.L.)
[2] Department of Biomedical Engineering, University of Michigan, Ann Arbor, MI 48109, USA; dxy@umich.edu
[3] Department of Mechanical Engineering, University of Michigan, Ann Arbor, MI 48109, USA; oldham@umich.edu
*   Correspondence: thomaswa@umich.edu; Tel.: +1-734-936-1228

Academic Editor: Huikai Xie
Received: 3 April 2017; Accepted: 10 May 2017; Published: 15 May 2017

Abstract: We present an electrostatic microelectromechanical systems (MEMS) resonant scanner with large out-of-plane translational stroke for fast axial-scanning in a multi-photon microscope system for real-time vertical cross-sectional imaging. The scanner has a compact footprint with dimensions of 2.1 mm × 2.1 mm × 0.44 mm, and employs a novel lever-based compliant mechanism to enable large vertical displacements of a reflective mirror with slight tilt angles. Test results show that by using parametrical resonance, the scanner can provide a fast out-of-plane translational motion with ≥400 μm displacement and ≤0.14° tilt angle over a wide frequency range of ~390 Hz at ambient pressure. By employing this MEMS translational scanner and a biaxial MEMS mirror for lateral scanning, vertical cross-sectional imaging with a beam axial-scanning range of 200 μm and a frame rate of ~5–10 Hz is enabled in a remote scan multi-photon fluorescence imaging system.

Keywords: MEMS scanner; axial scanning; multiphoton microscopy

---

1. Introduction

Optical imaging in vertical cross-sections with sub-cellular resolution is essential to biomedical research and clinical diagnosis because histology-like images can be provided to distinguish different features of tissue for early detection of cancer and other diseases. By combining axial and lateral scanning, this capability can be provided. Conventionally, axial scanning in confocal and multi-photon microscopes is achieved with the movement of either the objective or stage, and images in the vertical plane are reconstructed from a series of horizontal images. This approach is limited in speed and is prone to motion artifacts from vibrations introduced in the sample.

New methods to perform remote axial scanning have been developed to overcome these limitations, including group velocity dispersion (GVD)-based [1–4] and tunable lens-based temporal focusing [5,6]. Axial scanning with GVD modulation can achieve high speeds, but results in blurry images [4]. The use of a tunable lens for axial scanning has the advantages of high speed, low cost, and ease of integration, but suffers from changes in magnification and numerical aperture (NA) [6]. A remotely-located axial scan mirror that reflects the excitation beam has recently been demonstrated to move the focus [7–9]. High scan speeds can be achieved with mirrors that have minimal inertia. This method has been used successfully to collect aberration-free images at high speeds with a multi-photon microscope. Two high numerical aperture objectives are used to introduce equal but opposite aberrations in the excitation wavefront during scanning [8]. However, the scanning mechanisms used to move the mirror are based on bulky actuators, such as either a galvanometer or voice coil motor, which results in the difficulty of miniaturizing the system.

To realize axial scanning in a miniature instrument, a compact translational actuator that provides fast scan with large displacements and small tilting angles is needed. Microelectromechanical systems (MEMS) technology is well-suited to this application, and some recent advances have been made in MEMS translational actuators. In general, to achieve large displacements, thermoelectric, piezoelectric, and electromagnetic actuation mechanisms are usually used in MEMS actuators. Thermoelectric actuators can provide large displacements at low voltages, but have slow response times [10,11]. Thin-film piezoelectric scanners can achieve large displacements with high speeds, but require complex fabrication processes [12,13]. Electromagnetic actuators have been developed with fast response times and good displacement, but this technology has high power consumption and is difficult to scale down in size [14,15]. Compared with other actuation mechanisms, electrostatic approaches typically have small actuation force and the pull-in effect, but offer the advantages of low power consumption and complementary metal-oxide-semiconductor (CMOS)-compatible fabrication. Electrostatic MEMS actuators that use the principle of parametric resonance have achieved large axial displacements up to several hundreds of microns [16–19].

We have previously demonstrated an electrostatic MEMS scanner with axial scan capabilities to collect vertical cross-sectional images in a dual-axis confocal endomicroscope [17] and a multi-photon microscope [19]. Here we demonstrate a MEMS scanner with a smaller footprint and a higher speed to further extend the applicability of this axial scan technique in a multi-photon microscope system for real-time vertical cross-sectional imaging.

2. Scanner Design and Fabrication

2.1. MEMS-Based Remote-Scan Multi-Photon Imaging System

Figure 1a shows the schematic of a MEMS-based remote scan multi-photon imaging system. A bi-axial torsion MEMS mirror (M1, Figure 1b) and an out-of-plane translational MEMS scanner (M2) are used to perform lateral and axial scanning, respectively. Figure 1b shows the bi-axial torsional MEMS mirror [20]. This scanner employs a gimbal geometry that enables a 1.8 mm-diameter reflective mirror to rotate around the inner X- and outer Y-axes. The X-axis is defined as the fast axis with a resonant frequency of ~4.3 kHz, and the Y-axis is defined as the slow axis with a resonant frequency of ~1.05 kHz. Based on the optical design of the remote scan unit, the translation of M2 results in the axial displacement of the focus below the tissue surface with a magnification of ~2:1. That is, to achieve an imaging depth of 200 μm, the axial scanning device needs to provide a displacement of 400 μm.

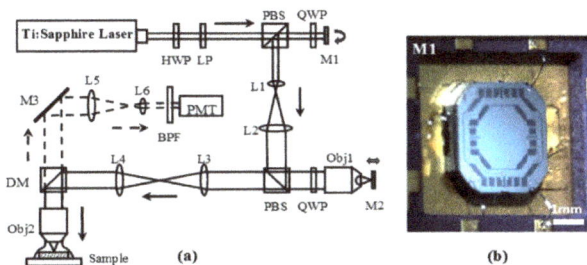

Figure 1. (**a**) Schematic for microelectromechanical systems (MEMS)-based remote scan multi-photon microscopic system. Key: HWP: half wave plate, LP: linear polarizer, L1-6: lenses, Obj1-2: objectives, M1: MEMS bi-axial torsional mirror for lateral scanning, M2: MEMS out-of-plane translational scanner, M3: fixed reflective mirror, PBS: polarizing beam splitter, QWP: quarter wave plate, DM: dichroic mirror, BPF: band pass filter, PMT: photomultiplier tube; (**b**) photo of MEMS bi-axial torsional mirror.

2.2. Design of the Out-of-Plane Translation MEMS Scanner

The basic structure for the out-of-plane translational MEMS device used for axial scanning in the multi-photon imaging instrument is shown in Figure 2A. A central reflective mirror is supported by four lever-based suspensions, and four comb-drives are used for actuation. The suspensions and the mirror form a compliant mechanism that can transfer the rotation of the lever into vertical translation of the mirror. This device is fabricated in a silicon on insulator (SOI) wafer with movable structures, comb-drives, and electrical pads formed in the silicon device layer. A cavity is opened in the silicon handle layer, and narrow trenches are opened in the silicon device layer for electrical isolation. The scanner has a dimension of 2.1 mm × 2.1 mm × 0.54 mm for integration into a miniature instrument. The mirror is designed with a diameter of 0.8 mm to cover the focused beam dimension over the expected 400 μm scan range. The lever-arm of the suspension is designed to have a spiral shape with a length of 1.33 mm to achieve a large vertical displacement. It also couples to the mirror and the anchor through two H-shaped torsional springs and one multi-turn folded-beam spring, respectively. The design of the H-shaped torsional spring is used to enable large rotations while providing high resistance to lateral bending, and the design of the multi-turn folded-beam spring is used to enable large deflections in its folding direction while providing a high resistance to lateral bending. The comb-drive has an in-plane structure, in which movable and stationary comb fingers with the same thickness are formed in the silicon device layer. Unlike conventional vertical staggered comb-drives, the in-plane comb-drive can only work in resonance to enable out-of-plane motions of the mirror. Based on the principle of parametric resonance, a driving voltage signal with a frequency near at 2 ω_0/n—where ω_0 is the natural frequency of the out-of-plane motion and n is an integer ≥ 1—should be used for actuation.

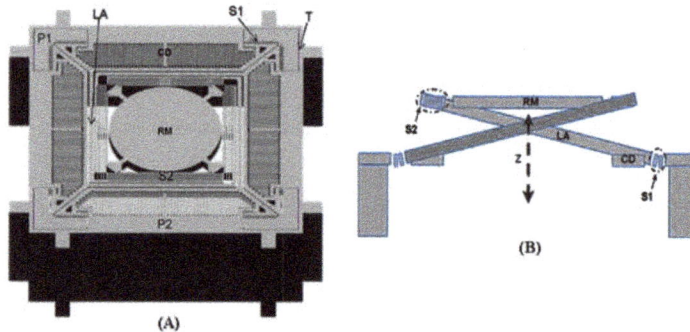

Figure 2. Schematics for out-of-plane translational MEMS scanner with a level-based compliant mechanism. (**A**) Front view of the basic structure; (**B**) cross-section view of the out-of-plane translational motion enabled by the level-based compliant mechanism. Key: CD: comb drive, LA: spiral-like level arm, P1-2: electrical pads, RM: reflective mirror, S1: H-shaped torsional spring, S2: multi-turn folded-beam spring, T: electrical isolation trench.

We optimized the geometry of springs to provide fast stable axial scanning with >400 μm displacement while avoiding spring failure. Figure 3 shows results for modal analysis of the optimized scanner using ANSYS software. The first mode is chosen as the desired out-of-plane translational motion with a resonant frequency of 1216.1 Hz, and the second mode is an in-plane translational motion with a resonant frequency of 4649.3 Hz. According to these results, the optimized structure will provide a stable out-of-plane translational motion with high resistance to parasitic vibrations. Figure 4 shows the stress distribution through the scanner, where there is a maximum value of ~638.2 MPa near at the fixed end of the H-shaped torsional spring with ±250 μm axial displacement. This value is well below the limit for fracture strength of single crystal silicon [21].

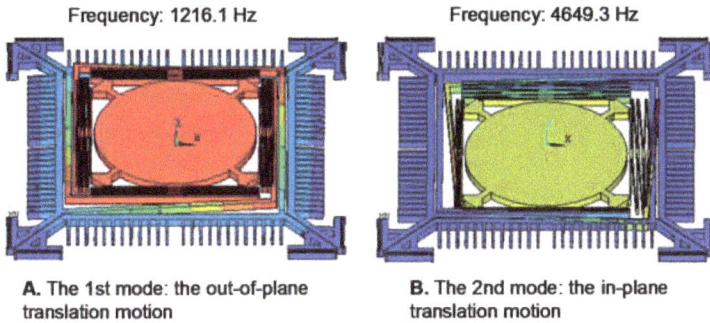

A. The 1st mode: the out-of-plane translation motion

B. The 2nd mode: the in-plane translation motion

Figure 3. Finite element method (FEM) modal analysis. (**A**) The first mode (1216 Hz): the desired out-of-plane translational motion; (**B**) the second mode (4649.3 Hz): the in-plane translation motion.

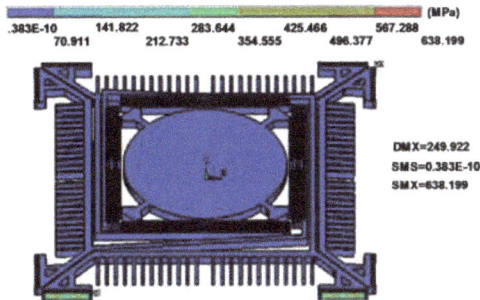

Figure 4. Stress distribution with a ±250 µm out-of-plane vertical displacement. Maximum stress of ~638.2 MPa is found near the fixed end of the H-shaped torsional spring.

2.3. Fabrication Process

A robust SOI micromachining process is developed for fabrication, which achieves a yield of >95%. Figure 5 shows the process flow, which starts with a 4-inch SOI wafer with a 40 µm silicon device layer, a 1 µm silicon dioxide (SiO_2) buried layer, and a 500 µm silicon handle layer. To avoid scratching and contaminating the reflective mirror surface, a 0.5 µm SiO_2 film was used as a hard mask, and was first deposited on the surface of the device layer by a plasma-enhanced chemical vapor deposition (PECVD) process. A 1 µm PECVD SiO_2 layer was also deposited on the backside surface to avoid photoresist burning during the deep reactive-ion etching (DRIE) process for removing backside silicon with a large open area (Figure 5a). Two masks (Figure 5b,c) and two DRIE silicon etching steps (Figure 5d,e) were used to define and form the scanner structures in the device and handle layers of the SOI wafer. The movable structures were released using a buffered hydrofluoric acid solution (BHF) to etch away the SiO_2 layers followed by an isopropyl alcohol (IPA) rinsing and drying. A 70 nm layer of aluminum (Al) film was coated on the device layer to provide >85% reflectivity over the visible and near-infrared spectrum (Figure 5f). This film was also used as the metal contact layer for electrical pads to perform wire bonding.

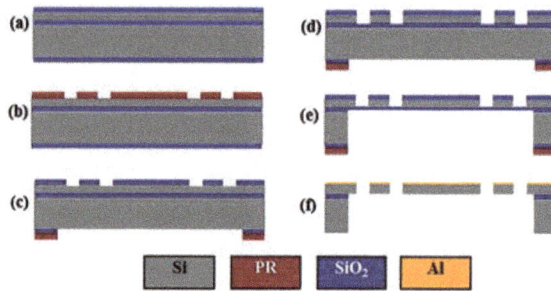

Figure 5. Process flow: (**a**) deposition of plasma-enhanced chemical vapor deposition (PECVD) SiO$_2$ hardmask layers; (**b**) patterning of the front side SiO$_2$ layer; (**c**) patterning of the backside SiO$_2$ layer; (**d**) deep reactive-ion etching (DRIE) of the device layer; (**e**) DRIE of the handle layer; (**f**) SiO$_2$ buffered hydrofluoric acid solution (BHF) release-etching, isopropyl alcohol (IPA) rinsing, IPA drying, and evaporation of Al layer.

3. Performance Characterization

Figure 6 shows scanning electron microscope (SEM) images of a fabricated device. The surface quality of the reflective mirror was characterized by an optical surface profiler (NewView 5000, Zygo, Berwyn, PA, USA). Measurements show that the mirror has a radius of curvature of ~2.6 m and a root mean square (RMS) roughness of ~2 nm.

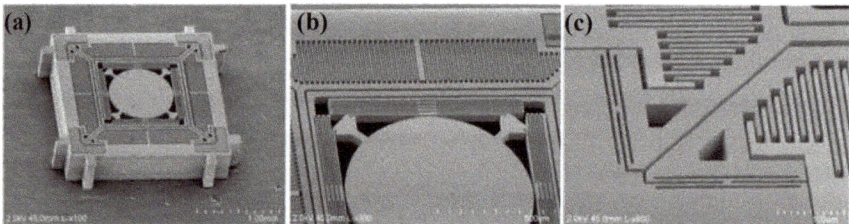

Figure 6. Scanning electron microscope (SEM) images of the fabricated scanner. (**a**) the complete device structure; (**b**) the multi-turn folded-beam spring; and (**c**) the H-shaped torsional spring.

We characterized the dynamic performance of the scanner using a displacement sensor to measure the out-of-plane translational displacement and a position sensing detector (PSD) to determine the tilt angle. Due to the compact geometry of the air damping, the squeeze film effect especially has a significant impact on out-of-plane translation. We reduced damping and achieved high-amplitude out-of-plane translation under ambient conditions by mounting the scanner onto a substrate with a ~0.5 mm-deep open-wall cavity. The scanner was driven into resonance by sweeping the drive frequency of a square-wave voltage at near twice the natural frequency of the out-of-plane translational mode. Figure 7 shows an image of the out-of-plane blur motion of the scanner.

Figure 8 shows that the dynamic response curves of the scanner exhibit a complex dynamic nonlinearity. We observed stiffness softening, mixed softening–hardening and hardening behaviors in the device by adjusting the voltage (Figure 8a) and the duty cycle (Figure 8b) of the square-wave drive signal. A relatively flat response region with large amplitudes (>400 μm) and wide adjustable frequency range (~390 Hz) was observed when forward sweeping the frequency of a drive signal with 80 V and 50% duty cycle, and a maximum amplitude of 480 μm was obtained at ~2.57 kHz. Measurements of the tilt angles about X and Y axes of the mirror over the frequency range for out-of-plane translational motion are also shown (Figure 9). The tilt angle is ≤0.14° for both X and Y axes when driven by a drive signal with 80 V and 50% duty cycle. The response curves for tilting and translation are similar and have the same frequency response range. This result suggests that tilting is not from vibration in other mechanical modes, but rather caused by process variations in the geometry of individual springs, resulting in the asymmetry of the scanner.

Figure 7. Blur motion image of the scanner in out-of-plane translational resonant mode.

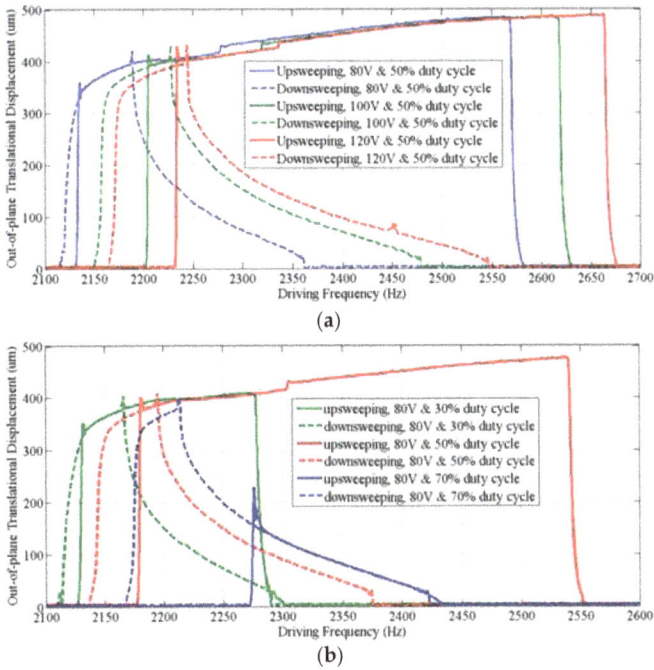

(a)

(b)

Figure 8. Frequency response curves of the out-of-plane translation motion. (**a**) Driven by square-wave signals with a 50% duty cycle at various voltages, the scanner exhibits a stiffness hardening behavior and a stiffness softening–hardening-mixed behavior during frequency upsweep and downsweep of the drive signals, respectively; (**b**) driven with square-wave signals that have different duty cycles at 80 V, the scanner exhibits not only stiffness hardening or softening–hardening-mixed behaviors, but also a stiffness softening behavior with upsweep of the frequency of a drive signal with a 75% duty cycle.

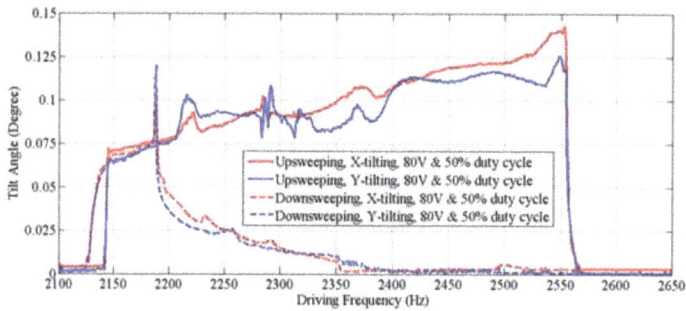

Figure 9. Frequency response curves for tilt angles over the frequency response range of the out-of-plane translational motion. These curves are similar to those for translation, and have the same response frequency range.

4. Imaging Result

Multi-photon excited fluorescence images of mouse colonic epithelium that express tdTomato were collected ex vivo at a frame rate of 5 Hz. Figure 10a shows a representative image obtained in the horizontal (XY) plane over a field of view (FOV) of 270 μm × 270 μm. For horizontal cross-sectional images, only the biaxial MEMS scanner was used to perform 2D Lissajous scanning in the XY plane. The inner X-axis and the outer Y-axis were respectively defined as the fast axis and the slow axis. Figure 10b shows an image of the same specimen obtained in the vertical (XZ) plane over a FOV of 270 μm × 200 μm by using the biaxial MEMS scanner and the out-of-plane translation scanner to perform 1D lateral scanning and axial-scanning, respectively.

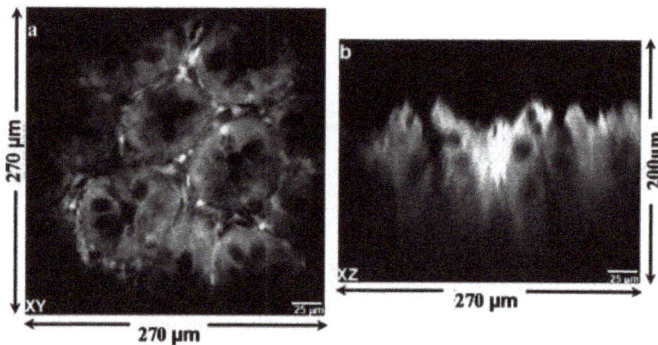

Figure 10. Multi-photon excited fluorescence images of mouse colonic epithelium that constitutively expresses tdTomato ex vivo reveal crypt structure. (**a**) The image in horizontal (XY) plane was collected using the biaxial MEMS scanner only; (**b**) the image in vertical (XZ) plane collected by using both the lateral and axial MEMS scanners.

5. Discussion and Conclusions

Axial scanning is needed to collect optical sections of tissue in the vertical plane, the direction for development of normal epithelium and invasion of disease. Standard objectives or stages that move in this dimension are slow, and collected images are prone to movement artifacts. Using a remotely located axial scanner/mirror is a promising technique that overcomes many limitations of conventional methods. A light-weight compact mirror that performs fast axial scanning may improve performance and extend applicability of this technique to miniature imaging instruments. This work presents a novel compact MEMS out-of-plane translational scanner developed to perform fast axial-scanning for a multi-photon microscopic system with a remote scan architecture. This scanner can achieve a fast (~1.27 kHz) out-of-plane translational motion with large axial displacements ($\geq$400 μm) and slight tilt angles ($\leq$0.14°) at ambient pressure. By employing this scanner and a biaxial MEMS mirror, vertical cross-sectional imaging with a beam axial-scanning range of 200 μm and a frame rate of ~5–10 Hz are enabled. The ability to acquire 3D images is limited because the scanner works in resonant mode only. Optical magnification in the current multi-photon system is ~2:1. This can introduce sensitivity to optical aberrations and difficulty for scanner design. The lever-based compliant mechanism demonstrated in this work can quasi-statically transfer small tilt angles into large pure axial displacements if vertically staggered comb drives are used. Future work will further optimize the structural design and modify the fabrication process to develop a scanner that can work in quasi-static mode.

Acknowledgments: Funding was provided in part by the National Institutes of Health (NIH) R01 EB020644. We thank the staff members of the University of Michigan Lurie Nanofabrication Facility for their support in our fabrication of MEMS scanners.

Author Contributions: H.L., K.R.O. and T.D.W. conceived and designed the experiments; H.L. and X.D. performed the experiments; H.L. analyzed the data; G.L. contributed reagents/materials/analysis tools; H.L. wrote the paper.

Conflicts of Interest: The authors declare no conflict of interest.

References

1. Durst, M.E.; Zhu, G.; Xu, C. Simultaneous spatial and temporal focusing for axial scanning. *Opt. Express* **2006**, *14*, 12243–12254. [CrossRef] [PubMed]
2. Du, R.; Bi, K.; Zeng, S.; Li, D.; Xue, S.; Luo, Q. Analysis of fast axial scanning scheme using temporal focusing with acousto-optic deflectors. *J. Mod. Opt.* **2008**, *56*, 99–102. [CrossRef]
3. Straub, A.; Durst, M.E.; Xu, C. High speed multiphoton axial scanning through an optical fiber in a remotely scanned temporal focusing setup. *Biomed. Opt. Express* **2011**, *2*, 80–88. [CrossRef] [PubMed]
4. Dana, H.; Shoham, S. Remotely scanned multiphoton temporal focusing by axial grism scanning. *Opt. Lett.* **2012**, *37*, 2913–2915. [CrossRef] [PubMed]
5. Grewe, B.F.; Voigt, F.F.; Hoff, M.V.; Helmchen, F. Fast two-layer two-photon imaging of neuronal cell populations using an electrically tunable lens. *Biomed. Opt. Express* **2011**, *2*, 2035–2046. [CrossRef] [PubMed]
6. Jiang, J.; Zhang, D.; Walker, S.; Gu, C.; Ke, Y.; Yung, W.H.; Chen, S.-C. Fast 3-D temporal focusing microscopy using an electrically tunable lens. *Opt. Express* **2015**, *23*, 24362–24368. [CrossRef] [PubMed]
7. Botcherby, E.J.; Juskaitis, R.; Booth, M.J.; Wilson, T. Aberration-free optical refocusing in high numerical aperture microscopy. *Opt. Lett.* **2007**, *32*, 2007–2009. [CrossRef] [PubMed]
8. Botcherby, E.J.; Smith, C.W.; Kohl, M.M.; Débarre, D.; Booth, M.J.; Juškaitis, R.; Paulsen, O.; Wilson, T. Aberration-free three-dimensional multiphoton imaging of neuronal activity at kHz rates. *Proc. Natl. Acad. Sci. USA* **2012**, *109*, 2919–2924. [CrossRef] [PubMed]
9. Rupprecht, P.; Prendergast, A.; Wyart, C.; Friedrich, R.W. Remote z-scanning with a macroscopic voice coil motor for fast 3D multiphoton laser scanning microscopy. *Biomed. Opt. Express* **2016**, *7*, 1656–1671. [CrossRef] [PubMed]
10. Wu, L.; Xie, H. A large vertical displacement electrothermal bimorph microactuator with very small lateral shift. *Sens. Actuators A Phys.* **2008**, *145*, 371–379. [CrossRef]
11. Zhang, X.; Zhou, L.; Xie, H. A fast, large-stroke electrothermal MEMS mirror based on Cu/W bimorph. *Micromachines* **2015**, *6*, 1876–1889. [CrossRef]
12. Qiu, Z.; Pulskamp, J.; Lin, X.; Rhee, C.; Wang, T.; Polcawich, R.; Oldham, K. Large displacement vertical translational actuator based on piezoelectric thin-films. *J. Micromech. Microeng.* **2010**, *20*, 075016. [CrossRef] [PubMed]
13. Zhu, Y.; Liu, W.; Jia, K.; Liao, W.; Xie, H. A piezoelectric unimorph actuator based tip-tilt-piston micromirror with high fill factor and small tilt and lateral shift. *Sens. Actuators A Phys.* **2011**, *167*, 495–501. [CrossRef]

14. Mansoor, H.; Zeng, H.; Chen, K.; Yu, Y.; Zhao, J.; Chiao, M. Vertical optical sectioning using a magnetically driven confocal microscanner aimed for in vivo clinical imaging. *Opt. Express* **2011**, *19*, 25161–25172. [CrossRef] [PubMed]

15. Zeng, H.; Chiao, M. Magnetically actuated MEMS microlens scanner for in vivo medical imaging. *Opt. Express* **2007**, *15*, 11154–11166.

16. Sandner, T.; Grasshoff, T.; Gaumont, E.; Schenk, H.; Kenda, A. Translatory MOEMS actuator and system integration for miniaturized Fourier transform spectrometers. *J. Micro Nanolithogr. MEMS MOEMS* **2014**, *13*, 011115. [CrossRef]

17. Li, H.; Duan, X.; Qiu, Z.; Zhou, Q.; Kurabayashi, K.; Oldham, K.R.; Wang, T.D. Integrated monolithic 3D MEMS scanner for switchable real time vertical/horizontal cross-sectional imaging. *Opt. Express* **2016**, *24*, 2145–2155. [CrossRef] [PubMed]

18. Li, H.; Duan, X.; Wang, T.D. An electrostatic MEMS scanner with in-plane and out-of-plane two-dimensional scanning capability for confocal endoscopic in vivo imaging. In Proceedings of the 2017 IEEE 30th International Conference on Micro Electro Mechanical Systems (MEMS), Las Vegas, NV, USA, 22–26 January 2017; pp. 514–517.

19. Duan, X.; Li, H.; Li, X.; Oldham, K.R.; Wang, T.D. Axial beam scanning in multiphoton microscopy with MEMS-based actuator. *Opt. Express* **2017**, *25*, 2195–2205. [CrossRef]

20. Duan, X.; Li, H.; Zhou, J.; Zhou, Q.; Oldham, K.R.; Wang, T.D. Visualizing epithelial expression of EGFR in vivo with distal scanning side-viewing confocal endomicroscope. *Sci. Rep.* **2016**, *6*. [CrossRef] [PubMed]

21. Chen, K.S.; Ayon, A.; Spearing, S.M. Controlling and testing the fracture strength of silicon on the mesoscale. *J. Am. Ceram. Soc.* **2000**, *83*, 1476–1484. [CrossRef]

micromachines

MDPI

Article

The Exploration for an Appropriate Vacuum Level for Performance Enhancement of a Comb-Drive Microscanner

Rong Zhao [1,2], Dayong Qiao [1,2,*], Xiumin Song [3] and Qiaoming You [4]

[1] Key Laboratory of Micro/Nano Systems for Aerospace, Ministry of Education,
   Northwestern Polytechnical University, Xi'an 710072, China; zhaorong@mail.nwpu.edu.cn
[2] Shaanxi Province Key Laboratory of Micro and Nano Electro-Mechanical Systems,
   Northwestern Polytechnical University, Xi'an 710072, China
[3] Xi'an Zhisensor Technologies Co. Ltd., Xi'an 710077, China; xiumin.song@zhisensor.com
[4] LeadMEMS Science and Technology Ltd., Xi'an 710075, China; qmyou@leadmems.com
* Correspondence: dyqiao@nwpu.edu.cn; Tel.: +86-138-9281-8414

Academic Editors: Kazunori Hoshino and Huikai Xie
Received: 17 February 2017; Accepted: 11 April 2017; Published: 16 April 2017

Abstract: In order to identify the influence of the vacuum environment on the performance of a comb-drive microscanner, and indicate the optimum pressure for enhancing its performance, a comb-drive microscanner fabricated on silicon-on-insulator (SOI) substrate was prepared and tested at different pressures, and the characteristics in vacuum were obtained. The test results revealed that the vacuum environment enhanced the performance in the optical scanning angle, and decreased the actuation voltage. With a 30 V driving voltage applied, the microscanner can reach an optical scanning angle of 44.3° at a pressure of 500 Pa. To obtain an enhancement in its properties, only a vacuum range from 100 to 1000 Pa is needed, which can be very readily and economically realized and maintained in a vacuum package.

Keywords: microscanner; optical scanning angle; vacuum operation; optimum pressure

1. Introduction

With the development of micro-electro-mechanical system (MEMS) technology, MEMS devices have been used in many fields, such as RF-MEMS, optical-MEMS, sensors, energy harvesters, and bio-MEMS. Among them, the microscanner, a promising optical-MEMS device, is widely used in LiDAR [1,2], pico-projectors [3], barcode readers [4], VR (Virtual Reality)/AR (Augmented Reality) applications [5], and so on. Mainstream actuation techniques used in microscanners are electrostatic [6], electromagnetic (EM) [7], piezoelectric [8], and electrothermal [9] actuations, and the drive forces are electrostatic forces, Lorentz or magneto-static forces, piezoelectric effects, and metal thermal effects, respectively. In the case of thermal actuation, two or more materials with different thermal expansion are used to achieve mechanical actuation. Although the thermal bimorph actuator provides a large static mechanical force at a relatively low driving voltage, the long thermal response time and non-resonant mode limit its application. Piezoelectric actuators can respond rapidly to driving signals, but the complicated fabrication of piezoelectric materials increases the difficulty in the development of these microscanners. Compared with thermal actuation and piezoelectric actuators, electrostatic and electromagnetic actuators are considered to be more suitable for microscanners because of the rapid response to the driving signal and their relatively high resonant frequency. Generally, electromagnetic actuators can offer a large driving force, but the deposited coils and permanent magnets they have result in a bulky packaging. In contrast, even though electrostatic actuators need a relatively high driving voltage, the simple and compact structure, the moderate scanning angle, and the comparatively simple fabrication process make them attract more interest in driving microscanners.

The application to display devices requires high-performance scanners, which should have high frequencies and large scanning angles to achieve good display quality. For all resonant microscanners based on different actuation mechanisms, the oscillation amplitude is determined by the input energy and loss. Air damping generates a large loss in all microscanners [10]. Especially for electrostatic microscanners, which have a structure of comb fingers and a mirror plate, they suffer from slide-film damping and squeeze-film damping. To achieve a large scanning angle, enhancing the input energy or reducing the loss energy (mainly caused by air damping) is expected. Therefore, two methods have been used to meet the requirement of a low driving voltage and a large scanning angle in its application: adopting hybrid actuation mechanisms to drive the microscanner [11] (enhancing input) and vacuum packaging [3,12] (reducing loss). However, the hybrid actuation combined electrothermal actuators and electromagnetic actuators, which is complex in fabrication and control. As for vacuum packaging, it would not change the device's structure. Additionally, it is effective for decreasing the driving voltage [13] and promoting the scanning angle and quality factor [14,15]. Therefore, vacuum packaging seems to be an ideal way to enhance the performance of the microscanner. Compared to other types of actuators, the amplitude of the electrostatic actuator is more obviously affected by vacuum packaging.

As stated by the description before, a vacuum-packaged electrostatic microscanner is appropriate for display devices. However, the earlier results showed that the oscillation frequency range has been decreased at high vacuums levels [13–15], which leads to an instability of the oscillation frequency, especially when temperature varies [16]. Although many reports agree that a high-level vacuum package will enhance the scanning angle of the microscanner, the narrow frequency range, the difficult sealing technology, and the high cost cannot be ignored. Furthermore, thermal management is a critical issue in high-vacuum packaging. Without a convection medium, such as air, to assist in heat dissipation, the thermal energy induced in the mechanical movement will accumulate on the device and affect its mechanical properties. Furthermore, the dynamic response and reliability could worsen, even though the vibration angle might increase. In order to obtain a relatively good performance, considering the leverage on stability, as well as on cost, in this paper we investigated the detailed characteristics of the microscanner in a vacuum to explore the appropriate vacuum level as used for packaging. The properties of the frequency response, excitation voltage, and the change of the stable and unstable regions are studied in atmospheric air and vacuum.

2. Materials and Methods

2.1. Device Description

The microscanner used in the vacuum test consists of a reflection mirror, a movable frame, and an electrostatic comb-drive actuator. The mirror and frame are connected with torsion beams. The scanning electron micrograph of the microscanner is shown in Figure 1. It is fabricated on SOI wafers with the process flow illustrated in Figure 2.

The fabrication process begins with an SOI wafer. A 454 μm-thick SOI wafer, which has a 4-μm-thick buried oxide layer and a 50-μm-thick device layer, is used. The first step of the process is making an isolation trench by photolithography and inductively-coupled plasma (ICP) etching with a polymer photoresist mask, and then the trench is filled with polysilicon by low pressure chemical vapor deposition (Figure 2a). Secondly, the photoresist polymer film is removed by O_2 plasma and the polysilicon over the wafer was then removed by chemical-mechanical polishing before depositing an aluminum film over the handle layer. After that, aluminum is deposited on the back of the substrate and patterned by wet etching. Then, ICP etching is used to form the back cavity using the aluminum mask (Figure 2b). Thirdly, the residual aluminum mask film on the substrate is removed and, to obtain the reflective film, aluminum is sputtered onto the device surface (Figure 2c). Fourthly, the photolithography proceeds and the device layer is etched to form the device structures, including the inner mirror, comb fingers, and outer frame (Figure 2d). Finally, the remaining photoresist polymer is cleaned, and the buried oxide underneath the moving parts is removed in a HF solution to form the movable mirror structure (Figure 2e).

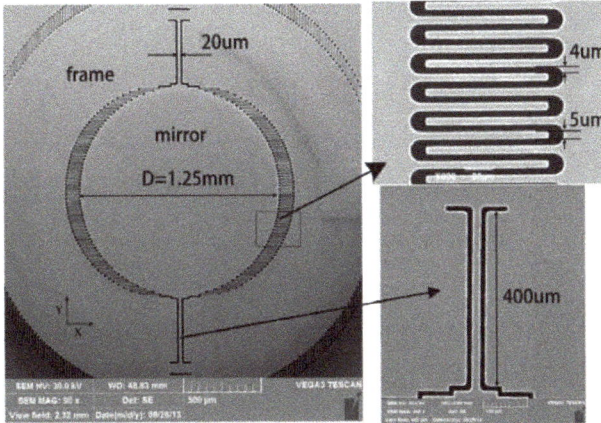

Figure 1. The scanning electron micrograph of the micro mirror structure. (**Top right**) The structure of comb finger; and (**bottom right**) the larger version of the torsion beam.

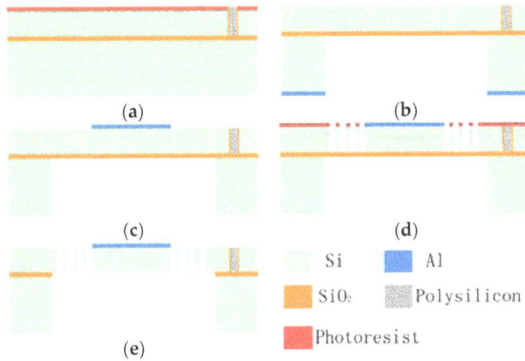

Figure 2. Fabrication process of the microscanner. (**a**) Isolation trench fabrication; (**b**) the back is etched to form the back cavity; (**c**) sputtering of the reflective film; (**d**) the front is etched to form the comb and structure; and (**e**) removal of the buried oxide to release the movable structures.

The torsion equation of motion of the microscanner is given by:

$$T_e = I\ddot{\theta} + b\dot{\theta} + K_\theta \theta \tag{1}$$

where θ is the torsion angle of the microscanner, b is the damping coefficient, T is the torque applied, K_θ is the elastic coefficient, and I is the moment of inertia, respectively.

The equation is a typical parametrically-excited system; thus, the comb-drive microscanner is a typical nonlinear parametric system [17]. The performance characteristics of the microscanner sweeping from different directions of frequency will demonstrate a hysteresis effect, and there will be two hopping frequencies, f_1 and f_2 ($f_2 > f_1$), in the frequency response curve, as shown in Figure 3a. The interval between the two jump frequencies f_1 and f_2 is the unstable region [18]. In the unstable region, oscillations can only be observed if the external frequency is swept down to this region from f_2, but when the frequency is swept up from f_1, no oscillation occurs. In contrast, in the stable region, the oscillation happens irrespective of the sweep direction.

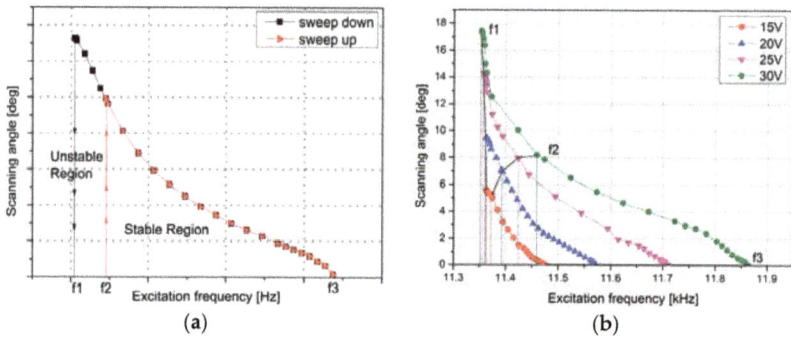

Figure 3. (a) Microscanner frequency response curves. (b) The frequency response curves with different driving voltages at atmospheric pressure of the microscanner used in this work.

The driving principle of the electrostatic microscanner is shown in Figure 4. A square wave is commonly used as the excitation signal, and the switch-off time of the driving signal coincides with the moment the mirror plates pass the resting position. When the voltage pulse ends, the plate swings back by inertia. The movement is then only guided by mechanical properties (spring stiffness and the moment of inertia of the plate). The next pulse starts at maximum deflection and ends again at the rest position. The microscanner can only work in resonance, and to achieve the largest scanning angle at a fixed driving voltage, it needs to be excited by a signal with a frequency near twice its natural frequency of the torsion mode [4]. Other signals, like triangular waves, sawtooth waves, or sine waves, can also be employed as the excitation signal for the microscanner, and the oscillation amplitudes of the microscanner excited by them are slightly smaller than the amplitude excited with a square wave at the same applied voltage [6]. In this work, the microscanner was excited by a square wave at 30 V.

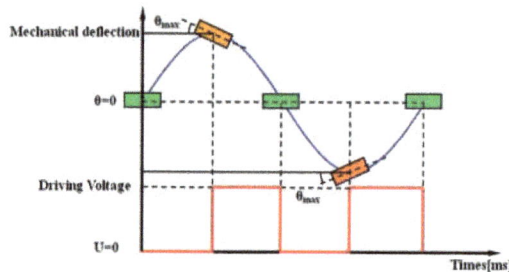

Figure 4. Driving principal of the microscanner. The frequency of the driving signal is double that of the microscanner.

The microscanner will reach the maximum scanning angle θ_{0r} when the resonance frequency is its natural frequency ω_r, with the following equation [18]:

$$\theta_{0r} = T_0/\omega_r b \tag{2}$$

where T_0 is the electrostatic torque generated by the comb fingers, and b is the damping coefficient. The optical scanning angle is a very important parameter to the microscanner, which directly determines the size of the image and the required driving voltage range. One of the properties to indicate superior performance of the device is that the device can obtain a larger scanning angle with a smaller actuation voltage. From Equation (2), it is showed that one way to enhance the scanning performance is to lower the damping coefficient.

Air damping is a significant factor influencing the performance of the microscanner [10]. Thus, theoretically, in a vacuum environment the device can reach a larger scanning scale than that at atmospheric pressure with the same actuation conditions.

2.2. Principles

For the resonant electrostatic microscanners driven by a square wave, the input energy comes from the driving voltage. At the rest position, the total energy provided by square wave is expressed as:

$$E = C_0 V^2 \tag{3}$$

where C_0 is the capacitance at the rest position (the maximum capacitance) and V is the voltage. In an oscillation cycle, the loss energy of the electric field is calculated by the following equation:

$$\Delta E^R = \begin{cases} C_0 V^2 \frac{\theta_0}{\theta_c}, \theta_0 \leq \theta_c \\ C_0 V^2, \theta_0 > \theta_c \end{cases} \tag{4}$$

where θ_0 is the torsion angle at voltage V, and θ_c is the angle where the fixed and movable comb fingers have no overlap. Since the quality factor Q is defined as the ratio between the kinetic energy of the mirror and the loss energy, Q is obtained by the equation:

$$Q = \frac{\pi I \omega^2 \theta_0^2}{\Delta E^R} \tag{5}$$

where ω is the oscillation frequency. In a low-damping system, the quality factor is the amplitude response at the resonant frequency, which can be estimated as:

$$Q = \frac{I \omega}{b} \tag{6}$$

From the above equations, the relationship between the voltage and the torsion angle can be roughly calculated as:

$$\theta_0 = \begin{cases} \frac{C_0 V^2}{b \pi \omega \theta_c}, \theta_0 \leq \theta_c \\ V \sqrt{\frac{C_0}{b \pi \omega}}, \theta_0 > \theta_c \end{cases} \tag{7}$$

A microscanner with large scanning angle and a low driving voltage is expected, which means a large ratio of θ_0/V is needed. Generally, the situation of $\theta_0 > \theta_c$ is considered; thus, the relationship between θ_0/V and b is expressed below: because the damping coefficient b is greater than zero, θ_0/V is monotonically decreased with b.

$$\frac{\theta_0}{V} = \sqrt{\frac{C_0}{b \pi \omega}}, \theta_0 > \theta_c \tag{8}$$

Equation (8) shows a nonlinear relationship between θ_0/V and b; as b is decreased very low, the increase of θ_0/V becomes imperceptible. Thus, an extreme vacuum degree may be unnecessary because the vibration angle would approach a constant value behind some certain vacuum degrees. For this reason, a balance between the vacuum level and the packaging cost needs to be explored.

In the case of a high vacuum (less than 1000 Pa), collisions with gas particles are the dominant damping mode, and the interaction between gas molecules is neglected [19]. The damping force is generated by the interaction between gas molecules and moving structures, and this includes the torque from the mirror plate and comb fingers. Figure 5 shows the schematic diagram of the microscanner used in this work. The torque generated by the pressure difference between the front and back surfaces of the mirror plate is estimated as [14]:

$$T_{plate} = \frac{\pi h R^3 P_i \dot{\theta}}{c} \left\{ \frac{R}{4h} \left[(2 - \sigma_n) \cdot \left(\frac{2}{\sqrt{\pi}} + 1 \right) + \sigma_n \sqrt{\frac{\pi T_w}{T_i}} \right] + \frac{1}{\sqrt{\pi}} \sigma_t \right\} \tag{9}$$

where h is the thickness of the mirror plate, R is the radius of the circular mirror plate, P_i is the environment pressure, $\dot{\theta}$ is the angular velocity ($d\theta/dt$), c is the thermal velocity of the gas molecules, σ_n and σ_t are the normal and tangential accommodation coefficients, respectively, T_w is the wall temperature, and T_i is the ambient temperature. The value of c is calculated from the gas molecules m, the ambient temperature T_i, and the Boltzmann constant k_B as:

$$c = \sqrt{2k_B T_i / m} \tag{10}$$

In addition, the torque generated by the molecular collisions on the comb fingers is calculated by the shear stress along the outside edges $\tau_{sidewall}$ and the front and rear edges τ_{edge} [19]. The shear stresses of the comb fingers are obtained by:

$$\tau_{sidewall} = N \frac{l_c \sigma_t P_i \dot{\theta}}{2c\sqrt{\pi}}, \tau_{edge} = N \frac{(l_c + R \sin \varphi) \sigma_t P_i \dot{\theta}}{c\sqrt{\pi}} \tag{11}$$

Then the torque T_{comb} is given by:

$$T_{comb} = \frac{N \cdot 4h}{2c\sqrt{\pi}} \int_0^{\pi/2} (3l_c \sigma_t P_i \dot{\theta} + 2R \sin \varphi \sigma_t P_i \dot{\theta}) R \sin \varphi d\varphi = \frac{2Nh\sigma_t P_i \dot{\theta}}{c\sqrt{\pi}} (3l_c R + \frac{\pi}{2} R^2) \tag{12}$$

In the above equations, the parameters not mentioned are denoted in Figure 5. Thus, the total damping torque T can be obtained by:

$$T = T_{plate} + T_{comb} = \frac{\pi h R^3 P_i \dot{\theta}}{c} \left\{ \frac{R}{4h} \left[(2 - \sigma_n) \cdot \left(\frac{2}{\sqrt{\pi}} + 1 \right) + \sigma_n \sqrt{\frac{\pi T_w}{T_i}} \right] + \frac{1}{\sqrt{\pi}} \sigma_t \right\}$$
$$+ \frac{2Nh\sigma_t P_i \dot{\theta}}{c\sqrt{\pi}} (3l_c R + \frac{\pi}{2} R^2) \tag{13}$$

Concerning an isothermal system ($T_w = T_i$), full momentum accommodation ($\sigma_n = 1$, $\sigma_t = 1$), the total torque T generated by the gas collisions is simplified to:

$$T = P_i \dot{\theta} \left[\frac{\sqrt{\pi} R^4}{4c} (2 + \pi + \sqrt{\pi} + 4h) + \frac{2Nh}{c\sqrt{\pi}} (3l_c R + \frac{\pi}{2} R^2) \right] \tag{14}$$

Since $T = b \times d\theta/dt$, and the quality factor Q at high vacuum is $Q = I\omega/b$, the quality factor in a high vacuum Q_v is expressed as:

$$Q_v = I\omega/P_i \left[\frac{\sqrt{\pi} R^4}{4c} (2 + \pi + \sqrt{\pi} + 4h) + \frac{2Nh}{c\sqrt{\pi}} (3l_c R + \frac{\pi}{2} R^2) \right] \tag{15}$$

In the case of a low vacuum (from 1000 Pa to 10^5 Pa), the friction damping generated by the viscous flow of ambient air becomes important. The quality factor in ambient air Q_a is calculated by Equation (5); thus, the loss energy is acquired.

The loss energy contains the loss of the mirror plate and comb fingers, which is estimated by [14,20]:

$$L_{plate} = \pi^2 \omega h R^3 \theta_0^2 \left(\frac{\omega \rho \eta}{2} \right)^{1/2} \left(1 + \frac{R}{2h} \right) \tag{16}$$

$$L_{comb} = \frac{2N\pi h\theta_0^2 \eta_{eff}\omega}{3g}\left(3l_c R + \frac{\pi}{2}R^2\right) \tag{17}$$

where ρ is the air density and η is the dynamic viscosity of air in Nsm^{-2}. At room temperature (300 K), the dynamic viscosity of air is 18.714×10^{-6} Nsm^{-2}. The η_{eff} is the effective dynamic viscosity of air, valued as $\eta/(1 + 9.658\, K_n^{1.159})$, where K_n is the Knudsen number [15].

Now, the quality factor in low vacuum is expressed as:

$$Q_a = \frac{I\omega}{\pi^2 h R^3 \left(\frac{\omega\rho\eta}{2}\right)^{1/2}\left(1 + \frac{R}{2h}\right) + \frac{2N\pi h\eta_{eff}}{3g}\left(3l_c R + \frac{\pi}{2}R^2\right)} \tag{18}$$

The equations above roughly analyzed the influence of air damping. Reviewing the analysis process, the impact of the vacuum is discussed in high vacuum (less than 1000 Pa) and in low vacuum (from 1000 Pa to 10^5 Pa) separately, owing to the different damping effect in each region. Therefore, the obtained equations can be useful for the rough explanation of the experimental results.

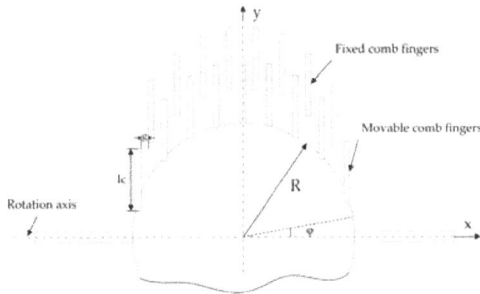

Figure 5. The geometry of the microscanner for calculations.

2.3. Experimental Procedure

To investigate the characteristics of the fabricated microscanner in a vacuum, a laser triangulation method is used in this paper, which includes changing the voltage and the pressure, testing the scanning amplitude, and transforming that into the optical scanning angle [21].

The specific steps are as follows: Firstly, the microscanner was mounted on a support base in the vacuum chamber, and illuminated by a 532 nm green semiconductor laser, while the reflected light was received by the indicated screen. Then, the AC excitation signal was applied to the microscanner to make it deflect at a certain frequency. Since the naked eye cannot distinguish the scanning spot, the trail of the reflected laser spot will produce a scanning line, which can be measured and used to calculate the optical scanning angle using the following equation:

$$\theta = \arctan\left(\frac{H + L/2}{S}\right) - \arctan\left(\frac{H - L/2}{S}\right) \tag{19}$$

where θ is the optical scanning angle of the mirror; L is the length of the laser scanning line after scanning; H is the distance between the fixed center point of the scanning line and the datum point; and S is the distance between the scanning mirror device and the receiving screen, respectively.

The measurement setup of the experiment is shown in Figure 6. When the excitation voltage and pressure are determined, the amplitude is tested through changing the frequency of the microscanner, and the corresponding amplitudes are measured and recorded. In this paper, two types of experiments were implemented. The first was carried out by sweeping the excitation frequency under given voltages and pressures to obtain the scanning angle θ_f. The second was carried out by sweeping the pressure under given voltages and frequencies to obtain the maximum scanning angle θ_p.

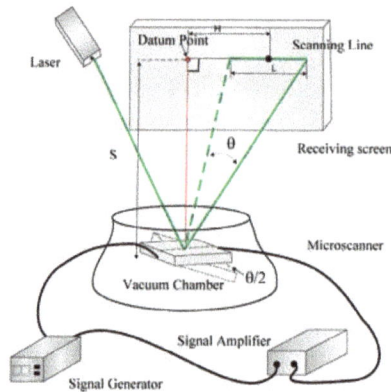

Figure 6. Schematic of the setup used in the vacuum test.

According to the positive correlation between the excitation voltage and the optical scanning angle shown in Figure 7, this experiment was performed to test microscanners at different excitation voltages, starting at 10 V, and ending at 30 V under different pressures.

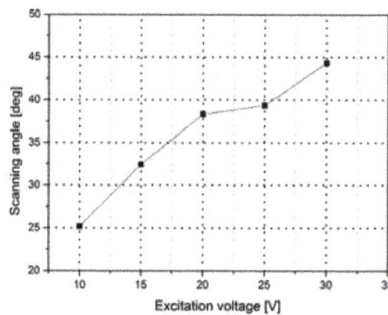

Figure 7. The maximum scan amplitude vs. excitation voltage.

3. Results

3.1. Time Response in a Vacuum

In a vacuum, when the excitation output signal is turned off, the scanning mirror does not immediately stop because, as the pressure reduces, the air damping decreases. The settling time in mode switching was tested in the experiment. Figure 8 shows the results of the settling time versus pressure at different excitation voltages. The settling time of the microscanner from dynamic to static states at 1 Pa is more than 10 s, longer than the time at atmospheric pressure. When the pressure is higher than 100 Pa, there is hardly any settling time from oscillation to stop states, and the settling time at higher pressure is less than 0.41 s.

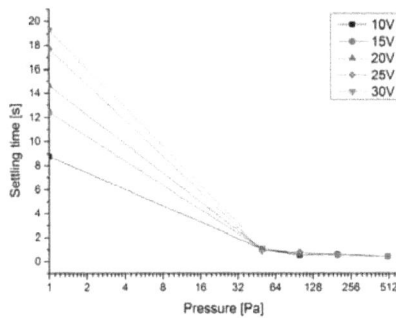

Figure 8. The settling time required for the microscanner to stop versus the pressure at different excitation voltages. The lower the pressure is, the longer time that is needed for the microscanner to stop.

The settling time is high in the case of the vacuum-packaged microscanner compared to the non-vacuum-packaged microscanner. The dynamic motion of a mechanical system is affected by the sum of the structural damping and the damping of the surrounding medium in which the structure moves. Therefore, the system takes a longer time to settle in a vacuum because of the absence of air damping.

3.2. The Minimum Actuation Voltage in a Vacuum

For the microscanner, a low actuation voltage is expected. A small required driving voltage means a larger driving voltage range that the device can operate in and lower energy consumption. In this experiment, the frequency versus actuation voltage was tested. Figure 9 shows that the test curve is similar to the macroscopic "tongue" shape. The "tip of the tongue" part represents the minimum voltage to drive the microscanner, and the driving frequency is applied irrespective of the sweep direction. The minimum driving voltage of the mirror is lower than 8 V in a vacuum environment at a pressure ranging from 1 to 1000 Pa. In contrast, the required driving voltage at atmospheric pressure is 12.5 V.

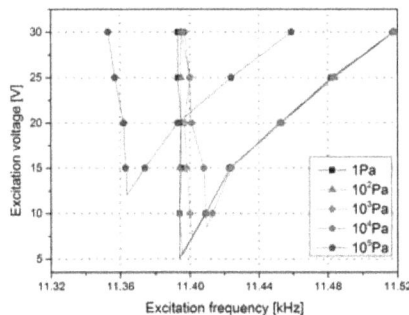

Figure 9. The excitation frequency versus the excitation voltage at different pressures.

3.3. The Range of Excitation Frequency in Vacuum

In a vacuum environment, Figure 10 shows that the range of the stable and unstable regions have been slightly influenced by the change of pressure, but it is broader than that at atmospheric pressure with the same excitation voltages applied. For the unstable region, when the pressure is higher than 1000 Pa, the range of the region becomes narrower with the increasing pressure.

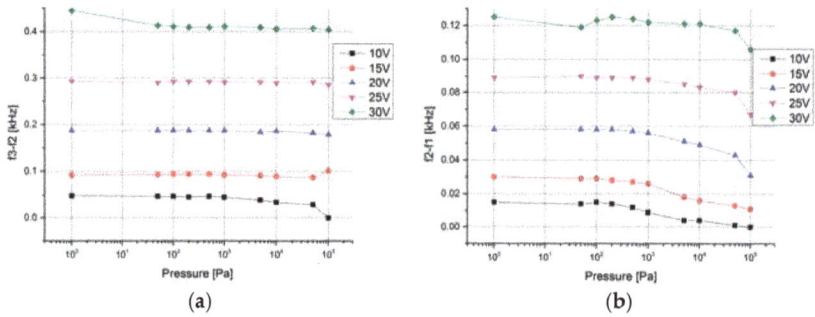

Figure 10. The change with pressure of the stable region and the unstable region. (**a**) The range change of the stable region. (**b**) The range change of the unstable region.

3.4. Characteristics in the Stable Region

In the stable region, as shown in Figure 11, the optical scanning angle has a negative correlation with pressure. Both the maximum optical scanning angle (θ_p) and the scanning angle at fixed frequencies (θ_f) have an improvement at lower pressure. Figure 11a shows the results of the tests under fixed frequencies: by applying the actuation voltage of 30 V, the scanning angle increased from 5.57° at atmospheric pressure to 8.74° at 500 Pa, raised by 56.96%; and with the voltage of 15 V applied, the angle increased by 352.13% from 1.93° to 8.74°. Figure 11b shows the change of the maximum optical scanning angle related to the pressure and actuation voltage. When the pressure ranges from 1 to 100 Pa, nearly all of the maximum scanning angles are above 8°, and with the decrease of the vacuum level, the influence of the actuation voltage on the scanning angle rises. The percentage of the change is inversely proportional to the actuation voltage. Here, a plateau zone has been found in Figure 11, ranging from 1 to 1000 Pa. When the vacuum level is in this zone, the microscanner will have an economical and expected scanning angle with the designed actuation voltage in the stable region.

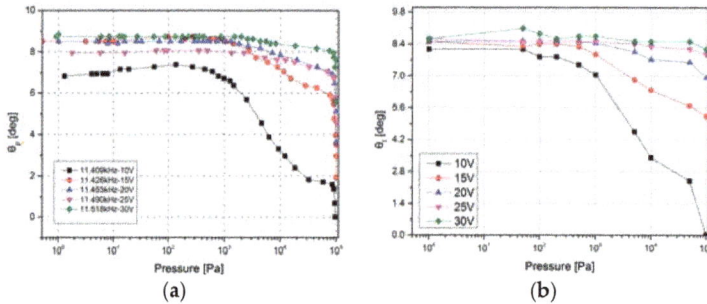

Figure 11. The optical scanning angle vs. pressure in the stable region. (**a**) The optical scanning angle vs. pressure with a fixed excitation frequency under different actuation voltages. (**b**) The maximum optical scanning angle vs. pressure under different actuation voltages.

3.5. Characteristics in the Unstable Region

In the unstable region, Figure 12 shows that the optical scanning angle does not change monotonically with the pressure. When the pressure is lower than a critical value P_c (P_c ranges from 100 to 1000 Pa, and is dependent on the excitation voltage), the optical scanning angle increases with the increasing air pressure, but when the pressure is higher than P_c, the optical scanning angle decreases with the increasing pressure. There is a peak value in the maximum optical scanning angle vs. pressure curve, and the higher the voltage is, the sharper the peak. Figure 12b represents the change of the maximum optical scanning angles in accordance with the change of pressure. At pressure P_c, the maximum optical scanning angle is 44.32° when 30 V is applied, 38.34° when 20 V is applied, and 32.44° when 15 V is applied, and at atmospheric pressure, the angles are 17.16°, 9.42°, and 5.57°, respectively.

Vacuum packaging is typically reasonable to increase the scanning angle of the microscanner working in the unstable region, and only a vacuum level ranging from 100 to 1000 Pa is needed to obtain the maximum optical scanning angle. This kind of vacuum level is readily achieved in vacuum packaging and is also economical to be maintained over the lifetime of the device.

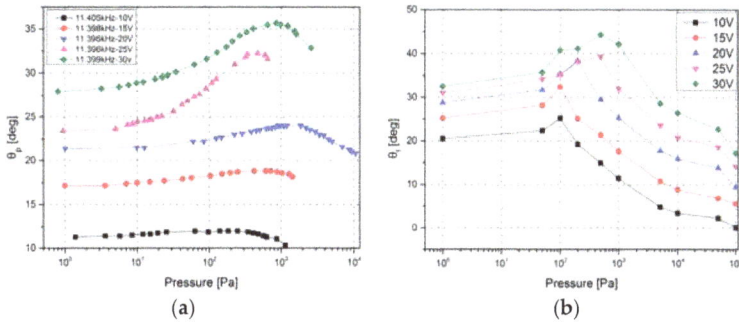

Figure 12. Theoretical scanning angle vs. pressure in the unstable region. (**a**) The optical scanning angle vs. pressure with a fixed excitation frequency under different actuation voltages. (**b**) The maximum optical scanning angle vs. pressure under different actuation voltages.

4. Discussion

The response time, the minimum actuation voltage, the optical scanning angle, and the range of the stable and unstable regions are indicators of the property values of the microscanner. An expected microscanner needs to have a larger projection scale, lower power consumption, and faster settling time. One method to optimize the performance and lower the power required is to reduce the air damping inside the packaged microscanner. According to the experiment results, at a pressure of 500 Pa, the maximum optical scanning angle is enhanced from 17.16° to 44.32°. From Figures 11 and 12, a plateau zone ranging from 100 to 1000 Pa has been obtained. In this zone, the behaviors of the scanning angle in both stable and unstable regions have a larger promotion than at atmospheric pressure. Additionally, the high vacuum also decreases the minimum actuation voltage from 12.5 V at atmospheric pressure to below 8 V at a pressure lower than 1000 Pa. A high vacuum level has a significant influence on the settling time: greater than ten seconds is needed for the microscanner to settle down at 1 Pa. However, at a pressure higher than 100 Pa, the settling time can be neglected.

On the other hand, the quality factor Q of the microscanner has been calculated from the experiment results (θ and ω) at different driving voltages, using Equations (15) and (18), shown in Figure 13. The measured values of Q are positively correlated with the driving voltages, but negative with the pressure. Additionally, the quality factor of the microscanner with a low driving voltage is more sensitive to pressure: the value of Q increases by about two orders of magnitude when compared with that in atmospheric air in the case of a 10 V driving voltage. With the pressure decreasing from 10^5 to 1000 Pa, the values of Q increase. The growing trends start to stabilize from at 1000 Pa and below. In comparison with the measured values, the theoretical Q values are calculated by Equations (15) and (18) with an average energy input in the high and low vacuum, respectively. Although the theoretical values of Q obtained by the equations are underestimated, the same segmentation point of the trend (1000 Pa) is valuable. The curve slope in the range from 100 Pa to 1000 Pa roughly fits with the measured values; thus, the principal based on molecular collision can explain the damping mechanism. When the pressure is lower than 10 Pa, the measured results show independence with pressure, and the loss is mainly caused by thermal dissipation or intrinsic damping.

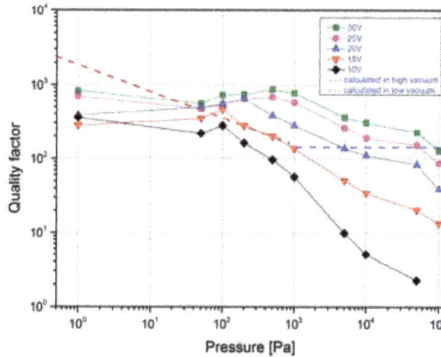

Figure 13. Quality factors of the microscanner measured at different driving voltages and calculated as a function of pressure.

Above all, the pressure of maintaining optimal properties is a medium vacuum level in the experiment, ranging from 100 to 1000 Pa, and providing a medium vacuum to meet the large-sized projection is particularly easy and economical to implement in an industrial package.

5. Conclusions

This paper emphatically expounds the performance testing of a microscanner at different pressures to explore the optimum pressure in improving the performance of the device. The test results revealed that the vacuum package is especially reasonable to enhance the scanning angle for the microscanner working in the unstable region. The device can achieve a 44.3° optical scanning angle at 30 V, requiring almost no settling time at pressures ranging from 100 to 1000 Pa, and the performance of the microscanner under these conditions is improved. Thus, a larger projection size is achieved in a medium vacuum environment to satisfy the demands of business projection, which is very readily and economically realized and maintained in the vacuum package. The results of the microscanner experiment in vacuum provide a reference for the vacuum packaging.

Acknowledgments: This research was sponsored by the National Natural Science Foundation of China (Grant Nos. 51375399 and 51375400), the Fundamental Research Funds for the Central Universities (3102014KYJD023) and NPU Foundation for Fundamental Research (Grant No. JCY20130119).

Author Contributions: Rong Zhao and Xiumin Song conceived and designed the experiments; Xiumin Song performed the experiments; Rong Zhao analyzed the data; Dayong Qiao contributed analysis tools; and Rong Zhao, Xiumin Song, and Qiaoming You wrote the paper.

References

1. Hofmann, U.; Senger, F.; Soerensen, F.; Stenchly, V.; Jensen, B.; Janes, J. Biaxial resonant 7mm-MEMS mirror for automotive LIDAR application. In Proceedings of the 2012 International Conference on Optical MEMS and Nanophotonics (OMN), Banff, AB, Canada, 6–9 August 2012; pp. 150–151.
2. Kasturi, A.; Milanovic, V.; Atwood, B.H.; Yang, J. UAV-borne lidar with MEMS mirror-based scanning capability. In Proceedings of the Conference on SPIE Defense+ Security, International Society for Optics and Photonics, Baltimore, MD, USA, 17–21 April 2016; Volume 9832.

3. Hofmann, U.; Senger, F.; Janes, J.; Mallas, C.; Stenchly, V.; von Wantoch, T.; Quenzer, H.J.; Weiss, M. Wafer-level vacuum-packaged two-axis MEMS scanning mirror for pico-projector application. In Proceedings of the SPIE 8977, MOEMS and Miniaturized Systems XIII, San Francisco, CA, USA, 1–6 February 2014.

4. Wolter, A.; Schenk, H.; Gaumont, E.; Lakner, H. MEMS microscanning mirror for barcode reading: From development to production. In Proceedings of the SPIE 5348, MOEMS Display and Imaging Systems II, San Jose, CA, USA, 24 January 2004; pp. 32–39.

5. Milanović, V.; Kasturi, A.; Yang, J.; Hu, F. A fast single-pixel laser imager for VR/AR headset tracking. In Proceedings of the SPIE 10116, MOEMS and Miniaturized Systems XVI, San Francisco, CA, USA, 28 January 2017.

6. Ataman, C.; Urey, H. Modeling and characterization of comb-actuatedresonant microscanners. *J. Micromech. Microeng.* **2006**, *16*, 9–16. [CrossRef]

7. Tenghsien, L.; Chingfu, T. Design, fabrication, and evaluation of vacuum testing of a novel electromagnetic microactuator. *J. Micro/Nanolith. MEMS MOEMS* **2011**, *10*, 043001.

8. Naono, T.; Fujii, T.; Esashi, M.; Tanaka, S. A large-scan-angle piezoelectric MEMS optical scanner actuated by a Nb-doped PZT thin film. *J. Micromech. Microeng.* **2014**, *24*, 015010. [CrossRef]

9. Morrison, J.; Imboden, M.; Little, T.D.C.; Bishop, D.J. Electrothermally actuated tip-tilt-piston micromirror with integrated varifocal capability. *Opt. Express* **2015**, *23*, 9555–9566. [CrossRef] [PubMed]

10. Xia, C.; Qiao, D.; Zeng, Q.; Yuan, W. The squeeze-film air damping of circular and elliptical micro-torsion mirrors. *Microfluidics Nanofluidics* **2015**, *19*, 585–593. [CrossRef]

11. How, K.K.; Lee, C. A two-dimensional MEMS scanning mirror using hybrid actuation mechanisms with low operation voltage. *J. Microelectromech. Syst.* **2012**, *21*, 1124–1135.

12. Hofmann, U.; Oldsen, M.; Quenzer, H.; Janes, J.; Heller, M.; Weiss, M.; Fakas, G.; Ratzmann, L.; Marchetti, E.; D'Ascoli, F.; et al. Wafer-level vacuum packaged resonant micro-scanning mirrors for compact laser projection displays. In Proceedings of the Conference on MOEMS and Miniaturized Systems VII, San Jose, CA, USA, 8 February 2008; p. 688706.

13. Tachibana, H.; Kawano, K.; Ueda, H.; Noge, H. Vacuum wafer level packaged two-dimensional optical scanner by anodic bonding. In Proceedings of the IEEE 22nd International Conference on Micro Electro Mechanical Systems, Sorrento, Italy, 25–29 January 2009.

14. Manh, C.H.; Kazuhiro, H. Vacuum operation of comb-drive micro display mirrors. *J. Micromech. Microeng.* **2009**, *19*, 105018. [CrossRef]

15. Chu, H.M.; Kazuhiro, H. Design, fabrication and vacuum operation characteristics of two-dimensional comb-drive micro-scanner. *Sens. Actuators A Phys.* **2011**, *165*, 422–430. [CrossRef]

16. Ishikawa, N.; Kentaro, I.; Renshi, S. Temperature dependence of the scanning performance of an electrostatic microscanner. *J. Micromech. Microeng.* **2016**, *26*, 035002. [CrossRef]

17. Ataman, C.; Urey, H. Nonlinear frequency response of comb-driven microscanners. In Proceedings of the SPIE on MOEMS Display and Imaging Systems II, San Jose, CA, USA, 24 January 2004; Volume 5348, pp. 166–174.

18. Conant, R.A. Micromachined Mirrors. Ph.D. Dissertation, UC Berkeley, Berkeley, CA, USA, 2002; pp. 9–11.

19. Martin, M.J.; Houston, B.H.; Baldwin, J.W.; Zalalutdinov, M.K. Damping Models for Microcantilevers, Bridges, and Torsional Resonators in the Free-Molecular-Flow Regime. *J. Microelectromech. Syst.* **2008**, *17*, 503–511. [CrossRef]

20. Klose, T.; Conrad, H.; Sandner, T.; Schenk, H. Fluidmechanical damping analysis of resonant micromirrors with out-of-plane comb drive. In Proceedings of the COMSOL Conference, Hannover, Germany, 4–6 November 2008.

21. Zhou, H.; Wang, D.; Huang, T. The study of measure model and optical path design of small angle measured by laser triangulation. *J. Yunnan Natl. Univ. Nat. Sci. Ed.* **2008**, *17*, 277.

micromachines

MDPI

Article

A Large-Size MEMS Scanning Mirror for Speckle Reduction Application [†]

Fanya Li [1,2], Peng Zhou [2], Tingting Wang [1,2], Jiahui He [1,2], Huijun Yu [2] and Wenjiang Shen [2,*]

[1] The School of Materials Science and Engineering, Xi'an Jiaotong University, Xi'an 710049, China; fyli2016@sinano.ac.cn (F.L.); ttwang2014@sinano.ac.cn (T.W.); jhhe2014@sinano.ac.cn (J.H.)
[2] Key Lab of Nanodevices and Applications, Suzhou Institute of Nano-tech and Nano-bionics, Chinese Academy of Sciences, Suzhou 215123, China; pzhou2015@sinano.ac.cn (P.Z.); hjyu2012@sinano.ac.cn (H.Y.)
* Correspondence: wjshen2011@sinano.ac.cn; Tel.: +86-512-6287-2688
† This paper is an extended version of our paper published in the 12th IEEE International Conference on Nano/Micro Engineered and Molecular Systems, 9–12 April 2017, Los Angeles, CA, USA.

Academic Editor: Huikai Xie
Received: 30 March 2017; Accepted: 24 April 2017; Published: 3 May 2017

Abstract: Based on microelectronic mechanical system (MEMS) processing, a large-size 2-D scanning mirror (6.5 mm in diameter) driven by electromagnetic force was designed and implemented in this paper. We fabricated the micromirror with a silicon wafer and selectively electroplated Ni film on the back of the mirror. The nickel film was magnetized in the magnetic field produced by external current coils, and created the force to drive the mirror's angular deflection. This electromagnetically actuated micromirror effectively eliminates the ohmic heat and power loss on the mirror plate, which always occurs in the other types of electromagnetic micromirrors with the coil on the mirror plate. The resonant frequency for the scanning mirror is 674 Hz along the slow axis, and 1870 Hz along the fast axis. Furthermore, the scanning angles could achieve ±4.5° for the slow axis with 13.2 mW power consumption, and ±7.6° for the fast axis with 43.3 mW power consumption. The application of the MEMS mirror to a laser display system effectively reduces the laser speckle. With 2-D scanning of the MEMS mirror, the speckle contrast can be reduced from 18.19% to 4.58%. We demonstrated that the image quality of a laser display system could be greatly improved by the MEMS mirror.

Keywords: microelectronic mechanical system (MEMS); speckle reduction; electromagnetic force; optical scanning

1. Introduction

Solid-state lasers can provide wider color gamut, longer lifetime, and higher brightness and contrast of images compared to light emitting diodes (LEDs), a popular light source for projection displays [1]. Laser display technology plays a significant role in our life, and can be applied in various fields such as movie theatres, home televisions and conference rooms. However, the existence of speckle degrades the images quality severely, which is an irregularly distributed pattern of light and dark particles caused by the interference of the reflective coherent laser beam from the rough screen comparable to optical wavelength [2–4]. One of the promising speckle reduction technologies in laser projection [5,6] is to employ MEMS scanning mirrors. At present, research on MEMS scanning mirrors are mostly focused on small diameter MEMS mirrors, while rarely on the larger size. Large-size mirrors can not only tolerate high optical power, but also ensure maximum utilization of light energy [7]. Microvision Company in the United States developed an electromagnetic two-dimensional scanning mirror, and successfully applied it to a laser Pico projection system, but the 1-mm diameter of the mirror was unable to meet the requirements of the high lumen imaging display [8]. Oliveira et al. were the pioneers who applied a MEMS scanning mirror to eliminate laser speckle. Limited to a 0.8-mm diameter, the MEMS mirror had issues when used in practical laser display systems [9]. Akram et al. in Vestfold University in Norway further improved the MEMS two-dimensional scanning mirror and improved the quality of the laser display images, but its diameter was only 2 mm, and also cannot be used in high power laser displays [10]. The large size and mass of a MEMS two-dimensional mirror limit the possibility of achieving a larger angle unless the driving moment is high enough. At the same time, the oscillating micromirror used to reduce the laser speckle should also have a high operating frequency [11].

Based on the above requirements, this paper proposes a 6.5-mm diameter, two-dimensional MEMS scanning mirror driven by the electromagnetic method. The efficient electromagnetic drive mode not only offered the driving moment of the large angle required, but also realized the high frequency. The mirror is used in a laser projection system to suppress laser speckle. This scanning mirror with a large diameter could be used in high power laser illumination for high lumen projection. Moreover, the high frequency of the scanning mirror could effectively reduce the speckle contrast and bring clearer and more comfortable images.

2. Design

This study employed the electromagnetic scanner in Figure 1 to demonstrate the proposed design concept. As indicated in Figure 1a, four external coils (A, B, C, and D) are symmetrically placed in corresponding positions and kept a certain distance for the mirror's free deflection. Specifically, the coils A and C are located beneath the ferromagnetic film on the outer frame, and coils B and D are placed beneath the rectangular ferromagnetic film on the back of the mirror. Coils A and C are responsible for the slow axis, while B and D coils are for the fast axis. When the A and C coils are driven by the square wave signals with the same frequency and 180° phase difference, the two coils, the bottom magnetic bar and the soft Ni film on the outer frame will compose a closed magnetic circuit, then an attractive force will drive the mirror to deflect a certain angle around the slow axis. Similarly, coils B, D, the bottom magnetic bar and the soft Ni film will compose another magnetic circuit when excitation signals are applied to B and D coils, so as to achieve the purpose of the two-dimensional scanning with our proposed model. Figure 1b,c shows the backside of mirror and the external coils, respectively. The two groups of coils and the signals are controlled separately, therefore, good independence, scanning linearity and accuracy can be achieved for the biaxial scanner. Detailed dimensions of the scanner and coils in our design are summarized in Table 1.

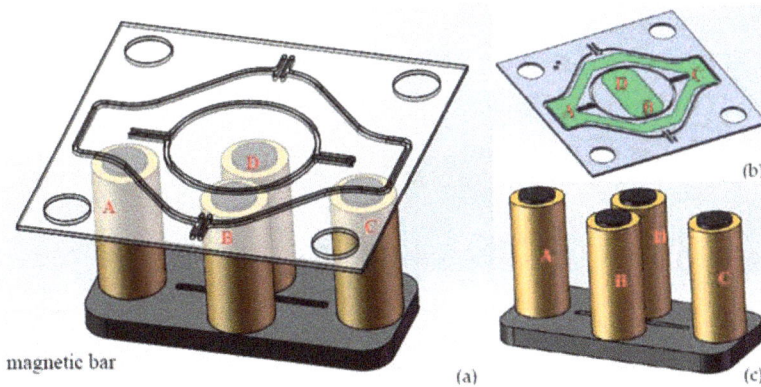

Figure 1. (a) Illustration of the microelectronic mechanical system (MEMS) mirror and the actuation coils underneath; (b) Back profile of mirror with nickel; (c) External coils.

Table 1. The dimensions of designed scanner and coils.

Parameter		Value	Unites
Diameter of the mirror		6.5	mm
Thickness of the mirror		200	μm
Width of the axis	Slow axis	100	μm
	Fast axis	160	
Length of the axis	Slow axis	1500	μm
	Fast axis	1750	
Thickness of the axes		200	μm
Thickness of the nickel film		20	μm
Outer diameter of the coil		3	mm
Inner diameter of the coil		2	mm
Height of the coil		10	mm
Number of turns for the coil		900	-
Resistence of the coil		30	Ω

We introduced a magnetic circuit model to solve the theoretical value of force shown in Figure 2. Based on the hypothesis [12] that all the magnetic fluxes pass through the core (no leakage except for the air gap), and according to the Maxwell's magnetic force formula, attractive force could be expressed by:

$$F = \frac{B_0{}^2 A_0}{\mu_0} \qquad (1)$$

In the equation, B_0 is defined as the magnetic flux density of the air gap, A_0 is the cross-sectional area of the gap, and μ_0 is the magnetic permeability of air. According to the Ampere's Law, we could derive that:

$$N \cdot i = \oint H ds = H_m l_m + H_c l_c + H_0 \cdot 2x , \qquad (2)$$

where N is the number of coil turns, and i is the current flowing through the coil. H_m, H_c, and H_0 respectively represent the strength of the magnetic field for the magnetic core, the clapper (ferromagnetic film) and the gap. l_m, l_c, and x represent the length of the magnetic core, the clapper and the air gap.

Because:

$$H = \frac{B}{\mu} = \frac{\Phi}{\mu A} \qquad (3)$$

the permeability of air can be negligible compared with the ferromagnetic materials' permeability, that is to say:

$$\mu_0 \ll \mu_m, \ \mu_0 \ll \mu_c$$

By assuming:

$$A_c = A_m = A_0 = A \qquad (4)$$

Then, substituting (2)–(4) into the Equation (1), the magnetic force can be expressed as follows:

$$F = k \frac{i^2}{x^2} \qquad (5)$$

where $k = \frac{\mu_0 N^2 A}{4}$. Obviously, the value of the driving force is proportional to the square of the current and inversely proportional to the square of the gap length.

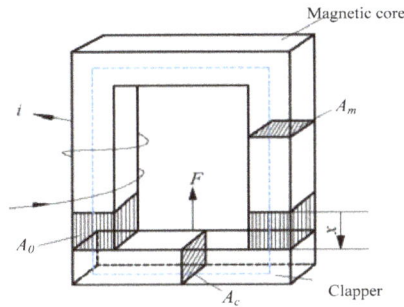

Figure 2. A magnetic circuit model consists of a magnetic core with copper winding and a clapper made out of ferromagnetic materials. The magnetic field created by the copper coil is concentrated in the magnetic core and clapper due to their high permeability. The magnetic circuit path is shown by dashed lines.

3. Fabrication

Figure 3 shows the detailed fabrication process. As in Figure 3a, we used a double sided polished *n*-type (100) 200-μm thick Si wafer as the starting substrate. A 20-nm Ti adhesion layer and a 100-nm Au coating were sputtered on the backside of Si substrate as the seed layer for the following electroplating step. A 20-μm thick film of photoresist (AZ4620) was patterned to selectively electroplate nickel. After the soft-magnetic Ni electroplating step, the photoresist and the Ti/Au thin films were removed by acetone solution and IBE (ion beam etching) technology, respectively. Then, a second photolithography step was used to define the window for bulk silicon etching, as shown in Figure 3d. After that, the structure of the mirror plate and axes were released by the deep reactive ion etching (DRIE) process. Lastly, the front side of the silicon was coated with a 120-nm aluminum layer to form the reflective mirror surface.

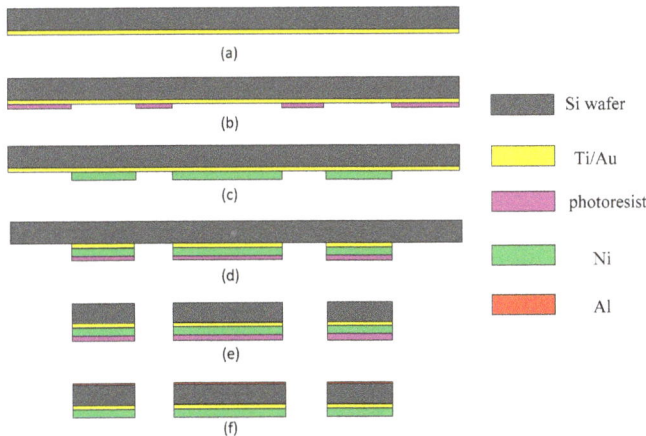

Figure 3. Fabrication flow for the MEMS mirror. (**a**) Ti/Au seed layer sputtering; (**b**) thick photoresist (AZ4620) spinning and exposure; (**c**) nickel electroplating and photoresist removal; (**d**) second photolithography and Ti/Au films removal; (**e**) Si etching (DRIE); (**f**) Al layer coating.

4. Characterization and Results

Figure 4 shows the package method and fully assembled prototype. The whole package is similar to the sandwich structure: The uppermost glass cover is to protect the mirror from the external environment; a 1.2-mm thick plastic top spacer above the mirror is chosen to ensure a large incident angle; the distance between the Ni film and the bottom coils is about 0.45 mm, which is the thickness of the bottom spacer. The thickness of the bottom spacer defines the maximum allowable rotation angle; coils are symmetrically placed in the coil holder, and the driving currents are applied through electrical connections to the printed circuit board (PCB) pads. The whole device size is only 16 mm × 16 mm × 12 mm.

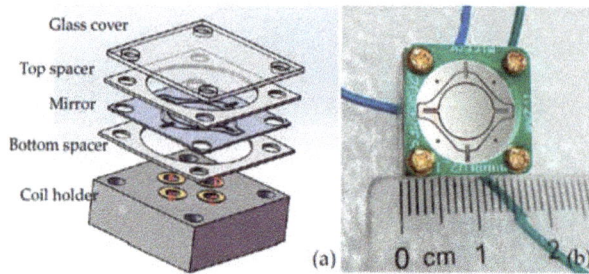

Figure 4. (**a**) Package method; (**b**) Fully assembled prototype.

After packaging, the device was tested with the setup shown in Figure 5. In the measurement, two Ampere meters were used to record the relation between the current and the mirror's scanning property. A function generator was employed to send square waves with a certain frequency to the coils as the excited signal, and to the oscilloscope for monitoring phase difference between input channels. The 2-D scanning mirror's vertical and horizontal axes were driven independently by signals from the function generator.

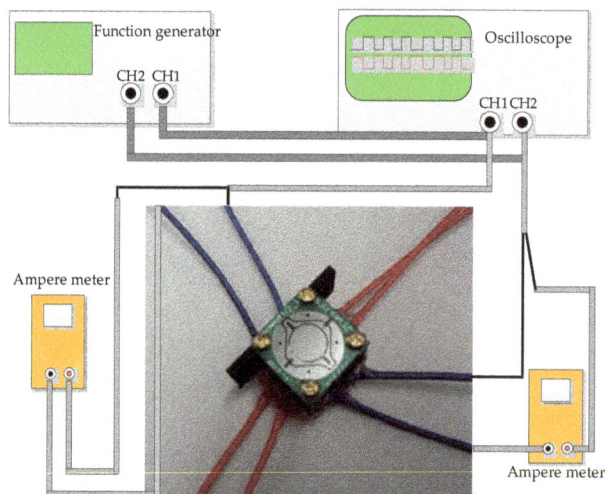

Figure 5. Schematic measurement setup: the 2-D scanning mirror was driven independently by function generators.

The results are plotted in Figure 6. The deflection angle can be calculated from the length of the scanning line and the distance between the screen and the scanning mirror. When the micro-mirror worked at the resonant vibration state, mechanical torsion angles increased with the AC (alternating current) driving currents (refer to root-mean-square values of the current flowing in the coils) for both the slow axis and fast axis. The slow scanning angle could achieve ±4.5° at the applied current of 21 mA and power consumption of 13.2 mW, and the fast scanning angle could reach ±7.6° at the applied current 38 mA and maximum power consumption of 43.3 mW. We used a 0.45-mm thick bottom spacer here, so the maximum allowable rotation angle for the slow axis and fast axis were ±4.5° and ±8° theoretically, matching well with the experimental data. By tuning the driving frequency, the frequency response of the scanning angle could be recorded. Figure 7 shows the frequency responses of the slow and fast axes when 20 mA was applied to a coil. According to the results, the resonant frequencies were 674 Hz and 1870 Hz for the slow and fast axes, respectively. The quality factor Q was calculated by the following equation [13]:

$$Q = \frac{f_0}{\Delta f} \tag{6}$$

where f_0 is the resonant frequency, and Δf is the half-power bandwidth. From the measured curve, we can derive the Q value; 122 for the slow axis and 623 for the fast axis. The large difference of Q values between the slow and fast axes was due to the dependence of damping on the resonant frequency. The Q value increases with the resonant mode and is proportional to $f^{0.5}$ according to Chu et al. [14]. As shown in Figure 1, the slow scanning of the mirror is driven by A and C coils, and the outer frame is scanning together with the mirror plate. So, the air damping is severer and the Q value is smaller for the slow axis scanning.

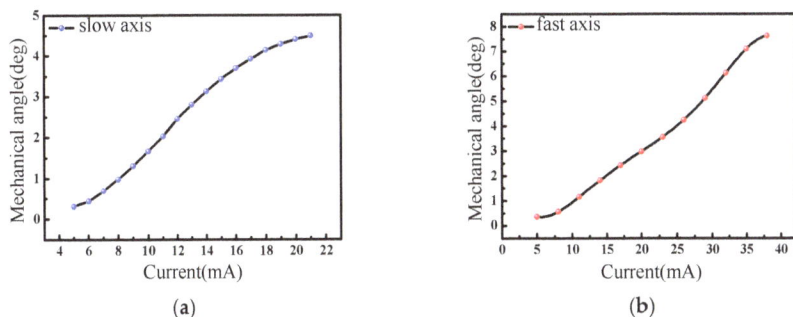

Figure 6. Current-angle relationship: (**a**) slow axis; (**b**) fast axis.

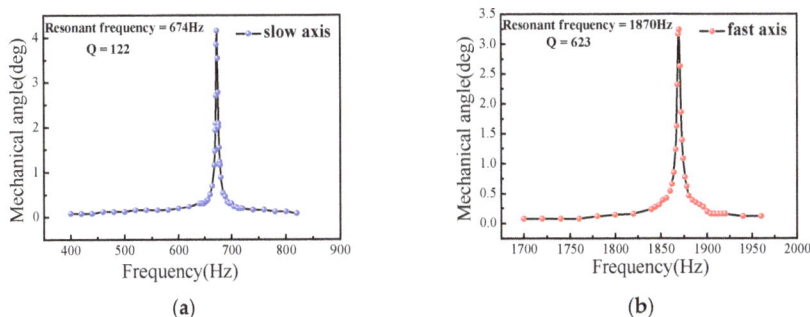

Figure 7. Frequency response: (**a**) slow axis; (**b**) fast axis.

Reliability is crucial to the successful application of MEMS devices when they reach commercialization [15,16]. To study the shock resistance of our fabricated devices, a shock test was performed in air at room temperature as follows [17]. A scanner prototype was fixed to a shock table by 3 M Epoxy Adhesive. Acceleration corresponded to the height of the table from which it was dropped. A piezoelectric transducer (PZT) sensor was mounted to the shock table to record actual acceleration. In the test, we changed the fixed direction so that the shock was applied in three (X, Y, Z) orientations, where the X and Y directions were along the fast axis and the slow axis respectively, and the Z direction was perpendicular to the mirror plane. The table was dropped to the floor three times in a row with three orthorhombic orientations, and half sine shock pulses with certain widths were produced. No significant fracture or electrical failures were observed until the prototype was tested at 900 g for X direction and 1500 g for Y, Z directions, where g is the gravity acceleration (g ≈ 9.8 m/s^2), demonstrating that the micro-mirror and applied package structure have good shock resistance. In addition, a vibration test was carried out on Electro Dynamic Shakers with varying frequencies (from 20 Hz to 2000 Hz) at a constant acceleration of 20 g [17]. After three periods of shaking, the device could still operate as before. These results show that the device is reliable and durable for practical applications.

5. Application to Speckle Reduction

In the application to laser projectors, the speckle phenomenon emerges by reflecting highly coherent laser beams with single wavelengths on random rough surfaces, resulting in a random spatial intensity distribution [18]. One of the criterion to describe speckle is speckle contrast ratio, which is defined as [19]:

$$C = \frac{\sqrt{\langle I^2 \rangle - \langle I \rangle^2}}{\langle I \rangle} \times 100\% \tag{7}$$

where $\langle I^2 \rangle$ and $\langle I \rangle$ represent the square mean value and the mean light intensity, $\sqrt{\langle I^2 \rangle - \langle I \rangle^2}$ denotes the standard deviation. The lower the C value, the clearer images could be derived, which is now of great concern.

As presented in Figure 8, the simplified speckle reduction system consists of a laser diode, a fabricated scanning mirror, a light pipe, optics elements such as a focusing lens and a diffuser with high transmittance, and a charge-coupled device (CCD) camera for acquiring pattern information. With the two-dimensional scanning of the mirror, the laser beam was reflected onto the diffuser placed at the entrance of the light pipe with angle diversity at different times. After multiple reflections inside the light pipe, the uniform illumination will be formed at the exit surface, which, with the existence of subsequent imaging optics, form the picture on the screen. The scanning area can be changed with different driving currents, however, we need to control the total reflected light entering into the light pipe. According to speckle suppression theory [20,21], once the speckle images at different times and different positions are uncorrelated, then these irrelevant speckle figures are finally superimposed on each other during a frame image formed on the screen. If N independent speckle patterns are overlapped on an intensity basis, and we assume that each pattern has an equal mean intensity, the speckle contrast C in the integrated image is reduced to [4]:

$$C = \frac{1}{\sqrt{N}} \tag{8}$$

where N is the number of independent speckle patterns. We could derive that the higher the value of N, the lower the value of speckle contrast results.

In our measurement, the focus length and aperture f-number of the CCD imaging lens are 25 mm and 8, respectively. The CCD has a pixel size of 3.75 μm × 3.75 μm and is located at 2 m away from the screen. We explored the mirror's stationary and vibrating conditions and their influence on the speckle contrast ratio. The calculated data by Matlab based on Equation (7) is presented in Table 2. When the mirror was working, the contrast was 4.58% during 50 ms integration time of the CCD camera, and when the mirror was turned off, the contrast was 18.19% at the same integration time. Figure 9 shows the speckle contrast images with and without the working mirror. With the 2-D scanning of the MEMS mirror, the speckle contrast for the laser projection display could be reduced from 18.19% to 4.58%. This result demonstrated that the scanning of the mirror can disturb the spatial and temporal coherences of the laser source and suppress the speckle pattern for laser projection images.

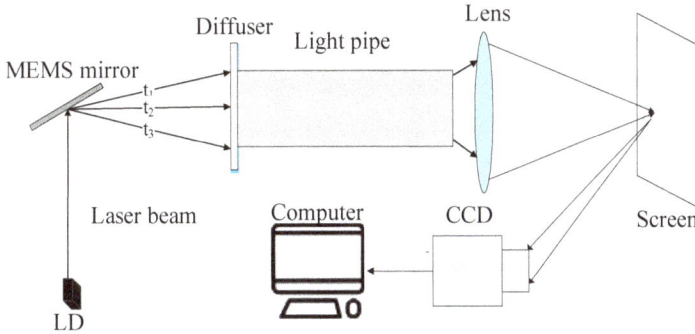

Figure 8. Speckle reduction system.

Table 2. The speckle contrast by different measurement systems.

Measurement System	Integration Time/ms	Maxmium Intensity	Minimum Intensity	Mean Intensity	Contrast Value/%
With mirror	50	157	108	126	4.58
Without mirror	50	255	94	179	18.19

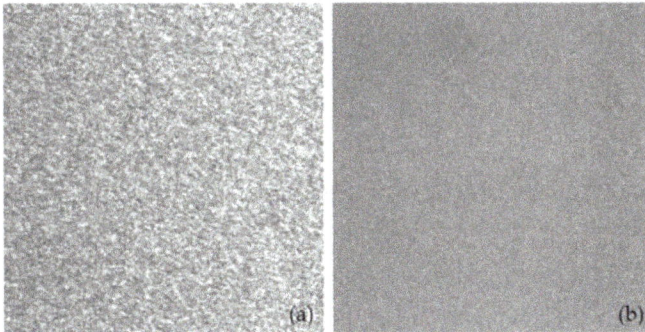

Figure 9. Speckle contrast images when CCD integration time was set to 50 ms: (**a**) Speckle pattern with inactive scanning mirror, $C = 18.19\%$; (**b**) Speckle reduction pattern with active mirror, $C = 4.58\%$.

6. Conclusions

We have proposed and fabricated a large-size MEMS scanning mirror. Our mirror can meet the requirements of high resonant frequency and large deflection angle used for speckle reduction application, and successfully reduces the laser speckle contrast from 18.19% to 4.58%. In addition, the fabricated devices have a shock resistance of more than 900 g and good vibration resistance, which is also crucial when used in commercial applications.

Acknowledgments: This research was supported by the National Key Research and Development Program of China (Grant No. 2016YFB0402003). The authors are grateful to SINANO's Nano Fabrication Facility (NFF) for their support in the fabrication process of micromirrors and related reliability tests.

Author Contributions: W.S. offered conceptual advices for finishing this article; J.H. and H.Y. gave technical support and offered assistance in the whole experiment process; F.L. mainly collected data about the performance for the mirror and wrote the manuscript; P.Z. and T.W. were responsible for the data of speckle reduction. All authors discussed the results and commented on the manuscript at all stages.

Conflicts of Interest: The authors declare no conflict of interest.

References

1. Tong, Z.; Shen, W.; Song, S.; Cheng, W.; Cai, Z.; Ma, Y.; Wei, L.; Ma, W.; Xiao, L.; Jia, S.; et al. Combination of micro-scanning mirrors and multi-mode fibers for speckle reduction in high lumen laser projector applications. *Opt. Express* **2017**, *25*, 3795–3804. [CrossRef] [PubMed]

2. Subramaniam, S.; Le, C.P.; Kaur, S.; Kalicinski, S.; Ekwinska, M.; Halvorsen, E.; Akram, M.N. Design, modeling, and characterization of a microelectromechanical diffuser device for laser speckle reduction. *J. Microelectromech. Syst.* **2014**, *23*, 117–127.

3. Young, E.J.; Kasdin, N.J.; Carlotti, A. Image Analysis with Speckles Altered by a Deformable Mirror. Available online: http://proceedings.spiedigitallibrary.org/proceeding.aspx?articleid=1744172 (accessed on 25 April 2017).

4. Tong, Z.; Chen, X.; Akram, M.N.; Aksnes, A. Compound speckle characterization method and reduction by optical design. *J. Disp. Technol.* **2012**, *8*, 132–137. [CrossRef]

5. Heger, A.; Schreiber, P.; Höfer, B. Development and Characterisation of a Miniaturized Laser Projection Display Based on MEMS-Scanning-Mirrors. Available online: http://proceedings.spiedigitallibrary.org/proceeding.aspx?articleid=1345482 (accessed on 25 April 2017).

6. Kim, S.; Han, Y.G. Suppression of speckle patterns based on temporal angular decorrelation induced by multiple beamlets with diverse optical paths. *J. Korean Phys. Soc.* **2014**, *64*, 527–531. [CrossRef]

7. Chen, X.; Svensen, Ø.; Akram, M.N. Speckle reduction in laser projection using a dynamic deformable mirror. *Opt. Express* **2014**, *22*, 11152–11166.

8. Sprague, R.B.; Montague, T.; Brown, D. Bi-Axial Magnetic Drive for Scanned Beam Display Mirrors. Available online: http://proceedings.spiedigitallibrary.org/proceeding.aspx?articleid=859733 (accessed on 25 April 2017).

9. Oliveira, L.C.M.; Barbaroto, P.R.; Ferreira, L.O.S.; Doi, I. A novel Si micromachined moving-coil induction actuated mm-sized resonant scanner. *J. Micromech. Microeng.* **2005**, *16*, 165. [CrossRef]

10. Akram, M.N.; Tong, Z.; Ouyang, G.; Chen, X.; Kartashov, V. Laser speckle reduction due to spatial and angular diversity introduced by fast scanning micromirror. *Appl. Opt.* **2010**, *49*, 3297–3304. [CrossRef] [PubMed]

11. Bayat, D. Large Hybrid High Precision MEMS Mirrors. Ph.D. Dissertation, École Polytechnique Fédérale de Lausanne, Lausanne, Switzerland, 2011.

12. Bleuler, H.; Cole, M.; Keogh, P.; Larsonneur, R.; Larsonneur, R.; Maslen, E.; Okada, Y.; Traxler, A. *Magnetic Bearings: Theory, Design, and Application to Rotating Machinery*; Springer Science & Business Media: Berlin, Germany, 2009.

13. Chen, M.; Yu, H.; Guo, S.; Xu, R.; Shen, W. An electromagnetically-driven MEMS micromirror for laser projection. In Proceedings of the 10th IEEE International Conference on Nano/Micro Engineered and Molecular Systems (NEMS), Xi'an, China, 7–11 April 2015; pp. 605–607.

14. Chu, H.M.; Hane, K. Design, fabrication and vacuum operation characteristics of two-dimensional comb-drive micro-scanner. *Sens. Actuators A Phys.* **2011**, *165*, 422–430. [CrossRef]

15. Tanner, D.M.; Walraven, J.A.; Helgesen, K.; Helgesen, K.; Irwin, L.W.; Brown, F.; Smith, N.F.; Masters, N. MEMS reliability in shock environments. In Proceedings of the 38th Annual 2000 IEEE International Reliability Physics Symposium, San Jose, CA, USA, 10–13 April 2000; pp. 129–138.

16. Naumann, M.; Dietze, O.; McNeil, A.; Mehner, J.; Daniel, S. Reliability of anchors at surface micromachined devices in shock environments. In Proceedings of the 2013 Transducers & Eurosensors XXVII: The 17th International Conference on Solid-State Sensors, Actuators and Microsystems (TRANSDUCERS & EUROSENSORS XXVII), Barcelona, Spain, 16–20 June 2013; pp. 570–573.

17. GB4590-84, Mechanical and Climatic Test Methods for Semiconductor Integrated Circuits. Available online: http://down.bzwxw.com/12/GB%204590-1984.pdf (accessed on 25 April 2017). (In Chinese)

18. Lee, J.Y.; Kim, T.H.; Yim, B.B.; Bu, J.U.; Kim, Y.J. Speckle reduction in laser picoprojector by combining optical phase matrix with twin green lasers and oscillating MEMS mirror for coherence suppression. *Jpn. J. Appl. Phys.* **2016**, *55*, 08RF03. [CrossRef]

19. Pan, J.W.; Shih, C.H. Speckle reduction and maintaining contrast in a LASER pico-projector using a vibrating symmetric diffuser. *Opt. Express* **2014**, *22*, 6464–6477. [CrossRef] [PubMed]

20. Pei, T.H.; Yeh, F.C.; Tsai, K.Y.; Li, J.H.; Liu, Z.R.; Hung, C.L. Simulation and experiment of speckle reduction by the beam splitting method on a pico-projection system. *Adv. Mater. Res.* **2014**, *933*, 572–577. [CrossRef]

21. Goodman, J.W. *Speckle Phenomena in Optics: Theory and Applications*; Roberts and Company Publishers: Greenwood Village, CO, USA, 2007.

micromachines

MDPI

Article

Large-Aperture kHz Operating Frequency Ti-alloy Based Optical Micro Scanning Mirror for LiDAR Application

Liangchen Ye, Gaofei Zhang * and Zheng You *

State Key Laboratory of Precision Measurement Technology and Instruments, Department of Precision Instrument, Tsinghua University, Haidian District, Beijing 100084, China; ylc12@mails.tsinghua.edu.cn
* Correspondence: zgf@mail.tsinghua.edu.cn (G.Z.); yz-dpi@mail.tsinghua.edu.cn (Z.Y.);
Tel.: +86-10-6277-6000 (G.Z. & Z.Y.)

Academic Editor: Huikai Xie
Received: 2 March 2017; Accepted: 7 April 2017; Published: 10 April 2017

Abstract: A micro scanning mirror is an optical device used to scan laser beams which can be used for Light Detection and Ranging (LiDAR) in applications like unmanned driving or Unmanned Aerial Vehicle (UAV). The MEMS scanning mirror's light-weight and low-power make it a useful device in LiDAR applications. However, the MEMS scanning mirror's small aperture limits its application because it is too small to deflect faint receiving light. In this paper, we present a Ti-alloy-based electromagnetic micro scanning mirror with very large-aperture (12 mm) and rapid scanning frequency (1.24 kHz). The size of micro-scanner's mirror plate reached 12 mm, which is much larger than familiar MEMS scanning mirror. The scanner is designed using MEMS design method and fabricated by electro-sparking manufacture method. As the experimental results show, the resonant frequency of the micro scanning mirror is 1240 Hz and the optical scanning angle can reach 26 degrees at resonance frequency when the actuation current is 250 mApp.

Keywords: large-aperture; micro scanning mirror; micro scanner; Ti-alloy; LiDAR

---

1. Introduction

LiDAR is widely used in many applications, such as space autonomous rendezvous docking, space target detection, UAV's navigation, Advanced Driver Assistance System (ADAS), automatic driving and so on. These applications have recently created a great demand for low-cost, low-power and low-weighted three-dimension imaging LiDAR.

Micro-mirror based LiDAR have drawn the attention of many researchers for the realization of 3D distance measurement [1–7]. Traditional laser scanners for LiDAR consist of heavy, expensive and large rotational optical devices. In comparison with traditional laser scanning sensors, MEMS scanners have the advantages of rapid scanning frequency, light-weight and low power [8]. Micro-scanners' advantages make it a promising technology for use in miniature LiDAR. However, current MEMS scanning mirrors' mirror plates are too small to be applied in LiDAR applications in order to detect receiving light. In LiDAR applications, large apertures are required for the measuring beam [9]. The range of the LiDAR is largely affected by the size of the mirror which reflects the received laser light. The size of mirror plate must be big enough to gather more light scattered by the target. Most MEMS scanners are designed in nearby applications like projectors or Optical coherent tomography (OCT) and cannot meet the requirements of LiDAR applications.

Efforts have been made to achieve large-aperture MEMS scanning mirrors. Sandner et al. [5] present a 1D-MEMS scanner module which has a resonant frequency of 250 Hz and a mirror plate size of 2.52×9.51 mm^2 per single mirror element. Lei et al. [10] present an electrothermal MEMS scanning mirror with a large aperture of 10×10 mm^2 which has a resonant frequency of 234 Hz. Milanovic et al. [11,12] present a gimbal-less Tip–Tilt–Piston MEMS scanner with a mirror plate size of 5 mm and a resonant frequency of 334 Hz.

However, current large-aperture MEMS scanner frequency is not rapid enough to achieve fast LiDAR images and the aperture is not large enough. Silicon is a material with a very high strength but its fatigue strength is much lower than its yield strength [13]. The fatigue strength decreases when the size of MEMS structure increases [14]. The metal based micro scanner is more robust than silicon-based MEMS devices due to its ductile properties in comparison with brittle silicon substrates [15]. Some metallic materials like stainless-steels have been researched to be able to replace the silicon substrate of a MEMS scanner due to the ductile properties in comparison with the brittle Si materials. Park et al. [15] presents a one-axis metal-based micro-scanner with a large mirror (3×3 mm^2), with a resonant frequency of 304 Hz and a scanning angle of 12 degrees. Youmin et al. [16] present a two-axis soft-magnetic stainless steel based micro scanner with a large-aperture of 4×5 mm^2 which has frequency of 112 and 1268 Hz in each axis.

In this paper, we present a novel one-axis Ti alloy-based electromagnetic micro scanning mirror with a very large-aperture and a high operating frequency. Ti-alloy substrate is used to achieve large aperture and fast scanning frequency. The size of the micro scanning mirror can reach 12 mm with kHz resonant frequency. The mirror substrate is fabricated by electro-sparking manufacture technology. A moving coil is attached to the back of the mirror and a pair of moon-like magnets are employed to achieve a large optical scanning angle a low actuation current. A Position Sensitive Device (PSD) is integrated to measure the rotation angle of the mirror plate. The design, simulation, fabrication and characterization of the micro-scanner is described in this article.

2. Micro Mirror Based LiDAR System

Figure 1a shows a concept for a one-axis micro scanning mirror based LiDAR. One-axis scan is realized by the motor's 360° rotation. The micro mirror scans the laser beam in another axis. The distance between the micro mirror and the target is determined by measuring the phase delay between the laser emitted and the laser received. To gather more reflected laser signal from the target, a larger-aperture micro mirror is required. To improve the horizontal resolution of the LiDAR image, the micro mirror needs to achieve fast operating frequency. Traditional scanners, like high-speed Galvano mirrors, are not suitable for this system because of their large size and slow, heavy weight. A kHz resonant frequency micro mirror with 12 mm aperture is required in this LiDAR system.

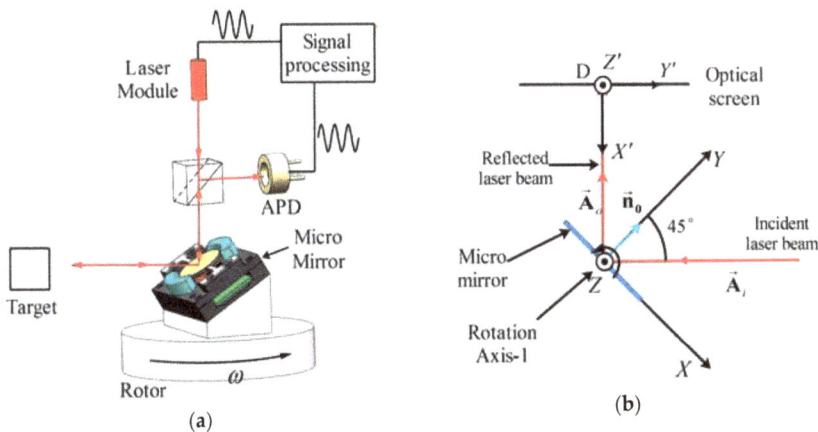

Figure 1. *Cont.*

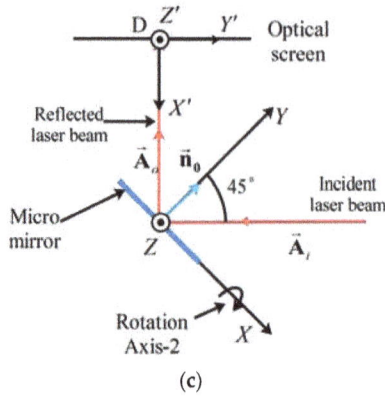

Figure 1. (a) Setup of two-axis Light Detection and Ranging (LiDAR) with 1D micro scanning mirror; (b) 1st arrangement of LiDAR system; (c) 2nd arrangement of LiDAR system.

Figure 1b,c show the normal arrangement of the LiDAR system. The rotational axis of the micro mirror can be set along the Z-axis (called the 1st arrangement) or along the X-axis (called the 2nd arrangement). The 1st arrangement is chosen in our LiDAR system for the curve-shaped field distortion caused by the 2nd arrangement.

The intersection point (x_D, y_D, z_D) of the reflected laser beam and the screen can be calculated by solving the below equation:

$$\begin{cases} (x_D, y_D, z_D) = a\vec{A}_o \\ y_D - x_D = \sqrt{2}l \end{cases} \tag{1}$$

where l is the distance between the micro mirror and the optical screen and $\vec{A}_o$ is the vector the of reflected laser beam, which can be calculated by solving the equation of reflection:

$$\vec{A}_o = \vec{A}_i - 2\vec{n}_0(\vec{n}_0 \cdot \vec{A}_i) \tag{2}$$

where $\vec{A}_i$ is the vector of the incident laser beam and $\vec{n}_0$ is the normal vector of the mirror plate.

The intersection point's coordinate $(x'_{D1}, y'_{D1}, z'_{D1})$ in the 1st arrangement and $(x'_{D2}, y'_{D2}, z'_{D2})$ in the 2nd arrangement can be calculated by coordinate transformation as:

$$\begin{cases} x'_{D1} = z'_{D1} = 0 \\ y'_{D1} = -l\tan 2\alpha \end{cases} \tag{3}$$

$$\begin{cases} x'_{D2} = 0 \\ y'_{D2} = l\tan^2\beta \\ z'_{D2} = \sqrt{2}l\tan\beta \end{cases} \tag{4}$$

where α and β are scanning angles of the mirror plate at the 1st arrangement and the 2nd arrangement separately.

From Equation (3), we can see that the laser beam length (y'_{D1}) in the 1st arrangement is about $\sqrt{2}$ times larger than the length (z'_{D2}) in the 2nd arrangement. In addition, the laser beam in the second arrangement is curve-shaped for the y-coordinate of the intersection $y'_{D2} \neq 0$. In our LiDAR, the 1st arrangement is occupied.

3. Design and FEM Simulation of the Micro Mirror

Figure 2a shows a sketch of the electromagnetic Ti-alloy based micro scanning mirror. The scanner consists of an Ag-coated mirror plate, Ti-alloy based mirror substrate, moving coils, permanent magnets and Al-alloy basement. The micro mirror is actuated by electromagnetic torque. Figure 2b shows the principle of the electromagnetic micro-scanner. The scanner is actuated by the magnetic torque along the torsional beam. The torque is generated by the interaction between the permanent magnets and the AC-excited coils on the back of the mirror plate. A pair of quarter-circle magnets (see Figure 2a) is applied to enhance the scanning angle of micro-scanner. Compared with flat magnets (see Figure 3a), quarter-circle magnets can achieve larger magnetic torque (see below discussion).

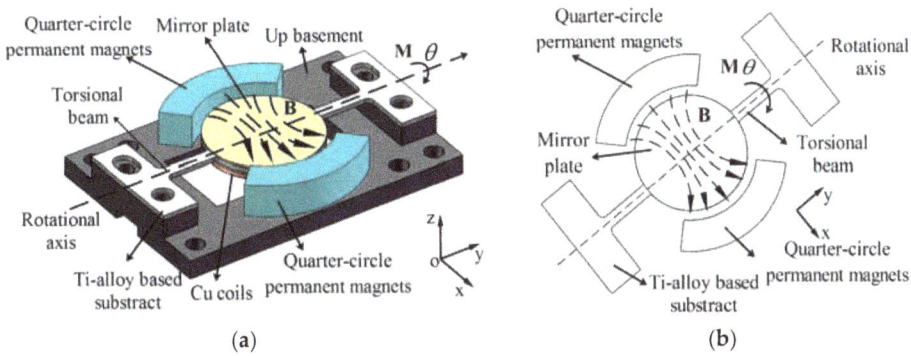

Figure 2. (a) Sketch of electromagnetic Ti-alloy based micro-scanner; (b) Principle of the electromagnetic micro-scanner.

When operating, the micro-scanner rotates along the rotational axis and the torsional angle (θ) is the only Degree of Freedom (DOF) in this dynamic system. The equation of the motion of the 1-DOF can be estimated as:

$$I\ddot{\theta} + D\dot{\theta} + K\theta = M \tag{5}$$

where I is the moment of inertia of the scanning mirror, D is the damping coefficient, K is the stiffness of the torsional beam and M is the torque generated by the coil's interaction with off-chip magnets.

The mirror of the micro-scanner contains a Ti-alloy substrate, a SiO_2 based mirror plate and a Cu based multi-turns coil. The moment of inertia can be written as:

$$I = \frac{1}{4}\rho_s \pi R_s^4 t_s + \frac{1}{4}\rho_m \pi R_m^4 t_m + \frac{1}{4}\rho_c \pi (R_{c1}^4 - R_{c2}^4) t_c \tag{6}$$

where ρ_s, R_s and t_s are the density, radius and thickness of mirror substrate, ρ_m, R_m and t_m are the density, radius and thickness of mirror plate, and ρ_c, R_{c1}, R_{c2} and t_c are the density, internal radius, external radius and thickness of the coil.

Followed by the formula reported in [17], the spring constant K is given as:

$$K = \frac{G_{yx}wh^3}{3l}(1 - \frac{192}{\pi^5}\frac{h}{\mu w}\tanh\frac{\pi w}{2h}), \mu = \sqrt{G_{yx}/G_{yz}} \tag{7}$$

where w, h and l are width, depth and length of torsional beam and G_{yx} and G_{yz} are shear moduli in different direction due to material anisotropy.

When operating, the micro-scanner is working on its resonant frequency which can be given as:

$$f = \frac{\sqrt{K/I}}{2\pi} \tag{8}$$

According to the parameters in Table 1, the resonance frequency of the micro scanner is 1141 Hz. The amplitude of the scanning angle at resonant frequency can be estimated as:

$$\theta(\omega_n) = \frac{M}{2\xi I} \tag{9}$$

where ξ is damping ratio which can be calculated as $\xi = D/2\sqrt{KI}$. Damping ratio is 0.0011 which can be calculated by the drag air damping model [18].

According to Equation (9), the amplitude of the scanning angle can be obviously enhanced by enlarging the torque generated by the magnetic field. The force acting on a current conductor in the magnetic field can be given as:

$$d\mathbf{F} = Id\mathbf{l} \times \mathbf{B} \tag{10}$$

where I is the current running through the path ($d\mathbf{l}$) and $\mathbf{B}$ is the external magnetic field.

Since the coil is a round loop, the torque ($\mathbf{M}$) generated by the coil's interaction with off-chip magnets can be estimated by integrating force along the coil:

$$\mathbf{M} = N \oint_L \mathbf{r} \times (Id\mathbf{l} \times \mathbf{B}) \tag{11}$$

where N is the number of coil turns and $\mathbf{r}$ is the vector from location of $d\mathbf{l}$ to the rotational axis.

When the permanent magnets are plate-type magnets, the magnetic field between the permanent magnets is constant (see Figure 2a). Then Equation (6) can be simplified as:

$$M = N \int_0^{2\pi} BIrd\theta \sin(\theta + \frac{\pi}{2})r \cos\theta = NBI\pi r^2 \tag{12}$$

where I is coil current, B is the external magnetic field, and r is radius of the coil, respectively.

To increase the torque generated by the magnetic field, quarter-circle magnets are applied in the micro-scanner. The distance between the coil and the magnets can be drawn closer and more torque can be gained on the coils. Accounting for the symmetry of the coil, Equation (11) can be simplified as:

$$M_y = N \oint_L r_c \sin\varphi I_c B_r(\varphi) \tag{13}$$

where M_y is the torque along the direction of torsional beam, I_c is the actuation current and B_r is component of magnetic flux density along the direction of the coil's radius.

A magnetic field-structure coupling Finite Element Method (FEM) simulation is conducted to analyze the interaction between the magnets and the micro-scanner with a coil (see Figure 3b). In the FEM simulation, the height, thickness and length of the two magnets is the same and the minimum distance between magnets and the mirror is the same. As the result of magnetic simulation show, the component of magnetic flux density along the direction of the coil's radius ($B_r(\varphi)$) in the micro-scanner with quarter-circle magnets is larger than that in the micro-scanner with plate-type magnets (see Figure 3c). In the magnetic field-structure coupling simulation, the same actuation current is applied to analyze the static rotational angle according to magnetic force. The micro-scanner with quarter-circle magnets can achieve a larger rotational angle than the scanner with plate-type magnets (see Figure 3d). More magnetic force can be generated when applying the quarter-circle magnets.

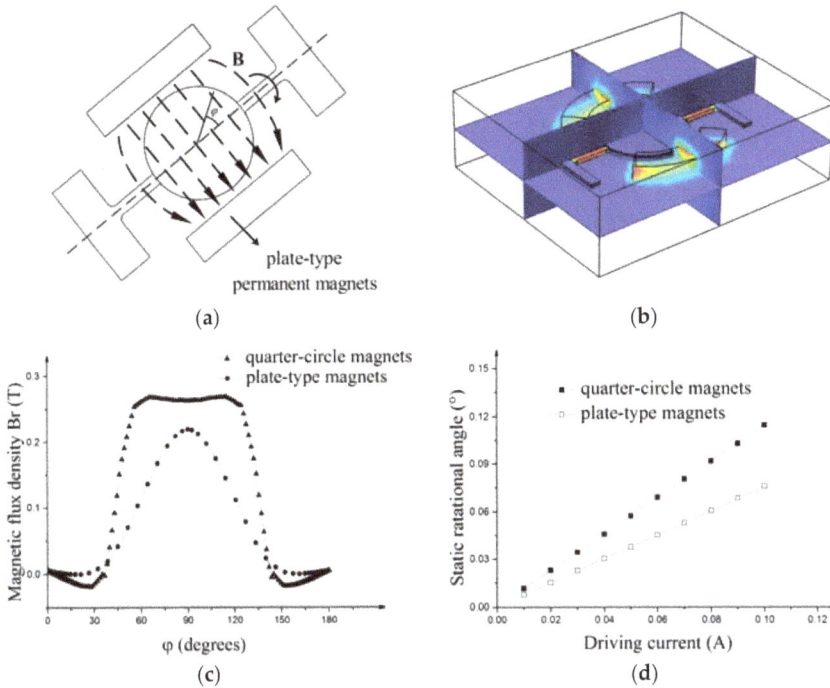

plate-type
permanent magnets

(a)

(b)

(c)

(d)

Figure 3. (**a**) Sketch of electromagnetic micro scanning mirror with plate-type magnets; (**b**) magnetic field-structure coupling Finite Element Method (FEM) simulation model of micro mirror; (**c**) magnetic flux density along direction of coil's radius ($B_r(\varphi)$) in micro mirror with quarter-circle magnets in comparison with one with plate-type magnets; (**d**) relationship between static rotational angle and actuation current of micro mirror with quarter-circle magnets against one with plate-shape magnets.

Table 1. Parameters of the Ti alloy based micro scanner.

Parameter	Symbol	Value	Parameter	Symbol	Value
Radius of mirror substrate	R_s	6 mm	Turns of coil	N	200
Thickness of mirror substrate	t_s	0.4 mm	Width of torsional beam	w	1 mm
Radius of mirror plate	R_m	6 mm	Depth of torsional beam	h	0.4 mm
Thickness of mirror plate	t_m	0.2 mm	Length of torsional beam	l	5.8 mm
Internal radius of coil	R_{c1}	6 mm	Internal radius of magnet	R_{mag1}	7 mm
External radius of coil	R_{c2}	5 mm	External radius of magnet	R_{mag2}	12.5 mm
Thickness of coil	t_c	0.5 mm	Thickness of magnet	t_{mag}	7 mm

Figure 4 shows the modal FEM simulation results of the 1-D micro scanning mirror. The torsional resonant mode and piston resonant mode are shown in Figure 3a,b. The frequency of torsional mode is 1199 Hz and the frequency of piston mode is1917 Hz. The torsional mode is selected as the operating mode. The frequency of the piston mode is designed much higher than that of torsional mode to avoid interference with scan mode. Figure 4c shows the frequency response of the optical scanning angle in the twisting mode. The optical scanning angle at the resonance frequency is 22.4° at an actuation current of 140 mApp.

(a) (b)

(c)

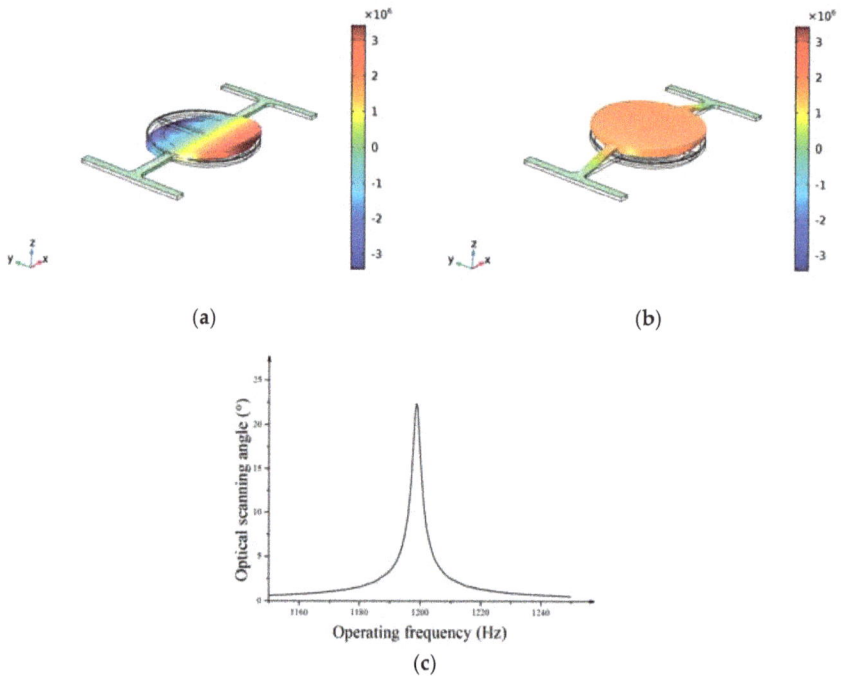

Figure 4. Modal analysis results from FEM simulation (a) Torsion mode of the micro mirror (scan mode); (b) piston mode of the micro mirror; (c) frequency response of the optical scanning angle in the twisting mode.

4. Packaging and Integration of Micro Scanning Mirror

Figure 5a shows the sketch of the electromagnetic Ti-alloy based micro scanning mirror module. The module consists of the Ag-coated mirror plate, the Ti-alloy based micro-mirror substrate, the cooper coils and the NdFeB-type permanent magnets (R11*90°, Jinchen Co. Ltd., Shenzhen, China), Position Sensitive Device (PSD, BS-PSD0018, Bosen Tech., Wuhan, China) and Al-alloy basement. The Ti-alloy based micro-mirror substrate is fabricated by electro-sparking manufacture method and the Al-alloy basement is fabricated by numerical control processing technology. The Ag-coated mirror plate is fabricated by spurting Ag on SiO_2 substrate. Cooper coils are winded and the diameter of cooper wires is 50 μm. Ag-coated mirror plate and cooper coils are fixed on the Ti-alloy substrate using 3M instant adhesive glue (CA40H Minnesota Mining and Manufacturing Company, St. Paul, MN, USA). The micro scanning mirror is actuated by electromagnetic torque. The electromagnetic actuation consists of a pair of permanent magnets which are fixed on the basement using CA40H glue and a moving coil attached to the back of the mirror substrate. When applying a sine waveform current, the coil yields a torque along the torsional beam and actuates the mirror plate rotating along the beam. Beneath the mirror plate, a PSD device fixed in the substrate is used to measure the rotational angle of the mirror plate by sensing the position of the laser point on PSD reflected by the mirror on the back of the mirror plate. When the mirror plate rotates at different angles, the laser point reflected by the mirror on the back will fix at a different position on the PSD which will lead to an output voltage linear to the position. Figure 5b shows the package of the micro scanning mirror. The whole chip size is $31.6 \times 21 \times 8.5$ mm^3 and the weight is 18 g.

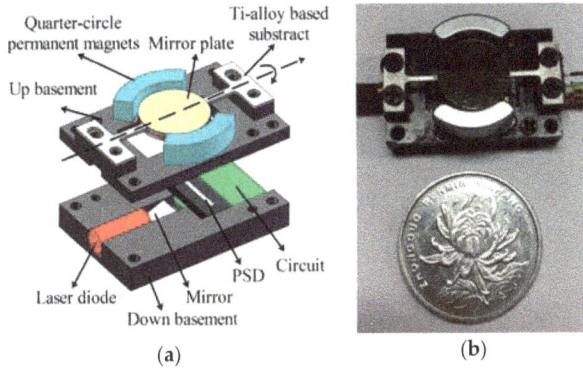

Figure 5. (a) Sketch of micro scanning mirror's structure; (b) package of the micro scanning mirror.

5. Experimental Results

In this section, features of the micro scanning mirror were measured. Operating frequency, optical scanning angle and angle measurement precision were the key characters of the micro scanning mirror. One-axis detection with the micro mirror was also achieved.

Figure 6 shows the experimental setup to measure the micro mirror's features. The laser beam reflected by the micro mirror radiates toward the optical screen and results in a laser line on the screen. The optical scanning angle of the micro mirror can be achieved by measuring the length of the laser line and the distance between the micro mirror and optical screen (see Equation (3)).

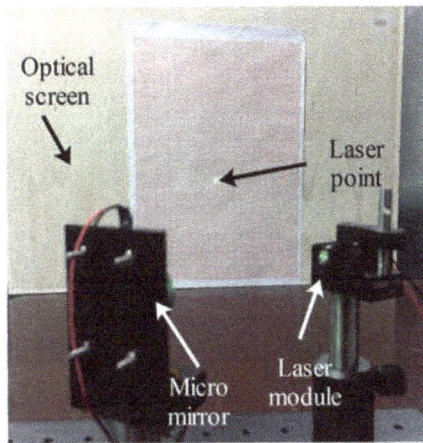

Figure 6. Experimental setup of micro mirror's character measurement.

Figure 7a illustrates the relationship between optical scanning angles with the operating frequency when the actuation current is 200 mA. The resonant frequency of the micro mirror was 1.24 kHz and the quality factor *Q* was 253. Figure 7b illustrates the relationship between the optical scanning angles with the actuation current of the coils. When the actuation current was 250 mApp, the micro mirror achieved the maximum scanning angle of 26 degrees (see Figure 7c).

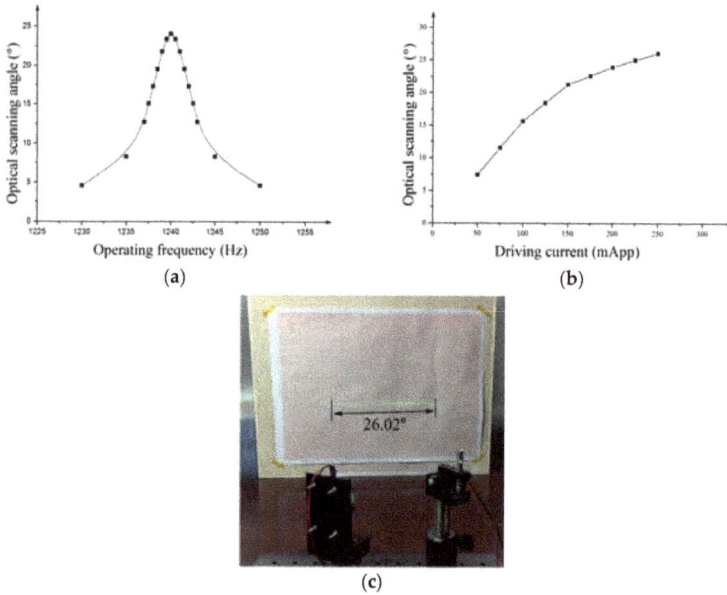

(a)

(b)

(c)

Figure 7. (**a**) Curve of optical scanning angle-versus-frequency; (**b**) curve of optical scanning angle versus actuation current; (**c**) maximum optical scanning angle of micro mirror.

The optical scanning angles were measured by the PSD optical device integrated in the package (see Figure 5a). An I-V conversion chip was applied to transform PSD's current signal into a voltage signal. Figure 8a shows the relationship between the PSD Voltages and the optical scanning angle. The relationship is linear and the correlation coefficient R^2 is 0.9959. The maximum amplitude of the scanning angle measurement in the experiment is 10 degrees while the whole optical scanning angle is 20 degrees. Amplitude of the scanning angle was measured by the PSD sensor and the precision was measured. Figure 8b shows the relationship between the measured angle and the amplitude of scanning angle with y-errors on the curve. Each scanning angle was measured nine times. The precisions of angle measurements at different scanning angles were calculated by achieving the 3σ error (thrice standard error) of each angle. Figure 8c shows the 3σ errors at each scanning angle and the maximum 3σ error (thrice standard error) of angle measurement is 0.07° at 10 degrees.

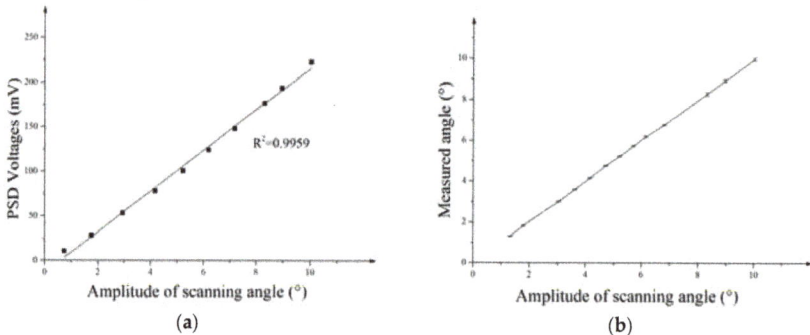

(a)

(b)

Figure 8. *Cont.*

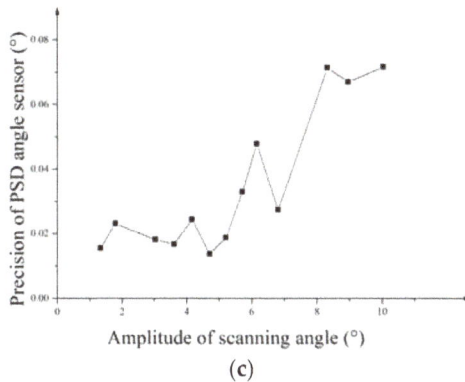

(c)

Figure 8. (**a**) The relationship between PSD Voltage and Optical scanning angle; (**b**) the relationship between measured angle and amplitude of scanning angle; (**c**) 3σ errors at each scanning angle.

The micro mirror was occupied in a one-axis LiDAR and one-axis detection was achieved. Figure 9a shows the setup of the LiDAR system. Laser was deflected by the micro mirror to the target and the angle of the target was measured by the mirror. The distance of the target was measured by the laser rangefinder module. The rangefinder module could detect the distance by measuring the phase delay between the output laser beam and the reflected laser beam. Figure 9c shows the one-axis detection picture of an optical post.

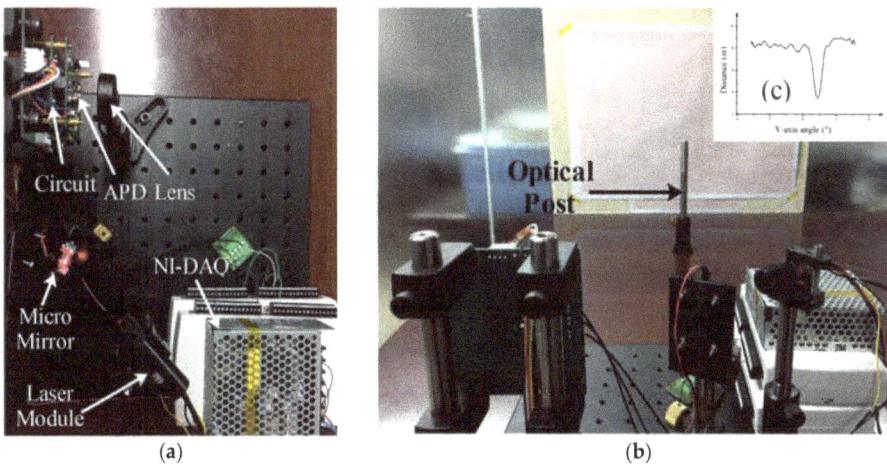

Figure 9. (**a**) Experimental setup of one-axis detection with micro mirror; (**b**) arrangement of LiDAR system and target; (**c**) one-axis detection picture of an optical post.

From the experimental results, we can see that in the first arrangement (see Figure 10a) the laser line scanned by the micro mirror is curve-shape distorted which will make our imaging solution complex (see Equation (4)). However, in the 2nd arrangement (see Figure 10b), the laser line is straight so that the imaging solution is simple (see Equation (3)). The experimental results fit our simulation results well.

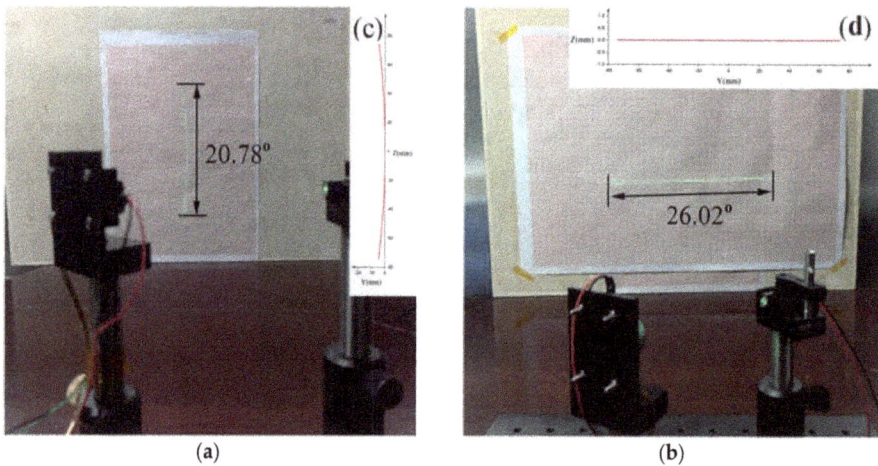

Figure 10. (**a**) Curve-shaped field distortion caused by arrangement 1 of micro mirrors; (**b**) scanned laser line on the screen in the 2nd arrangement; (**c**) simulation of curve-shaped field distortion caused by arrangement 1; (**d**) simulation of scanned laser line on the screen in the 2nd arrangement.

6. Discussion

It's very difficult to design a large-aperture and fast operating frequency micro mirror. From Equations (6)–(8), the resonance frequency can be calculated as:

$$f = k_{size,1}\sqrt{\frac{G}{\rho}} \tag{14}$$

where $k_{size,1}$ is a coefficient which is only related to the size of the structure. The ratio of shear modulus and density ($\sqrt{G/\rho}$) of the material itself determines the mechanical resonance frequency. It is similar to the effect of the ratio of Young's modulus and density ($\sqrt{E/\rho}$) [15], because Young modulus (E) and shear modulus (G) has the relationship of $G = E/(2 + 2\nu)$ and poison ratios of most common materials are between 0.2 and 0.4.

On the other hand, the maximum shear stress of the micro scanner should be maintained to be less than the strength of the material. The maximum shear stress of the micro-scanner can be given as [19]:

$$\tau_{max} = \frac{\beta G w}{2\alpha l}\theta_{max} \leq [\tau] \tag{15}$$

where α and β are coefficients related to the width (w) and depth (h) of the beam [20] and $[\tau]$ is the shear strength of material.

When designing a larger micro-scanner, the moment of inertia becomes larger and the resonance frequency decreases. From Equation (7), to increase the resonance frequency of the large-aperture micro-scanner, the width (w) and the depth (h) should be increased and the length l should be decreased. From Equation (14) we can see that the shear stress will be increased which will lead to a decrease in maximum scanning angle. It is difficult to design a micro-scanner with large-aperture and large resonance frequency at the same time.

Table 1 lists the parameters of some common materials. The ratio $\sqrt{E/\rho}$ of silicon is about two larger than other materials. However, from the realized large-aperture Si based micro scanner [5,10–12], mirror size of silicon based micro-scanner is below 1 cm and the frequency is less than 350 Hz. The metal based micro scanner is more robust than MEMS devices due to its ductile properties in comparison with brittle silicon substrates [15].

For most metal material, the ratio $\sqrt{E/\rho}$ is approximately equal. Therefore, micro-scanners with the same size but different metal material have approximately the same resonance frequency. From Equation (5), we can see that for the micro-scanner with the same size, the shear stress is proportional to shear modulus G. From Table 2, we can see that Young modulus of stainless-steel is about 1.7 times larger than Ti-alloy. Compared with the stainless-steel based micro scanner, the Ti-alloy based micro scanner has lower shear stress and higher tensile strength. The maximum scanning angle of the Ti-alloy based micro scanner is significantly larger than the stainless-steel based micro mirror.

Table 2. Parameters of some common materials [21].

Material	Young Modulus (GPa)	Density (kg·m^{-3})	Poison Ratio	$\sqrt{E/\rho}$ (m/s)	Tensile Strength (MPa)	$[\sigma]/E$ (1×10^{-3})
Silicon	150–170	2.33×10^3	0.28	8023–8541	1320 [1]	8.3–8.8
SUS304 Stainless steel	193	7.9×10^3	~0.3	4943	520	2.7
TC4 Ti-alloy	116	4.5×10^3	0.34	5077	895	7.7
7050 Al alloy	68.5	2.7×10^3	~0.3	5036	485	7.1

[1] The size of the Silicon specimen is $3000 \times 250 \times 24.5$ μm^3 [14].

Relative shear strength can be defined here which can be given as:

$$[\tau]_r = \frac{[\tau]}{G} \tag{16}$$

From Equation (15), we can calculate the relationship between maximum scanning and shear strength:

$$\theta_{max} \leq k_{size,2}[\tau]_r \tag{17}$$

where $k_{size,2}$ is a coefficient which is only related to size.

For most material, tensile strength can be obtained from research and we calculate the relative tensile strength $[\sigma]/E$ instead. Table 1 demonstrates that Ti-alloy has the maximum relative strength. Ti-alloy substrate is chosen to design the micro scanning mirror.

Our device has the advantages of mirror plate size and scanning frequency. In comparison to the Galvo Scanner (see Table 3) which is usually occupied in the LiDAR system, the Ti-alloy-based micro-mirror has comparable large apertures while the operating frequency is about an order of magnitude larger. Compared with large-aperture MEMS scanning mirror, the aperture, operating frequency and scanning angle are much larger. The MEMS mirror has the advantages of small size and low power consumption while it does not meet traditional LiDAR scanner's demand like the large aperture. Our device is designed using the MEMS design method and it can link the gap between the traditional scanner and the MEMS scanning mirror. The Ti-alloy based micro mirror can meet the demand of the LiDAR system while it retains some advantages of the MEMS mirror like small size and low power consumption.

Table 3. Features of our device in comparison with Galvo Scanner and large-aperture MEMS scanning mirror.

Type	Diameter/mm	Frequency/Hz	FOV/Deg	Driving Voltage/Current	Angle Measurement
Our device	12	1240	26	250 mA	PSD
Fraunhofer [5]	2.5×9.5	250	30	180 Vpp	Photodiodes
Huikai Xie [10]	10×10	234	10	7 Vpp	NO
Mirrorcle tech. [11,12]	5	330/334	10×10	157 V	NO
Galvo Scanner [22]	10	150	40	1.25 A rms	YES

7. Conclusions

A novel, one-axis Ti alloy-based electromagnetic micro scanning mirror with a very large-aperture and rapid resonant frequency is presented in this paper. The micro mirror is designed for the demand of a MEMS based LiDAR system. Ti-alloy substrate is used to achieve larger aperture and faster scanning frequency and a pair of moon-like magnets are used to achieve larger optical scanning angle with low actuation current. The Ti alloy-based electromagnetic micro-scanner has very large-aperture (12 mm) and rapid scanning frequency (1.24 kHz). The optical scanning angle can reach 26 degrees when the actuation current is 250 mApp. In comparison with the Galvo Scanner which is usually occupied in the LiDAR system, the Ti-alloy-based micro-mirror has comparable large apertures while the operating frequency is about an order of magnitude larger. The size of the micro-scanner reached 12 mm which is much larger than similar MEMS scanning mirrors. The Ti alloy-based large-aperture micro scanner will speed up the micro mirror's application in LiDAR.

Acknowledgments: The work was supported by the Research Foundation of Tsinghua University.

Author Contributions: Liangchen Ye is responsible for the research. Gaofei Zhang and Zheng You gave advice on the experiment. Liangchen Ye performed the experiments and analyzed the data.

Conflicts of Interest: The authors declare no conflict of interest.

References

1. Kasturi, A.; Milanovic, V.; Atwood, B.H.; Yang, J. UAV-Borne LiDAR with MEMS Mirror-Based Scanning Capability. In Proceedings of the SPIE 9832, Laser Radar Technology and Applications XXI, Baltimore, MD, USA, 19–20 April 2016; Volume 9832.
2. Ataman, C.; Lani, S.; Noell, W.; de Rooij, N.; Abe, K. A dual-axis pointing mirror with moving-magnet actuation. *J. Micromech. Microeng.* **2013**, *23*, 25002. [CrossRef]
3. Lee, X.; Wang, C. Optical design for uniform scanning in MEMS-based 3D imaging LiDAR. *Appl. Opt.* **2015**, *54*, 2219–2223. [CrossRef] [PubMed]
4. Zhang, X.; Koppal, S. J.; Zhang, R.; Zhou, L.; Butler, E.; Xie, H. Wide-angle structured light with a scanning MEMS mirror in liquid. *Opt. Express* **2016**, *24*, 3479–3487. [CrossRef] [PubMed]
5. Sandner, T.; Wildenhain, M.; Gerwig, C.; Schenk, H.; Schwarzer, S.; Wolfelschneider, H. Large aperture MEMS scanner module for 3D distance measurement. In Proceedings of the SPIE 7594, MOEMS and Miniaturized Systems IX, San Francisco, CA, USA, 25–27 January 2010; Volume 7594, p. 75940D.
6. Hu, Q.; Pedersen, C.; Rodrigo, P.J. Eye-safe diode laser Doppler LiDAR with a MEMS beam-scanner. *Opt. Express* **2016**, *24*, 1934–1942. [CrossRef] [PubMed]
7. Hofmann, U.; Senger, F.; Soerensen, F.; Stenchly, V.; Jensen, B.; Janes, J. Biaxial resonant 7mm-MEMS mirror for automotive LIDAR application. In Proceedings of the 2012 International Conference on Optical MEMS and Nanophotonics, Banff, AB, Canada, 6–9 August 2012; pp. 150–151.
8. Holmstrom, S.T.S.; Baran, U.; Urey, H. MEMS Laser Scanners: A Review. *J. Microelectromech. Syst.* **2014**, *23*, 259–275. [CrossRef]
9. Wolter, A.; Hsu, S.-T.; Schenk, H.; Lakner, H.K. Applications and requirements for MEMS scanner mirrors. In Proceedings of the SPIE 5719, MOEMS and Miniaturized Systems V, San Jose, CA, USA, 25–26 January 2005; pp. 64–75.
10. Wu, L.; Xie, H. Large-aperture, rapid scanning MEMS micromirrors for free-space optical communications. In Proceedings of the IEEE/LEOS International Conference on Optical MEMS and Nanophotonics 2009, Clearwater, FL, USA, 17–20 August 2009; pp. 131–132.
11. Milanovic, V.; Matus, G.A.; McCormick, D.T. Gimbal-less monolithic silicon actuators for tip-tilt-piston micromirror applications. *IEEE J. Sel. Top. Quantum* **2004**, *10*, 462–471. [CrossRef]
12. Mirrorcle. Support. Available online: http://www.mirrorcletech.com/support.php (accessed on 3 March 2017).
13. Wolter, A.; Schenk, H.; Korth, H.; Lakner, H.; Abe, K. Torsional stress, fatigue and fracture strength in silicon hinges of a micro scanning mirror. In Proceedings of the SPIE 5343, Reliability, Testing, and Characterization of MEMS/MOEMS III, San Jose, CA, USA, 24 January 2004; pp. 176–185.

14. Namazu, T.; Isono, Y. Fatigue Life Prediction Criterion for Micro—Nanoscale Single-Crystal Silicon Structures. *J. Microelectromech. Syst.* **2009**, *18*, 129–137. [CrossRef]

15. Park, J.H.; Akedo, J.; Sato, H. High-speed metal-based optical microscanners using stainless-steel substrate and piezoelectric thick films prepared by aerosol deposition method. *Sens. Actuators A Phys.* **2007**, *135*, 86–91. [CrossRef]

16. Wang, Y.; Gokdel, Y.D.; Triesault, N.; Wang, L.; Huang, Y.Y.; Zhang, X. Magnetic-Actuated Stainless Steel Scanner for Two-Photon Hyperspectral Fluorescence Microscope. *J. Microelectromech. Syst.* **2014**, *23*, 1208–1218. [CrossRef]

17. Urey, H.; Kan, C.; Davis, W.O. Vibration mode frequency formulae for micromechanical scanners. *J. Micromech. Microeng.* **2005**, *15*, 1713–1721. [CrossRef]

18. Davis, W.O. Empirical analysis of form drag damping for scanning micromirrors. In Proceeding of the Conference on MOEMS and Miniaturized Systems VIII, San Jose, CA, USA, 27–28 January 2009; Volume 7208.

19. Zhang, C.; Zhang, G.; You, Z. A Two-Dimensional Micro Scanner Integrated with a Piezoelectric Actuator and Piezoresistors. *Sensors* **2009**, *9*, 631–644. [CrossRef] [PubMed]

20. Ren, W.; Chen, Y.; Fan, Q. *Mechanics of Materials*; Tsinghua University Press: Beijing, China, 2005; pp. 54–55.

21. Wen, B.; Wang, B.; Lu, X. *Manual of Metal Material*; Electronic Industry Press: Beijing, China, 2013; pp. 157, 526, 585, 625, 636.

22. Thorlabs. GVS411—1D Large Beam (10 mm) Diameter Galvo System, UV Enhanced Aluminum Mirror, PSU Not Included. Available online: https://www.thorlabs.com/thorproduct.cfm?partnumber=GVS411 (accessed on 3 March 2017).

micromachines

MDPI

Article

An Enhanced Robust Control Algorithm Based on CNF and ISM for the MEMS Micromirror against Input Saturation and Disturbance

Jiazheng Tan [1], Weijie Sun [1] and John T. W. Yeow [2,*]

[1] College of Automation Science and Engineering, South China University of Technology, Guangzhou 510000, China; tanjiazheng1992@outlook.com (J.T.); auwjsun@scut.edu.cn (W.S.)

[2] Advanced Micro-/Nano-Devices Lab, Department of Systems Design Engineering, Waterloo Institute for Nanotechnology, University of Waterloo, 200 University Avenue West, Waterloo, ON N2L 3G1, Canada

* Correspondence: jyeow@uwaterloo.ca

Received: 14 September 2017; Accepted: 2 November 2017; Published: 3 November 2017

Abstract: Input saturation is a widespread phenomenon in the field of instrumentation, and is harmful to performance and robustness. In this paper, a control design framework based on composite nonlinear feedback (CNF) and integral sliding mode (ISM) technique is proposed for a MEMS micromirror to improve its performance under input saturation. To make the framework more effective, some essential improvements are supplied. With the application of the proposed design framework, the micromirror under input saturation and time-varying disturbances can achieve precise positioning with satisfactory transient performance compared with the open-loop performance.

Keywords: micromirror; input saturation; disturbance rejection; composite nonlinear feedback; integral sliding mode

---

1. Introduction

Since the first micromirror for scanning applications was reported in 1980 [1], the research of microelectromechanical systems (MEMS) micromirrors has become increasingly popular. Early MEMS micromirrors focused on imaging applications such as confocal microscopy [2] and fingerprint sensing [3]. Later, it was extended to some other different areas such as optical coherence topography [4], optical switches [5], and high-resolution displays [6]. Recently, some new applications include response surface method [7] and ghost imaging [8]. On the basis of actuation method, torsional micromirrors can be classified into different types: electrothermal [9], electrostatic [10], electromagnetic [11,12], piezoelectric [13], and so forth. Recently, the electromagnetic micromirrors have attracted special attention owing to their ability to produce large scan angles with low voltage and remote actuation (meaning that the mirrors can be controlled by the non-contact force at a distance).

In this paper, we study the electromagnetic micromirror reported in [11]. which was made by Pallapa and Yeow. The structure of the mirror is shown in Figure 1. In their original work, it is reported that the mirror was fabricated by the application of hybrid MEMS fabrication using polymers and magnets [11]. In the following the fabrication process is briefly provided [6]. In the first place, adopting the lithography method, the mould of the micromirror and two torsional bars are fabricated with a silicon wafer used as the mould substrate. The mould is fabricated by standard photolithography. Next, polydimethylsiloxane (PDMS) is filled into the mould as a basement of the micromirror and two torsional bars, and the MQFP-12-5 isotropic magnetic powder (Nd-Fe-B) (Magnequench International Inc., Pendleton, IN, USA) with a particle size D50 of 5 μm is doped into the polydimethylsiloxane (PDMS) at the weight percentage of 80%. Subsequently, an ultrasonic horn tip probe is immersed into the composite, therefore leading to uniform dispersion. Then, using the e-beam evaporation approach, the micromirror is plated with a layer of gold with a thickness of 1.0 mm as a reflective layer. Finally, the micromirrors are magnetized under a field of 1.8 Tesla, resulting in that the micromirror contains a hard magnetic feature. Besides, the rectangle coils are fabricated by the standard printed circuit board (PCB) manufacturing technique and constructed under the micromirror. The detailed design parameters of the mirrorare shown in Table 1.

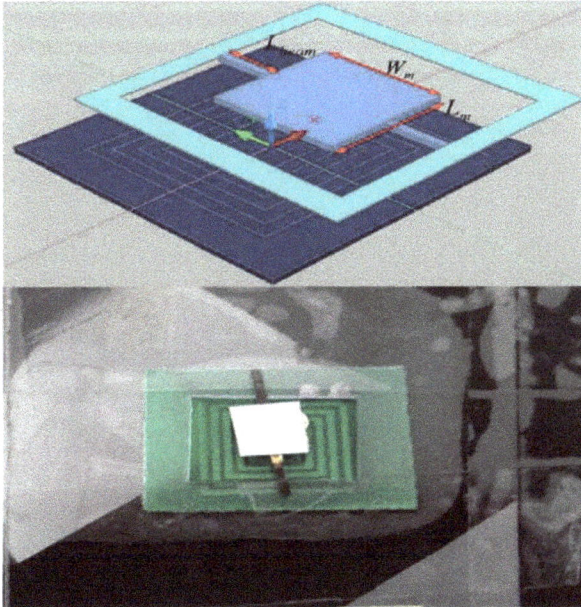

Figure 1. The structure of the micromirror. The micromirror consists of the micromirror-plate, the torsional beam, the driving microcoil, a support structure and other parts. Two straight torsion micromirror bars are fixed on the frame at a distance position 1 mm from the left side center of the mirror to coil center. The axis of rotation of the scanner is along Y vector. Rectangular spiral coil are assembled to the bottom of the frame in the X, Y plane and the center of the coil is the origin in Cartesian Coordinate.

Table 1. Paramenters of the micromirror.

Symbol	Parameter	Value
W_{beam}	width of the torsion bar	250 μm
L_{beam}	length of the torsion bar	2 mm
t_{beam}	thickness of the torsion bar	250 μm
W_m	width of the mirror	4 mm
L_m	length of the mirror	4 mm
t_m	thickness of the mirror	250 μm

To rotate an electromagnetic micromirror, a peripheral driving circuit based on a microcoil is essential. Clearly, the amplitude of the driving circuit's output is limited, meaning that the plant's input is subject to input saturation. Therefore, it is essential to propose a suitable method to guarantee the performance of the micromirror under input saturation. Because the mathematical model of the micromirror is linear [12], we will search for a solution to the input saturation problem in the field of linear system control theory.

The input saturation problem has been extensively investigated in the field of linear system control theory. For this problem, transient performance of the closed-loop system should not be neglected. Settling time and overshoot are the two most important evaluation indices of transient performance. However, it is unfortunate that short settling time and negligible overshoot are usually contradictory aspects when facing input saturation, which means that most of the existing control techniques cannot help but make a trade-off between them. As a kind of nonlinear control scheme, the composite nonlinear feedback (CNF) technique has been presented to tackle such a dilemma [14]. The CNF technique is composed of two components; i.e., a linear part and a nonlinear part. The linear feedback part is introduced to yield a closed-loop system with small damping ratio, while the nonlinear feedback part is designed to increase the damping ratio when the output signal reaches the reference asymptotically. Therefore, it is possible for the CNF technique to simultaneously achieve quick response and negligible overshoot. The CNF technique has also been implemented in various of engineering applications such as hard disk drive servo systems [15–17], helicopter flight control systems [18,19], position servo systems [20,21], and grid-connected voltage source inverters [22].

The normal CNF technique is able to improve the transient performance of controlled systems in the absence of external disturbances [15,16]. However, disturbances are non-negligible in practical environments, especially for the problems in high-precision control fields. Thus, both the enhanced composite nonlinear feedback and robust composite nonlinear feedback techniques are formulated to deal with constant disturbances [17,21]. Without considering input saturation, robust design for CNF technique has also been achieved by improving the nonlinear part to deal with unmodelled dynamic effects [23]. Furthermore, due to the easy implementation property, the integral sliding mode technique has been combined with the CNF technique for the rejection of time-varying disturbances under input saturation [24,25].

Our motivation is to achieve good transient performance and disturbance rejection for the micromirror simultaneously under input saturation. Previous work of control system design for the electromagnetic micromirror can be seen in [26,27]. However, input saturation was not considered in those works. As shown in Figure 2, the open-loop performance of the micromirror is far from favourable (long settling time and large overshoot), meaning that further improvement is essential (the difference between the two responses is mainly because of the unmodelled dynamics and the air damping effect). Inspired by the aforementioned discussion, in this paper, we apply the integral sliding mode (ISM)-based CNF design framework considered by Bandyopadhyay, Deepak, and Kim [25]. Furthermore, to improve this framework, some improvements are introduced. With the proposed improvement, it is expected that the proposed control scheme would have much better robustness under input saturation than the existing work [25], while the transient and steady-state performance become much more satisfying. The effectiveness of the proposed design framework is illustrated by experimental results, and it would show that the proposed framework forces the micromirror to perform well under input saturation and disturbance.

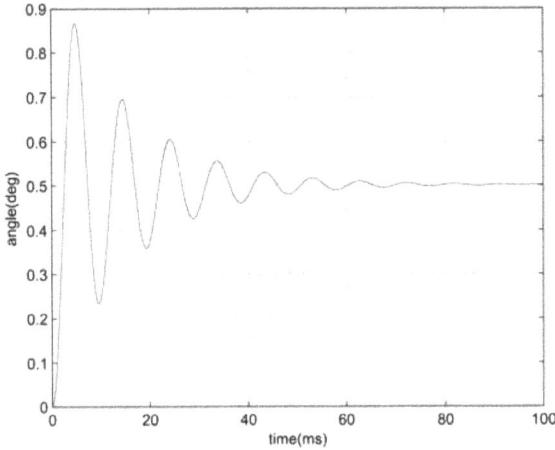

Figure 2. Open-loop positioning performance under input saturation (Theoretical model simulation).

This paper is organized as follows. In Section 2, the whole control design framework and the relevant improvements are proposed. Then, the effectiveness of the enhanced control scheme is examined by experimental results in Section 3. In Section 4, conclusions are drawn.

2. ISM-Based CNF Control Design

Consider a linear plant described in the following form with input saturation and disturbance:

$$\dot{x} = Ax + Bsat(u) + Bw(x,t), x(0) = x_0$$

$$y = Cx \qquad (1)$$

where $x \in R^n$, $u \in R$, $y \in R$, and $w \in R$ are the state, control input, output, and disturbance input of the system. A, B, C are constant matrices with appropriate dimensions.

The function, $sat(\cdot) : R \rightarrow R$, represents the input saturation defined as

$$sat(u) = sgn(u)min\{u_{max}, |u|\} \qquad (2)$$

where u_{max} is the saturation level of the input.

To introduce the whole control design framework, the following standard assumptions on the given system are made

A1: (A, B) is stabilizable.

A2: (A, C) is detectable.

A3: (A, B, C) is invertible with no invariant zero at $s = 0$.

A4: $w(x, t)$ represents the bounded matched uncertainty or disturbance and

$$|w(x,t)| \leq w_{max} \qquad (3)$$

where w_{max} is the maximum amplitude of w.

In the following, we will introduce an ISM-based CNF scheme for system (Equation (2)) such that the controlled output y can asymptotically achieve the positioning of reference input r under constant or time-varying disturbances within limits.

The ISM-based CNF scheme can be constructed by the following a step-by-step design procedure.

(1) Design the part of CNF control law [17].

The linear feedback part is first given as

$$u_L = Fx + Gr \tag{4}$$

r denotes the constant reference signal, F is the feedback gain of the states, and G is the feedforward gain of the reference. F is chosen such that $(A + BF)$ is a Hurwitz matrix and the closed-loop system $C(sI - A + BF)^{-1}B$ has a small damping ratio [15].

Besides, G is defined as

$$G = -[C(A + BF)^{-1}B]^{-1} \tag{5}$$

Next, the nonlinear feedback part is formulated as

$$u_N = \rho(y - r)B^T P(x - x_e) \tag{6}$$

where

$$x_e = -(A + BF)^{-1}BGr \tag{7}$$

and $\rho(y - r)$ is a non-positive nonlinear function locally Lipschitz in $(y - r)$. u_N is introduced to change the system closed-loop damping ratio as the output approaches the step command input [15]. In this paper, we select

$$\rho(y - r) = -\beta e^{-\alpha \alpha_0 |y - r|} \tag{8}$$

where α and β are positive parameters to be tuned, and

$$\alpha_0 = \begin{cases} \frac{1}{|h_0 - r|}, & r \neq h_0 \\ 1, & r = h_0 \end{cases} \tag{9}$$

For simplicity, we set $h_0 = 0$ in this paper. With the discontinuous coefficient α_0, $\rho(y - r)$ can suit the amplitude variation of constant reference [16].

In (6), $P > 0$ is the solution to the following Lyapunov equation:

$$(A + BF)^T P + P(A + BF) = -W \tag{10}$$

where W is a positive-definite matrix. It is known that the solution P always exists if $(A + BF)$ is Hurwitz.

For convenience, we set W as

$$W = 10^\theta \cdot \hat{E} \tag{11}$$

where $\hat{E}$ represents the appropriate dimensional identity matrix and $\theta \in R$.

In this respect, the linear feedback part and nonlinear feedback part can be combined to form the CNF control law

$$u_{CNF} = u_L + u_N = Fx + Gr + \rho(y - r)B^T P(x - x_e) \tag{12}$$

The following theorem shows that the closed-loop system comprising the given plant (Equation (2)) with and the CNF control law of Equation (12) is asymptotically stable. It also determines the magnitude of r that can be tracked by such a control law without exceeding the control limit [17].

Theorem 1. *Under assumptions A1–A4, for the closed-loop system composed of Equations (2) and (12), if the following conditions are satisfied*

(1) For any $\delta \in (0, 1)$, there exists a largest positive scalar $c_\delta > 0$ such that

$$\forall \tilde{x} \in X(F, c_\delta) := \{\tilde{x} : \tilde{x}^T P \tilde{x} \leq c_\delta\} \Rightarrow |F\tilde{x}| \leq (1 - \delta)u_{max}, \tag{13}$$

where $0 < \delta < 1$ *and*

$$\tilde{x} = x - x_e \tag{14}$$

(2) The initial condition x_0 satisfies

$$x_0 - x_e \in X(F, c_\delta) \tag{15}$$

(3) The amplitude of the reference satisfies

$$|Hr| \leq \delta_1 u_{max} \tag{16}$$

where $0 < \delta_1 < \delta$, $H = FG_e + G$.

Then, for any non-positive function $\rho(y-r)$ locally Lipschitz in $(y-r)$, the control law (Equation (12)) can stabilize the system (Equation (2)) and drive the output of Equation (2) to asymptotically track the command reference in the absence of disturbances w as well as in the presence of input saturation [15].

(2) Design the part of integral sliding mode law.

Define the following notations:

$$x_d = \int_0^t (Ax(\tau) + Bu_{CNF}(\tau))d\tau$$

$$e = C_0(x - x_d) \tag{17}$$

where C_0 is an appropriate matrix satisfying $C_0 B > 0$.

x_d represents the trajectory of the nominal plant (the original plant that does not suffer from disturbance). It is clear that the boundedness of x_d is equivalent to the stability of the closed-loop system (Equation (2)) with the CNF control law (Equation (12)) in the absence of disturbances w and input saturation, and it has been proved in [17].

The origin of e is the existence of w, and the goal of the integral sliding mode law is to eliminate the influence of w, forcing x to follow x_d.

Further, choose the following sliding manifold

$$s(e, t) = k_1 e + k_2 \int_0^t e(\tau)d\tau \tag{18}$$

where k_1, k_2 should be properly chosen to ensure $\dot{s}(e, t) = 0$ is strictly Hurwitz.

Different from the work of Bandyopadhyay, Deepak, and Kim [25], an additional integral term $\int_0^t e(\tau)d\tau$ is supplied to the sliding manifold (Equation (18)). The effect of the integral term is twofold. Firstly, it can bring to faster responses, which is also proved by [28]. Furthermore, it can better force the trajectory to stay on the sliding surface $s = 0$ in the presence of parameter perturbation (because it is made of polymer, the mirror is soft, meaning that it is likely for it to suffer from parameter perturbation).

Construct the following Lyapunov function:

$$V = \frac{1}{2}s(e, t)^T s(e, t) \tag{19}$$

It can be calculated that

$$
\begin{aligned}
\dot{V} &= [k_1 C_0(\dot{x} - \dot{x}_d) + k_2 C_0(x - x_d)]s(e, t) \\
&= \{k_1(u_{ISM} + w) + k_2 \int_0^t [u_{ISM}(\tau) + w(\tau)]d\tau\}C_0 Bs(e, t)
\end{aligned}
\tag{20}
$$

Accordingly, the integral sliding mode control law is constructed as

$$u_{ISM} = u_{eq} + u_{sw} \tag{21}$$

where u_{eq} is the equivalent control and u_{sw} is the switched control.

Let

$$u_{eq} = -w(x,t) \tag{22}$$

and

$$u_{sw} = -Msign[C_0Bs(e,t)] - N[C_0Bs(e,t)] \tag{23}$$

where M, N are positive parameters and $M \geq w_{max}$.

The term $-Msign[C_0Bs(e,t)]$ is discontinuous, and it probably causes high-frequency unmodeled dynamics. The high-frequency unmodeled dynamics can lead to the chattering phenomenon, which seriously harms the actuator and the direct application of the mirror. The application of the continuous term can reduce the discontinuous effect. Therefore, the term $-N[C_0Bs(e,t)]$ is introduced to reduce chattering, which is different from the work of Bandyopadhyay, Deepak, and Kim [25].

Applying the ISM law (Equation (21)), it is obvious that

$$\dot{V} \leq 0, \quad (\dot{V} = 0, s = 0)$$

Thus

$$\lim_{t \to \infty} e = 0 \Rightarrow \lim_{t \to \infty} x(t) = x_d(t)$$

So

$$\lim_{t \to \infty} y(t) = \lim_{t \to \infty} Cx(t) = Cx_e = -C(A + BF_x)^{-1}BGr = r \tag{24}$$

Now, the whole control scheme composed of (Equations (12) and (21)) could be expressed as

$$u = u_{CNF} + u_{ISM} \tag{25}$$

and it can stabilize the system (Equation (2)) and drive the output of Equation (2) to asymptotically track the command reference under the above conditions (Equations (13), (15) and (16)) in the presence of disturbances w.

When we design the ISM part, the input saturation is not taken into account. Clearly, to ensure that the amplitude of Equation (25) will not go beyond the limit, the constraint of the maximum amplitude of w is essential.

Therefore, we need the following theorem.

Theorem 2. *Under assumptions A1–A4 and the conditions Equations (13), (15) and (16), for the closed-loop system composed of Equations (2) and (25), if the maximum amplitude of disturbance satisfies*

$$|(\delta - \delta_1)u_{max}| = w_{max} \tag{26}$$

then, by the application of Equation (25), the output of Equation (2) can asymptotically track the command reference and the amplitude of Equation (25) will not go beyond the limit [25].

The block diagram of the proposed control design is shown in Figure 3.

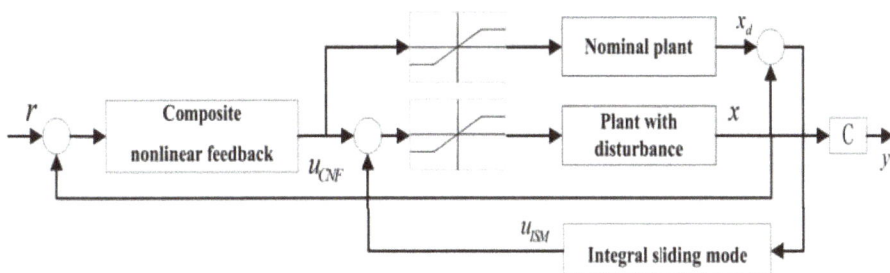

Figure 3. The integral sliding mode-composite nonlinear feedback (ISM-CNF) controller diagram.

3. Experimental Results

In this section, we validate the performance of the aforementioned control scheme composed of composite nonlinear feedback and integral sliding mode technique. The validation is performed on a torsional micromirror-based experimental platform. The platform is composed of a He-Ne laser, a micromirror together with coils, a PSM2-10 position-sensitive detector (PSD), a voltage-controlled current amplifier (VCCA) circuit, and an NI PXI-7852R field-programmable gate array (FPGA) system (see Figure 4). The structure of the micromirror is depicted in Figure 1. The controller is programmed in the FPGA card. The output of the FPGA card, which is the output of the controller, is in the form of voltage. The voltage signal is converted to current by the VCCA circuit proportionally, and the ratio is one to one. A magnetic field sets up when the currents are flowing in the coils, resulting the Lorentz force to rotate the mirror. The amplitude of current is restricted within 1.0 A because the coils have a restriction on maximum passing current.

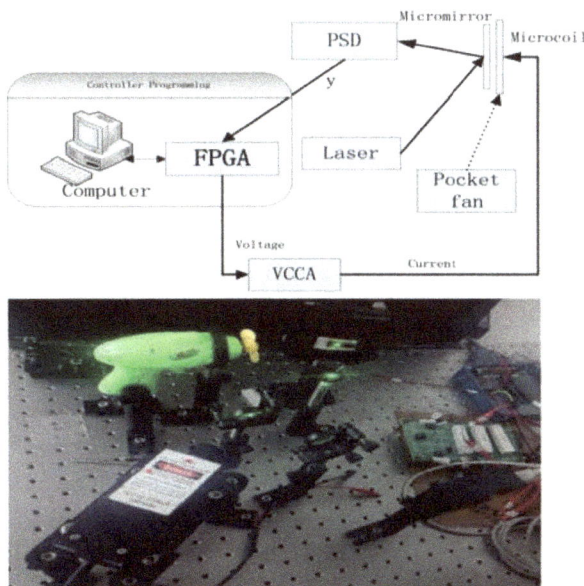

Figure 4. The experimental platform. FPGA: field-programmable gate array; PSD: position-sensitive detector; VCCA: voltage-controlled current amplifier.

The model of the micromirror under external disturbance and input saturation is shown as follows [12]

$$
\begin{bmatrix} \dot{x}_1 \\ \dot{x}_2 \end{bmatrix} = \begin{bmatrix} 0 & 1 \\ -\frac{k}{J} & -\frac{b}{J} \end{bmatrix} \begin{bmatrix} x_1 \\ x_2 \end{bmatrix} + \begin{bmatrix} 0 \\ \frac{1}{J} \end{bmatrix} (sat(u) + w)
$$

$$
\hat{y} = \begin{bmatrix} 1 & 0 \end{bmatrix} \begin{bmatrix} x_1 \\ x_2 \end{bmatrix} = x_1 \tag{27}
$$

where x_1, x_2, u are respectively the angular position of the micromirror, the angular velocity, and the driving torque. $J = 6.26 \times 10^{-12} (\mathrm{kg \cdot m^2})$ is the moment of inertia, $b = 0.82 \times 10^{-9} (\mathrm{N \cdot m \cdot s})$ is the damping coefficient, and $k = 2.96 \times 10^{-6} (\mathrm{N \cdot m})$ is the spring coefficient of the torsional bars. The practical open-loop performance is shown in Figure 5. One may have noticed some differences between the theoretical open-loop performance and practical open-loop performance (settling time and overshoot). The possible reasons are shown as follows. Firstly, the mirror's practical model is not totally linear. In other words, the nonlinear unmodelled dynamics can influence the performance. Secondly, in the theoretical model of the mirror, air frication is not considered, which can lead to smaller overshoot and shorter settling time.

The controller parameters are chosen as follows. For the CNF part, $F = [-0.58 \quad -6.2 \times 10^{-4}]$, $\alpha = 8.7 \times 10^{-5}$, $\beta = 2.6 \times 10^{-5}$, $\theta = 0.268$. For the ISM part, we select $M = 1$, $N = 2.5$, $k_1 = 450$, $k_2 = 2$, $C_0 = [1 \quad J]$. To summarize, the proposed scheme is achieved as the sum of two parts.

In the following experiment, to state the results more clearly, we also provide the performance of the controller proposed by Bandyopadhyay, Deepak, and Kim. The controller is shown as follows:

$$
\begin{aligned}
u_{BDK} &= u_{CNF} + \bar{u}_{ISM} \\
\bar{u}_{ISM} &= u_{eq} + \bar{u}_{sw} \\
\bar{u}_{sw} &= -\bar{M}sign[\bar{G}B\bar{s}(e,t)] \\
\bar{s} &= \bar{G}(x - x_d)
\end{aligned} \tag{28}
$$

Obviously, the difference between the two controllers is in the part of ISM. Therefore, in the part of CNF, we select the same parameters. In the part of ISM, for the above controller, we select $\bar{G} = [0 \quad J]$, $\bar{M} = 500$.

The positioning experimental results are shown in Figure 5, in which the output responses of the closed-loop system and open-loop system are given. The figure has shown that the torsional micromirror can achieve the expected precise positioning with the distinguished features (i.e., high-speed seeking performance and negligible overshoot under control input saturation). Clearly, the positioning performance of the proposed control scheme under input saturation is validated.

To examine the robustness, additional introduced disturbances are imposed on the closed-loop system. The whole disturbances are divided into two categories (i.e., w_1 and w_2, where $w_1 = 0.3\sin(1200t)$ V is presented in the form of voltage and w_2 is the wind interference generated by a pocket fan, as shown in Figure 4). The level of the disturbance from the pocket fan is equivalent to a voltage disturbance with the amplitude of 0.34 V. The regulation performance of the torsional micromirror under the following three cases are studied: (1) $r = 0.4°$, $w = w_1$; (2) $r = 0.4°$, $w = w_2$; (3) $r = 0.4°$, $w = w_1 + w_2$. The experimental results are shown in Figures 6–8. Compared with the controller proposed by Bandyopadhyay, Deepak, and Kim [25], the proposed scheme with the improvement forces the outputs to settle into the target asymptotically. It indicates that the whole design framework is workable and the proposed improvement is essential.

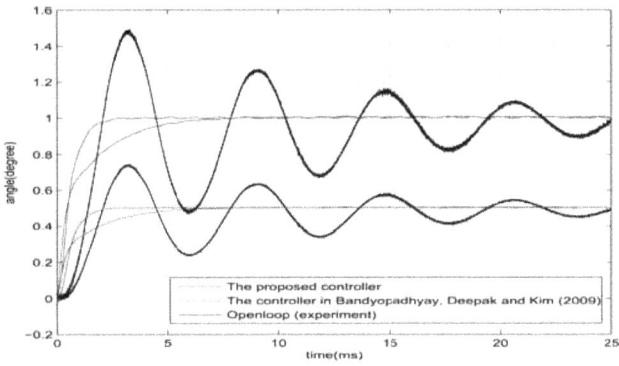

Figure 5. Positioning performance under input saturation.

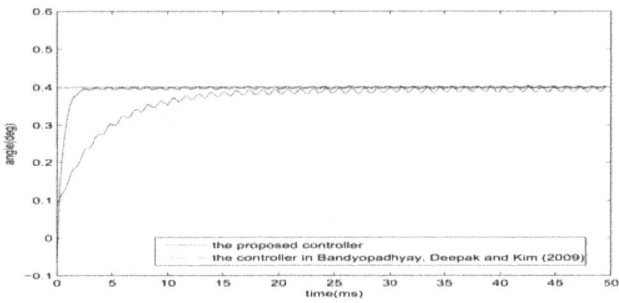

Figure 6. $r = 0.4°$, $w = w_1$.

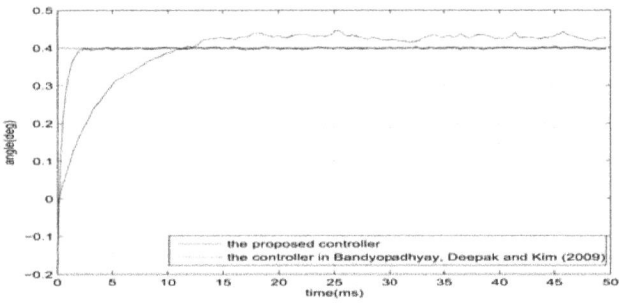

Figure 7. $r = 0.4°$, $w = w_2$.

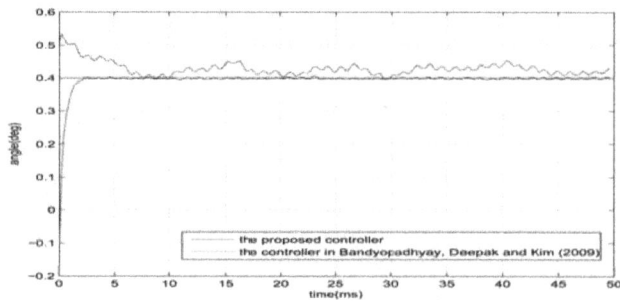

Figure 8. $r = 0.4°$, $w = w_1 + w_2$.

In general, the overall results state clearly that the proposed scheme is effective in improving the transient and steady-state performance and robustness under input saturation.

4. Conclusions

In this paper, to deal with input saturation and achieve disturbance rejection for the MEMS micromirror, a robust control design framework based on composite nonlinear feedback and integral sliding mode is proposed, and some essential improvement is supplied to further enhance the framework. The effectiveness of the proposed scheme is verified by experimental results. Applying the proposed scheme, the micromirror can deal with input saturation and time-varying disturbances at the same time, while providing much better transient and steady-state performance compared with the open-loop performance.

Acknowledgments: This work is supported by National Natural Science Foundation of China under Grants 61374036, Natural Science Foundation of Guangdong Province under Grants 2015A030313200, and Science & Technology Project of Guangdong Province under Grants 2017A010101009.

Author Contributions: Jiazheng Tan and Weijie Sun conceived and designed the proposed controller; Jiazheng Tan performed the experiments; John T. W. Yeow contributed experimental conditions; Jiazheng Tan wrote the tpaper.

Conflicts of Interest: The authors declare no conflict of interest.

References

1. Petersen, K.E. Silicon torsional scanning mirror. *IBM J. Res. Dev.* **1980**, *24*, 631–637.
2. Dickensheets, D.L.; Kino, G.S. Micromachined scanning confocal optical microscope. *Opt. Lett.* **1996**, *21*, 764–766.
3. Neukermans, A.P.; Slater, T.G. Micromachined Torsional Scanner. U.S. Patent 5,629,790, 13 May 1997.
4. Yeow, J.T.W.; Yang, X.D.; Chahwan, A.; Gordon, M.L.; Qi, B.; Vitkin, A.; Wilson, C.; Goldenberg, A. Micromachined 2-D scanner for 3-D optical coherence tomography. *Sens. Actuators A Phys.* **2005**, *117*, 331–340.
5. Chu, P.B.; Lee, S.S.; Park, S. MEMS: The path to large optical crossconnects. *IEEE Commun. Mag.* **2002**, *40*, 80–87.
6. Yalcinkaya, A.D.; Urey, H.; Brown, D.; Montague, T.; Sprague, R. Two axis electromagnetic microscanner for high resolution displays. *J. Microelectromech. Syst.* **2006**, *15*, 786–794.
7. Martowicz, A.; Klepka, A.; Uhl, T. Analysis of static and dynamic properties of micromirror with the application of response surface method. *Int. J. Multiphys.* **2016**, *6*, 115–127.
8. Chen, J.; Gong, W.; Han, S. Sub-Rayleigh ghost imaging via sparsity constraints based on a digital micro-mirror device. *Phys. Lett. A* **2013**, *377*, 1844–1847.
9. Samuelson, S.R.; Xie, H. A large piston displacement MEMS mirror with electrothermal ladder actuator arrays for ultra-low tilt applications. *J. Microelectromech. Syst.* **2014**, *23*, 39–49.

10. Hao, Z.; Wingfield, B.; Whitley, M.; Brooks, J.; Hammer, J.A. A design methodology for a bulk-micromachined two-dimensional electrostatic torsion micromirror. *J. Microelectromech. Syst.* **2003**, *12*, 692–701.

11. Pallapa, M.; Yeow, J.T.W. Design, fabrication and testing of a polymer composite based hard-magnetic mirror for biomedical scanning applications. *J. Electrochem. Soc.* **2013**, *161*, B3006–B3013.

12. Isikman, S.O.; Urey, H. Dynamic modeling of soft magnetic film actuated scanners. *IEEE Trans. Magnet.* **2009**, *45*, 2912–2919.

13. Koh, K.H.; Kobayashi, T.; Lee, C. Investigation of piezoelectric driven MEMS mirrors based on single and double S-shaped PZT actuator for 2-D scanning applications. *Sens. Actuators A Phys.* **2012**, *184*, 149–159.

14. Lin, Z.; Pachter, M.; Banda, S. Toward improvement of tracking performance nonlinear feedback for linear system. *Int. J. Control* **1998**, *70*, 1–11.

15. Chen, B.; Lee, T.; Peng, K.; Venkataramanan, V. Composite nonlinear feedback control for linear systems with input saturation: theory and an application. *IEEE Trans. Autom. Control* **2003**, *48*, 427–439.

16. Lan, W.; Thum, C.K.; Chen, B.A. hard disk drive servo system design using composite nonlinear feedback control with optimal nonlinear gain tuning methods. *IEEE Trans. Ind. Electron.* **2010**, *57*, 1735–1745.

17. Peng, K.; Chen, B.; Cheng, G.; Lee, T. Modeling and compensation of nonlinearities and friction in a micro hard disk drive servo system with nonlinear feedback control. *IEEE Trans. Control Syst. Technol.* **2005**, *13*, 708–721.

18. Cai, G.; Chen, B.; Peng, K.; Dong, M.; Lee, T. Modeling and control of the yaw channel of a uav helicopter. *IEEE Trans. Ind. Electron.* **2008**, *55*, 3426–3434.

19. Peng, K.; Cai, G.; Chen, B.; Dong, M.; Lum, K.; Lee, T. Design and implementation of an autonomous flight control law for a uav helicopter. *Automatica* **2009**, *45*, 2333–2338.

20. Cheng, G.; Peng, K.; Chen, B.; Lee, T. Improving transient performance in tracking general references using composite nonlinear feedback control and its application to high-speed xy table positioning mechanism. *IEEE Trans. Ind. Electron.* **2007**, *54*, 1039–1051.

21. Cheng, G.; Peng, K. Robust composite nonlinear feedback control with application to a servo positioning system. *IEEE Trans. Ind. Electron.* **2007**, *54*, 1132–1140.

22. Eren, S.; Pahlevaninezhad, M.; Bakhshai, A.; Jain, P. Composite nonlinear feedback control and stability analysis of a grid-connected voltage source inverter with lcl filter. *IEEE Trans. Ind. Electron.* **2013**, *60*, 5059–5074.

23. Mobayen, S.; Majd, V.J. Robust tracking control method based on composite nonlinear feedback technique for linear systems with time-varying uncertain parameters and disturbances. *Nonlinear Dyn.* **2012**, *70*, 171–180.

24. Mondal, S.; Mahanta, C. Composite nonlinear feedback based discrete integral sliding mode controller for uncertain systems. *Commun. Nonlinear Sci. Numer. Simul.* **2012**, *17*, 1320–1331.

25. Bandyopadhyay, B.; Deepak, F.; Kim, K. Integral sliding mode based composite nonlinear feedback control. In *Sliding Mode Control Using Novel Sliding Surfaces*; Springer: Heidelberg, Germany, 2009.

26. Chen, H.; Pallapa, M.; Sun, W.J.; Sun, Z.D.; Yeow, J.T.W. Nonlinear control of an electromagnetic polymer MEMS hard-magnetic micromirror and its imaging application. *J. Micromech. Microeng.* **2014**, *24*, 045004.

27. Tan, J.; Sun, W.; Yeow, J.T.W. Internal Model-Based Robust Tracking Control Design for the MEMS Electromagnetic Micromirror. *Sensors* **2017**, *17*, 1215.

28. Stepanenko, Y.; Cao, Y.; Su, C.Y. Variable structure control of robotic manipulator with PID sliding surfaces. *Int. J. Robust Nonlinear Control* **1998**, *8*, 79–90.

micromachines

MDPI

Article

PZT-Actuated and -Sensed Resonant Micromirrors with Large Scan Angles Applying Mechanical Leverage Amplification for Biaxial Scanning

Shanshan Gu-Stoppel *, Thorsten Giese, Hans-Joachim Quenzer, Ulrich Hofmann and Wolfgang Benecke

Micro System Technology, Fraunhofer Institute for Silicon Technology, Itzehoe 25524, Germany; thorsten.giese@isit.fraunhofer.de (T.G.); hans-joachim.quenzer@isit.fraunhofer.de (H.-J.Q.); ulrich.hofmann@isit.fraunhofer.de (U.H.); wolfgang.benecke@isit.fraunhofer.de (W.B.)
* Correspondence: shanshan.gu-stoppel@isit.fraunhofer.de; Tel.: +49-4821-171-424

Received: 9 April 2017; Accepted: 2 July 2017; Published: 6 July 2017

Abstract: This article presents design, fabrication and characterization of lead zirconate titanate (PZT)-actuated micromirrors, which enable extremely large scan angle of up to 106° and high frequency of 45 kHz simultaneously. Besides the high driving torque delivered by PZT actuators, mechanical leverage amplification has been applied for the micromirrors in this work to reach large displacements consuming low power. Additionally, fracture strength and failure behavior of poly-Si, which is the basic material of the micromirrors, have been studied to optimize the designs and prevent the device from breaking due to high mechanical stress. Since comparing to using biaxial micromirror, realization of biaxial scanning using two independent single-axial micromirrors shows considerable advantages, a setup combining two single-axial micromirrors for biaxial scanning and the results will also be presented in this work. Moreover, integrated piezoelectric position sensors are implemented within the micromirrors, based on which closed-loop control has been developed and studied.

Keywords: micromirror; PZT; piezoelectric; position sensors; biaxial scanning

---

1. Introduction

Either for biomedical [1], automotive [2] or for entertainment uses like pico-projector [3], micromirrors are attract increasing interest due to the miniaturized size, low power consumption and low production cost compared to conventional scanning devices. To drive a microelectromechanical systems (MEMS) mirror there are basically four actuating principles: thermal, magnetic, electrostatic and piezoelectric principles. Thermal micromirrors reach large deflections driven by low driving voltages [4]. Yet the power consumption is high compared to the other three driving principles. Meanwhile, the actuation frequencies are limited by the thermal response time. Electromagnetic actuation delivers high force and requires low driving voltage [5], but the needed external magnets and the electromagnetic interference impair the integration and the availabilities of the micromirrors for many applications. Electrostatic microscanners are realized by established manufacturing technology and provide good mechanical performance [6]. However, for the actuation high driving voltage is required and the comb finger capacitors cause high lateral air damping [7], which demands vacuum packaging to improve the mechanical efficiency [6]. In contrast, piezoelectric materials [8–10], for example lead zirconate titanate (PZT), deliver high driving force at low driving voltage, so that piezoelectrically driven micromirrors achieve large deflection even operated under ambient conditions. Therefore, together with the improvement of processing technology of piezoelectric materials, piezoelectrically driven micromirrors are showing obvious advantages.

For most applications of micromirrors, biaxial scanning is needed to sense surface of objects, environments or displaying images in two dimensions, which can be realized by raster or Lissajous scanning principles [11]. Two-dimensional raster micromirrors have been reported by [3,5,12], where the micromirrors are driven resonantly in one dimension and quasi-statically in the other. Thereby the quasi-statically driven axes have been designed to possess frequencies of 1–2 kHz for lowering the stiffness and reaching large displacement, which can affect the mechanical robustness of the devices severely. The second option for micromirrors reflecting laser spots to fill a rectangle field is Lissajous scanning, where both axes are driven resonantly. Depending on designed frequency difference of the two axes, different Lissajous figures will repeat and fill the field. Previous works show that very good results regarding high light density can be realized, if the two axes have a frequency difference of 60 Hz [13]. However, such low frequency difference causes inevitably so strong mechanical coupling of the two axes, that complex controlling is needed to decouple the biaxial motions [11]. Therefore, despite great integration and sophistication, biaxial micromirrors have yet vulnerabilities. By comparison, the approach of applying two single-axial micromirrors to realize biaxial scanning shows significant advantages of high flexibility and no crosstalk of the two axial motions, which simplifies the controlling [10]. Thus, a setup combining two single axial micromirrors for biaxial scanning and the results will be presented in this work.

Generally high frequency and large displacement of resonant micromirrors are required for many applications like pico-projector [14]. Also, quasi-statically driven vector micromirrors request high-resonant frequency for better mechanical robustness and low settling time [15]. However, these two requirements are contradictory to each other, since high frequency demands high stiffness, while large displacement requires low stiffness of the devices. So, to achieve these two targets simultaneously is the challenge of constructing micromirrors. One of the focuses of this work is to utilize analytic modelling, where the micromirror plate and actuators are considered as an entire system, to improve the mechanical efficiency for realizing high frequency, large displacement and low consumption. Piezoelectric micromirrors are usually driven by beam actuators, which also influence resonant frequencies and consume power due to their bending motions. Hence, mechanical leverage amplification has been applied for ensuring that the power is primarily consumed for mirror torsion than actuator bending. On the other hand, previous works showed that the maximum achievable scan angles of piezoelectric micromirrors are strongly affected by the breaking strength of the used material, for example poly-Si [16]. Thus, fracture strength of poly-Si has been studied for optimizing the mirror designs.

At the end of the work, closed-loop control based on PZT position sensors are studied and presented. In [17], closed-loop control for piezoelectric micromirrors based on capacitive sensors was reported. Also integrated piezoresistive sensors were described in [18]. Despite the great sensor sensitivities, hybrid integration of capacitive sensors and expensive fabrication of piezoresistive sensors remain their tradeoffs. In contrast to them, integrated piezoelectric sensors cost no extra fabrication steps and deliver measuring signals with large signal-to-noise-ratio (SNR) as well.

2. Modelling and Analysis

2.1. Dynamic Leverage Amplification

The design developments of this work have been strongly supported by finite element method (FEM) simulations. Generally, it is difficult to simulate realistic resonant behaviors of one micromirror except its resonance frequency, since results like achievable displacements are affected by factors like air damping, dielectric and mechanical loss of piezoelectric material, which are complicated to predict accurately. Therefore, static simulations of the designs have been performed for the assessment of the achievable scan angles, since the static behavior equals the border case of strongly damped dynamic behavior. For comparing the achievable displacements of different designs and assessing the efficiency of these designs, a same driving voltage was applied for the static simulations, while the resonant frequencies were calculated by dynamic simulations to evaluate resonant behaviors comprehensively. Before the individual designs were compared with each other, a basic design concept had been developed. Since the focus of the work lies on single-axis micromirrors, the mirror plate is placed in the center of the device, which is linked by torsion bars and connecting bars to two symmetric, surrounding actuators. Figure 1 shows the top view of such a micromirror and the cross-sectional view, which reveals also the principle of such designs: The surrounding piezoelectric actuators are activated by turns. The torque delivered by the actuator is transferred by the connecting bars and torsion bars to rotate the mirror plate. The geometry of the connecting bars has been designed to amplify the displacement of actuators, so that the mirror plate reaches much larger displacements.

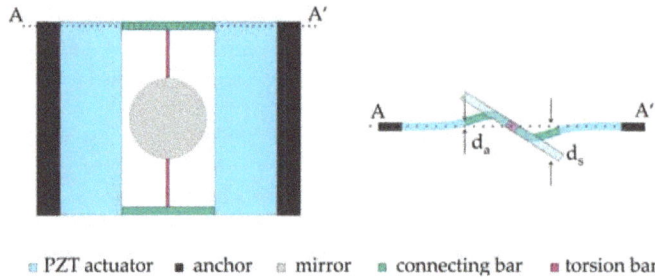

□ PZT actuator ■ anchor ▢ mirror ■ connecting bar ■ torsion bar

Figure 1. Top view and cross-sectional view of the basic design S. AA': Cross-section of actuators and connecting bars, d_a stands for displacement of actuators and d_m stands for displacement of mirror plate.

Actuators, mirror plate, torsion bars and connecting bars can be sorted into two groups: The torsion bars and mirror plate constitute the torsion group (Group T), while the actuators and connecting bars constitute the bending group (Group B). These two groups behave as two coupled oscillators, so that the total energy of the entire system is divided into two parts. Since the actuators possess considerably higher moment of inertia and stiffness than the mirror plate and torsion bars, the mirror plate will show much larger displacement than the actuators in the torsional mode. This is the abovementioned leverage amplification effect and has been proven by FEM simulations using frequency-domain study (Figure 2).

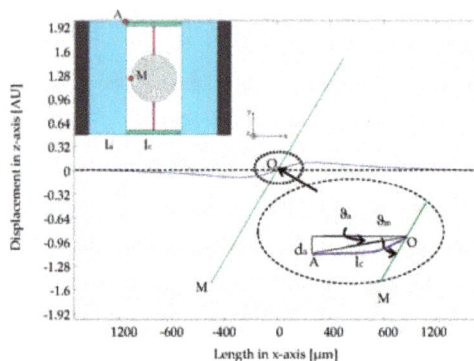

Figure 2. Top view of design S1 with two reference points A and M (A stands at the end of the actuator and M stands on the edge of the micromirror.) and cross-sectional view of a micromirror in the torsional mode, which is calculated by finite element method (FEM) simulations: Deflected actuators and connecting bars (blue) and the torsional mirror plate (green) (O: Center of the mirror plate; l_a: Length of the actuators; l_c: Length of the connecting bars; ϑ_a: Angle of actuators and ϑ_m: Angle of mirror plate).

The angle ϑ_a is formed by the zero line and OA describing the deflection of the actuator, as Figure 2 shows, while ϑ_m stands for the torsion angle of the mirror plate. Thereby, the displacement amplification is obviously observed and the ratio n_a of ϑ_m to ϑ_a indicates the amplification efficiency. To calculate this key indicator, n_a model has been built by using Euler–Lagrange equation [19]:

$$L = T - V \tag{1}$$

In Equation (1) L stands for the Lagrange function, T stands for the kinetic energy and V stands for the potential energy. Furthermore, there is a following equation system, which is derived from Equation (1) and applies for the Group T and Group B of the mirror device.

$$\frac{d}{dt}\left(\frac{\partial L}{\partial \dot{\vartheta}_a}\right) - \frac{\partial L}{\partial \vartheta_a} = M_a = M_{PZT} \tag{2}$$

$$\frac{d}{dt}\left(\frac{\partial L}{\partial \dot{\vartheta}_m}\right) - \frac{\partial L}{\partial \vartheta_m} = M_m = 0 \tag{3}$$

In Equations (2) and (3) M_a stands for bending moment of actuators equaling the moment delivered by the PZT M_{PZT}. Since there is no external force influencing the torsion motion of mirror plate, torsion moment M_m equals 0. Given the states of kinetic energy T and potential energy V, torsion angle ϑ and angular acceleration $\ddot{\vartheta}$ are depending on the effective mass and effective stiffness of the actuator and the moment of inertia of the mirror plate, Equations (2) and (3) can also be described as [11]:

$$\frac{33}{140}(m_1 + m_2 K^2)l_c^2 \ddot{\vartheta}_a + [(k_1 + k_2 K^2)l_c^2 + k_m]\vartheta_a - k_m \vartheta_m = M_{PZT} \tag{4}$$

$$I_{effm}\ddot{\vartheta}_m + k_m \vartheta_m - k_m \vartheta_a = 0 \tag{5}$$

In the above equations, the symbols depict the following variables:

ϑ_a:	Torsional angle of actuators;	ϑ_m:	Torsional angle of mirror plate;
m_1:	Mass of actuators;	m_2:	Mass of connecting bars;
k_1:	Stiffness of actuators;	k_2:	Stiffness of connecting bars;
k_m:	Stiffness of mirror;	l_c:	Length of connecting bars;
I_{effm}:	Effective moment of inertia of mirror;	M_{PZT}:	Torque delivered by PZT;
K:	Geometric matching factor.		

According to Fourier Transformation, Equations (4) and (5) can be further described as:

$$\frac{33}{140}(m_1 + m_2 K^2)l_c^2 \dot{\vartheta}_a \cdot j\omega + [(k_1 + k_2 K^2)l_c^2 + k_m]\dot{\vartheta}_a/(j\omega) - k_m \dot{\vartheta}_m/(j\omega) = M_{PZT} \tag{6}$$

$$I_{effm}\dot{\vartheta}_m \cdot j\omega + k_m \dot{\vartheta}_m/(j\omega) - k_m \dot{\vartheta}_a/(j\omega) = 0 \tag{7}$$

The Equation system of (6) and (7) can be presented as an equivalent circuit (Figure 3), where $\dot{\vartheta}$ can be considered as the current, the effective mass or moment of inertia, like I_{effm}, is demonstrated by inductivity L and the inverse of stiffness, like $1/k_m$, is demonstrated by capacitance C. The electric resistance R demonstrates the damping. The values of the circuit components are obtained based on geometry and material parameters of the micromirror design (Table 1). It is to emphasis, using this model the relative energy distribution of different parts (torsional micromirror and bending actuators) can be calculated, the absolute energy dissipation caused by air damping and mechanical losses is not considered. Therefore, the values of R are set as 0.

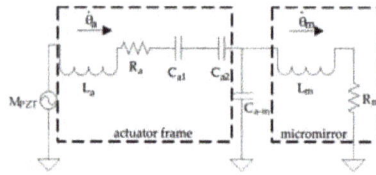

Figure 3. Equivalent electric circuit describing mechanical behavior of micromirrors and actuators.

Table 1. The electric components and the equivalent mechanical parameters.

Electric Component	L_a	L_m	C_{a1}	C_{a2}	C_{a-m}	R_a, R_m
Equivalent mechanical parameter	$I_{effa} = \frac{33}{140}(m_1 + m_2K^2)l_c^2$	I_{effm}	$\frac{1}{(k_1+k_2K^2)l_c^2+k_m}$	$-\frac{1}{k_m}$	$\frac{1}{k_m}$	Damping = 0

The analytic model using the equivalent circuit has been proven by FEM simulations (frequency-domain study), since both results show the identical amplitude spectrums (Figure 4a,b).

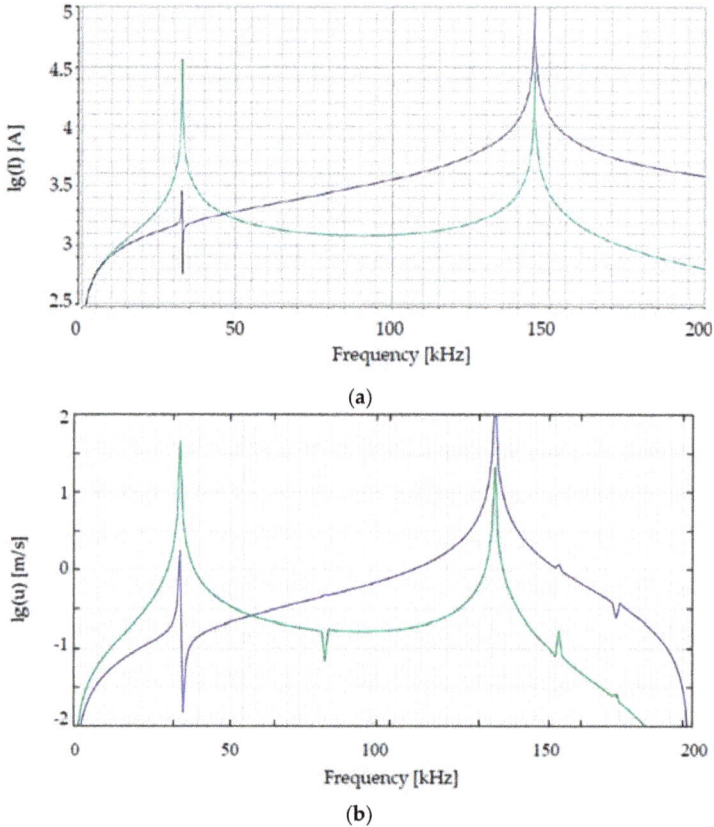

(a)

(b)

Figure 4. Amplitude spectrums of $\dot{\vartheta}_a$ (blue) and $\dot{\vartheta}_m$ (green) of design S1 calculated by (a) analytic modelling and (b) FEM simulations (I: current of the equivalent circuit and u: velocity of actuator and mirror plate.)

Although no damping mechanisms have been taken into account for the analytic modelling and FEM simulation, bandwidths are apparent at the resonance peaks, as Figure 4a,b show, which means damping. It is caused by different reasons: The damping of analytic modeled results comes from the electric components of the equivalent electric circuit, which has been used for calculation of analytic modelling. FEM simulation results show also damping because of the intrinsic material damping due to the material properties.

The advantage of using this analytic model is to give a clear tendency of influence of every single geometric parameter of the designs, while FEM simulations are accurate, comprehensive but time-consuming. Both the analytic modelling and FEM simulations in Figure 4 have verified the dynamic leverage amplification effect of the design concept. At the resonant frequency of the torsional mode the simulated design, where the first amplitude peak in the amplitude spectrum appears, the displacement of the mirror plate is 32 times as large as that of the actuators, which has been later proven by characterization results.

2.2. Von Mises Stress and Fracture Strength

Due to the dynamic leverage amplification, the PZT delivered torque can be very efficiently used for rotating the mirror plate, so that the micromirror reaches very large scan angles already at low driving voltages and low power consumption. Hence the limitation for reaching larger scan angles for the micromirrors is the fracture strength of poly-Si, of which the micromirrors primarily consist. To investigate the fracture behavior of poly-Si in different micromirrors, these mirrors have been deflected in FEM simulations to a certain rotating angle, which is the maximum achievable rotating angle of these mirrors proven by the characterization results. The observed failure behavior means the calculated maximum Von Mises Stress now equals the fracture strength of this micromirror. The dependence of the fracture strength of polycrystalline materials on their Young's modules was reported by [20]. Furthermore, in [20–23] fracture behavior of different polycrystalline materials has been investigated and the fracture strengths lie between 0.6% and 3% of the Young's moduli of these materials. According to them failure strength of the micromirrors was expected at a mechanical stress level of 1.5 GPa, which approximates 1% of the Young's modules of used poly-Si.

First of all, the simulation results disclose an important influence of the geometry of micromirrors, especially shapes of the springs, on the maximum Von Mises Stresses. The following pictures in Figure 5 show such a comparison: Different maximum Von Mises Stresses (1.8 GPa, 1.6 GPa and 1.4 GPa) appear within different micromirrors, even if these micromirrors achieve a same mechanical tilting angle of 15°.

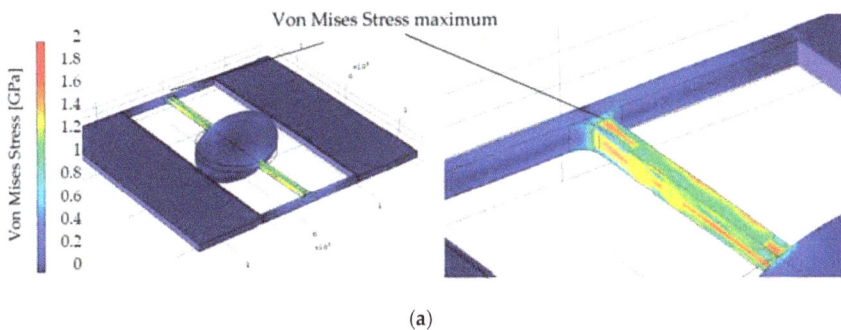

(a)

Figure 5. *Cont.*

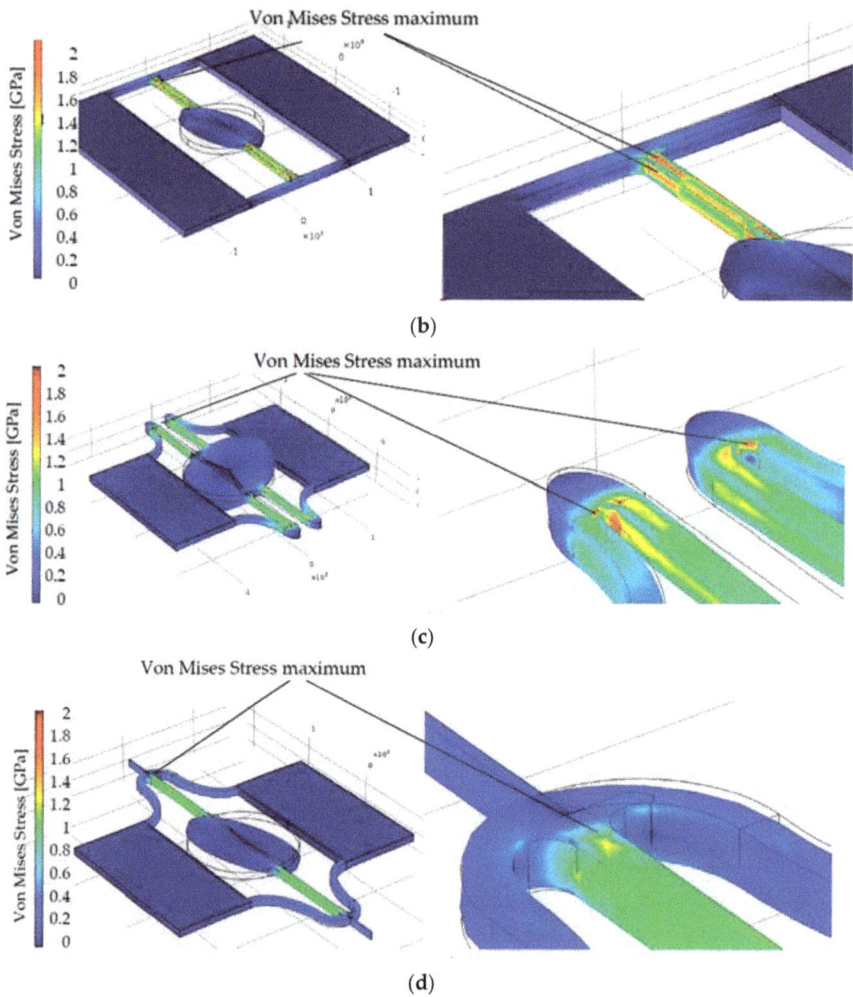

Figure 5. Different maximum Von Mises Stresses in different mirrors, when these mirrors achieve a same mechanical tilting angle of 15°: (**a**) Von Mises stress maximum of 1.8 GPa in design S1; (**b**) Von Mises stress maximum of 1.8 GPa in design S2; (**c**) Von Mises stress maximum of 1.6 GPa in design E5; (**d**) Von Mises stress maximum of 1.4 GPa in design E4.

This comparison shows a clear correlation of the maximum Von Mises Stress of a micromirror with its design, for example, meandering and rounding springs can reduce the maximum mechanical stress severely. Additionally, a second finding has been revealed by the later characterization: Although the bearable mechanical stress level was assumed as 1.5 GPa, the micromirrors bear mechanical stress of up to 3.4 GPa and the measurement results show a strong dependence of the fracture strength on the geometry of the designs. Analysis on such correlations and characterization results of these three type micromirrors will be shown in Section 4.2.

3. Fabrication

For manufacturing of the 1D micromirrors, wafers of 725 μm silicon, 1 μm SiO_2 and 80 μm epitaxial poly-Si are used as substrates. An additional 1 μm thick SiO_2 layer on top of the poly-Si is followed by an evaporated thin Ti/Pt layer acting as bottom-electrode and PZT-seedlayer. Subsequently, 2 μm PZT is hot magnetron sputtered featuring a high piezoelectric modulus. On top of the PZT layer, a thin Cr/Au layer serves as top-electrode. After the deposition of all functional layers the Cr/Au layer is wet-etched, while PZT and Ti/Pt layers are dry-etched. Before the 80 μm polysilicon is deep reactive-ion etching (DRIE)-patterned to define the mirrors and actuators, a 100 nm Al layer is deposited as the reflection surface. Finally, the 725 μm silicon and the 1 μm SiO_2 are etched using DRIE from the rear side to release the mirror. The process flow is illustrated in Figure 6 [10].

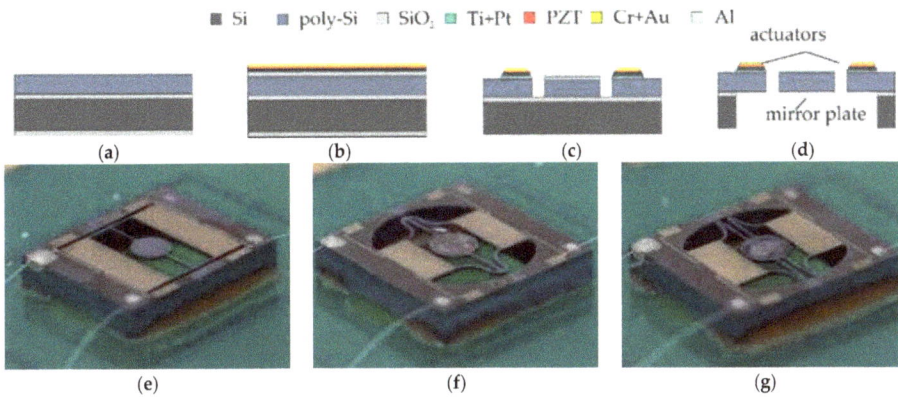

Figure 6. Cross-sectional process flow and device photos: (**a**) Substrate made of poly-Si, SiO_2 and Si layer; (**b**) Deposition of functional layers; (**c**) Structuring by wet and dry etching from the front side; (**d**) Release by dry etching from the rear side; (**e**) Device photo of design S1; (**f**) Device photo of design E4; (**g**) Device photo of design E5.

4. Characterization

4.1. Dynamic Behavior

To verify the dynamic leverage amplification effect of the design concept, as an example the micromirror with design S1 has been measured by a Polytec® Laser-Doppler-Vibrometry (LDV, Polytech Ophthalmologie AG, Zuzwil, Switzerland). As the measurement results in Figure 7a,b show, the mirror plate has much larger displacement than the actuators, when they are driven in the torsional mode. The amplification factor approximates 30, which is identical to the analytic and FEM modelling results shown in Section 2.1. It should be noticed that the LDV measurements have been performed for motion of the mirror plate with a mechanical scan angle of less than about 2.5°, as Figure 7a shows, since the laser spots will be reflected by the rotating mirror plate with larger scan angles out of measurable range of the used LDV lenses.

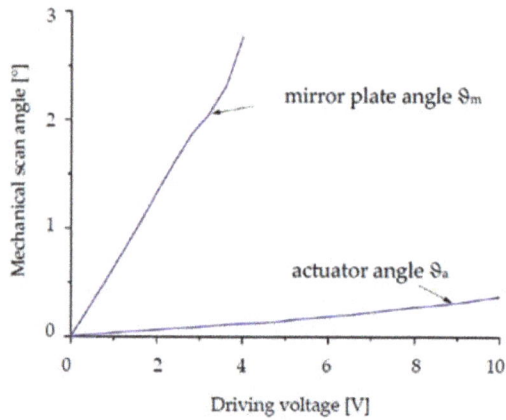

(a)

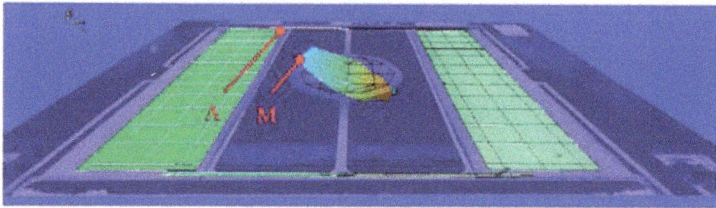

(b)

Figure 7. Laser-Doppler-Vibrometry (LDV) measurement results of micromirror with design S1 (measure points are A and M): (**a**) Mechanical scan angles of the mirror plate and the actuators; (**b**) Measurement recording of a deflected micromirror with scan angle of 2.5° measured at point M and the actuators with scan angle of about 0.08° measured at point A.

4.2. Fracture Strength

Besides the dependence of the maximum Von Mises Stress of micromirrors on the design, which has been shown by the FEM simulations, the dependence of the fracture strength on the design has been also studied by the characterization results. Two similar designs have been compared in Figure 8a,b, which are based on the basic design S shown in Figure 1. First of all, the cracking origins within the two designs appear both on the torsion bars adjacent to the connecting bars, which can be recognized in Figure 8c.

Additionally, the only geometry difference between the two designs is the shape of torsion bars. While design S1 has a rectangular torsion bar with a smaller width b than its height t ($b/t = 0.75$), design S2 has a square torsion bar with the same width b as its height t ($b/t = 1$) (Figure 8a,b). This only difference results in different distribution of the maximum Von Mises Stress in these micromirrors: The maximum Von Mises Stress of designs S1 appears on the side wall of the torsion bar, whose surface is rough due to the dry etching process, while the maximum Von Mises Stress of designs S2 appears on the polished and smooth top of the torsion bar. Figure 8d shows a SEM image of the torsion bar, where the smooth top surface and the rough side wall can be seen. Due to the notch effect, such rough surface of the side wall benefits the crack formation and crack growth. So, the material bears significantly lower mechanical stress, if the crack origin appears on such tough surfaces [21–23]. It is the reason for this phenomenon, that one of the designs can withstand mechanical stress of 3.4 GPa, while the torsion bar of the second micromirror breaks already at 3 GPa, even though both designs are very similar and the main mechanical functional structures (torsion bars, connecting bars, mirror plate and actuators) consist of the same material, poly-Si.

These findings revealed by the FEM simulations and the characterization give a clear picture of the fracture behavior of the micromirrors. Firstly, the maximum Von Mises Stress is strongly dependent on the designs, and determines at which scan angles the mechanical structures will break. Secondly, also the fracture strength of the micromirrors is dependent on the designs, since the designs have a large influence on the distribution of Von Mises Stress (i.e., where the maximum will occur).

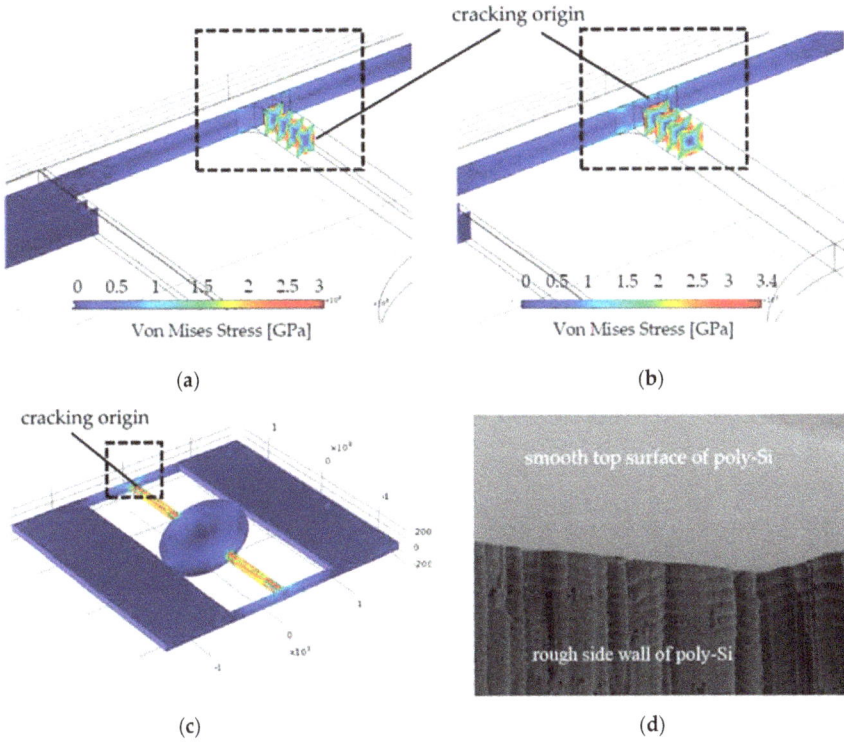

Figure 8. Sectional view of simulation of Design S: (**a**) Design S1 with a rectangular torsion bar ($b/t = 0.75$) and the maximum bearable Von Mises Stress is 3 GPa; (**b**) Design S2 with a square torsion bar ($b/t = 1$) and the maximum bearable Von Mises Stress is 3.4 GPa; (**c**) Top view of Design S; (**d**) SEM picture of a torsion bar with smooth top surface and rough side wall of poly-Si.

The characterization results of different four design types in Figures 5 and 8 are shown in the following table (Table 2).

Table 2. The characterization results of Design S1, S2, E4 and E5.

Design	Aperture Diameter [mm]	Driving Voltage Peak-to-Peak [V]	Full Optical Scan Angle [°]	FEM Simulated Frequency [kHz]	Measured Frequency [kHz]
S1	1	22	106.1	35.4	33.3
S2	1	22	106.3	47.1	45.1
E4	1.2	25	86.7	28.1	26.8
E5	1.2	20	104.2	34.1	31.2

4.3. Setup of Biaxial Scanning Using Two 1D Micromirrors

The above characterization has been conducted to study the mechanical behavior of single-axial micromirrors. Since another focus of this work is to realize biaxial scanning by using two single-axis micromirrors, Figure 9a,b show the construction for this purpose, where two resonant micromirrors have been connected in series [10]. Both micromirrors are parallel to each other with a small distance. The torsion axis of micromirror 1 is inclined to the incident laser beam with an angle of 45°. While micromirror 1 possesses a 1 mm diameter circular aperture, micromirror 2 has a rectangular aperture of 1.4 mm × 4 mm. Both single-axial micromirrors rotate about the own torsion axes, which are perpendicular to each other. The laser beam has been pointed at micromirror 1 and the linear laser beam trajectory of micromirror 1 has been further reflected by micromirror 2 delivering a rectangular light screen. Figure 9c shows such a realized rectangular light screen, which, for example, reaches optical scan angles of 30.7° × 34.5°.

A rectangle scanned by two independent single-axial micromirrors can possess various length ratios between the horizontal and vertical edges depending on the possible scan angle of the two mirrors. Also, the picture resolution of the displayed pictures, which relates to the frequency ratio [11], can be arbitrarily defined. The flexibility of combing two independent single-axial micromirrors and non-mechanical-crosstalk are the most important advantages of this approach, which simplify the complexity of design, manufacturing and controlling of micromirrors greatly.

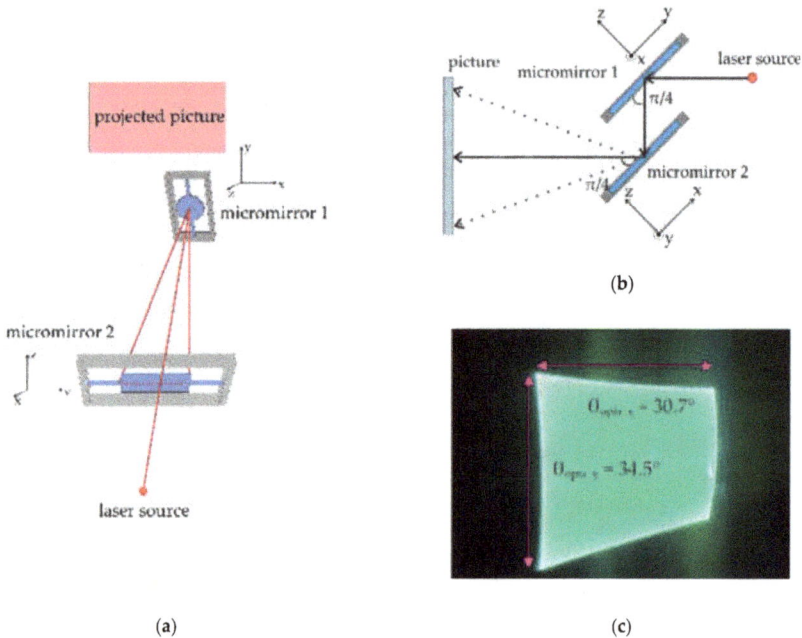

Figure 9. Setup and result of two single-axial micromirrors for 2D scanning: (**a**) Front view; (**b**) Side view; (**c**) Illuminated rectangle.

4.4. Position Sensing and Closed-Loop Control

The last point of this work is to investigate position sensing and closed-loop control of the micromirror. Since the piezoelectric material has the property of converting mechanical energy to electrical energy, one of the position-sensing approaches is to use one of the PZT cantilever as the position sensor, while the second one serves as the actuator, as Figure 10 shows [24].

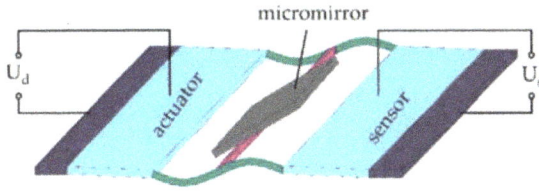

Figure 10. Schematic drawing of realizing micromirror position sensing applying one PZT cantilever as position sensor based on the direct piezoelectric effect (U_d stands for the driving voltage and U_s stands for the sensing voltage.)

Figure 11a shows the comparison of the driving voltage, position signal of the micromirror measured by a Position-Sensitive-Device (PSD) and a PZT sensing signal in time domain, which are measured on design S1. The result proves a great amplitude correlation between the PZT sensing signal and the PSD signal, which represents the exact mirror position signals. The minor phase shift between the PZT sensing signal and the PSD signal is caused by the capacitance of the PZT actuator and the electrical supply cables of the measurement setup. Additionally, Figure 11b shows the PZT sensing signal in frequency domain. After a simple signal processing of 64 averaging this signal shows already a good signal quality of 45 dB SNR, while the mechanical scan angle was only 0.7°. Since this micromirror can achieve a mechanical scan angle of 26.5°, meaning a full field of view of 106°, the total sensing resolution n is larger than 12 bit, as Equation (8) shows, which can enable the controlling of projecting picture of 1920 pixels.

$$\text{SNR} = 1.76 + 6.02 \cdot n \tag{8}$$

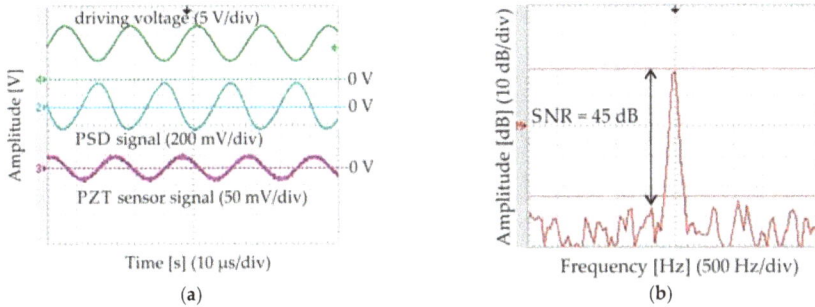

Figure 11. Measurement results of the PZT sensing signal of design S1: (**a**) Comparison of driving voltage (green), PSD position signal (blue) and PZT position signal (violet) in time domain; (**b**) PZT position signal in frequency domain with a SNR of 45 dB after 64 averaging process at a mechanical scan angle of 0.7°. PSD = Position-Sensitive-Device; SNR = signal-to-noise-ratio.

Based on this great signal quality, closed-loop control has been developed, as Figure 12 demonstrates. The first driving signal is delivered by a controller, which is converted by a Digital-Analog-Converter (DAC) and amplified by a booster, before it reaches the micromirror. Then analog sensing signals from the micromirror are processed by an Analog-Frond-End (AFE) and ADC (Analog-Digital-Converter), before they are demodulated, processed and feed to a Phase-Locked-Loop (PLL), which compose the closed-loop control.

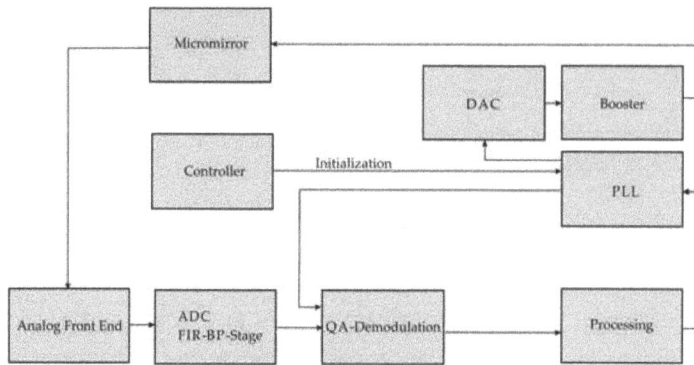

Figure 12. Block diagram of the closed-loop control for the piezoelectrically actuated and sensed micromirror. DAC = Digital-Analog-Converter; ADC = Analog-Digital-Converter; PLL = Phase-Locked-Loop; FIR = Finite-Impulse-Response, BP = Band-Pass.

5. Conclusions

This work demonstrates the good performance of piezoelectric micromirrors regarding the achievable scan angles and the resonant frequencies. To reach these two targets simultaneously, the mechanical efficiency of the entire system should be raised, which is enabled by the dynamic leverage amplification. This work shows an analytic model to calculate and enlarge the amplification factor by adapting the material and geometrical parameters of micromirrors to increase the mechanical efficiency of the system, whose effect has been confirmed by FEM simulations and characterization results. Furthermore, dependences of the maximum Von Mises Stress and the fracture strength of the micromirrors on the designs have been extensively investigated, to clarify the mechanical fracture behavior and adapt the designs for reaching larger scan angles.

For demonstrating the advantages of using single-axial micromirrors for biaxial scanning, like high flexibility and no crosstalk, combination setup, integrated piezoelectric position sensors and closed-loop control have been developed and presented. In the future, works for design improvements are projected to achieve larger scan angles, different frequency ratios of the two combined micromirrors for enhancement of the displayed picture quality. Finally, further technology developments for processing piezoelectric materials are also foreseen, to improve the material properties, like the linearity and long-term stability.

Acknowledgments: This study was a self-funded project by Fraunhofer institute for silicon technology in Germany. The authors want to thank Joachim Janes for the LDV measurements and thank Amit Kulkarni for the SEM photos.

Author Contributions: Shanshan Gu-Stoppel conceived and designed the experiments, performed the experiments, analyzed the data and wrote the paper; Thorsten Giese conceived the concept of the closed-loop control; Hans-Joachim Quenzer, Ulrich Hofmann and Wolfgang Benecke contributed the supervision and gave valuable advices to this work.

Conflicts of Interest: The authors declare no conflict of interest.

References

1. Pengwang, E.; Rabenorosoa, K.; Rakotondrabe, M.; Andreff, N. Scanning micromirror platform based on MEMS technology for medical application. *Micromachines* **2016**, *7*, 24. [CrossRef]
2. Chao, F.; He, S.; Chong, J.; Mrad, R.B.; Feng, L. Development of a micromirror based laser vector scanning automotive HUD. *IEEE Int. Conf. Mechatron. Autom.* **2011**, 75–79. [CrossRef]
3. Ikegami, K.; Koyama, T.; Saito, T.; Yasuda, Y.; Toshiyoshi, H. A biaxial piezoelectric MEMS scanning mirror and its application to pico-projectors. *Int. Conf. Opt. MEMS Nanophoton.* **2014**, 95–96. [CrossRef]

4. Pal, S.; Xie, H. A curved multimorph based electrothermal micromirror with large scan range and low drive voltage. *Sens. Actuators Phys.* **2011**, *170*, 156–163. [CrossRef]

5. Yalcinkaya, A.D.; Urey, H.; Brown, D.; Montague, T.; Sprague, R. Two-axis electromagnetic microscanner for high resolution displays. *J. Microelectromech. Syst.* **2006**, *15*, 786–794. [CrossRef]

6. Hofmann, U.; Janes, J.; Quenzer, H.-J. High-Q MEMS resonators for laser beam scanning displays. *Micromachines* **2012**, *3*, 509–528. [CrossRef]

7. Kaajakari, V. *Practical MEMS*; Small Gear Publishing, Louisiana Tech University: Ruston, LA, USA, 2009.

8. Weinberger, S.; Hoffmann, M. Aluminum nitride supported 1D micromirror with static rotation angle >11°. *Proc. SPIE* **2013**, *8616*. [CrossRef]

9. Baran, U.; Brown, D.; Holmstrom, S.; Balma, D.; Davis, W.O.; Muralt, P.; Urey, H. Resonant PZT MEMS Scanner for High-Resolution Displays. *J. Microelectromech. Syst.* **2012**, *21*, 1303–1310. [CrossRef]

10. Gu-Stoppel, S.; Janes, J.; Quenzer, H.J.; Hofmann, U.; Benecke, W. Two-dimensional scanning using two single-axis low-voltage PZT resonant micromirrors. *Proc. SPIE* **2014**, *897706*. [CrossRef]

11. Gu-Stoppel, S. *Entwicklung, Herstellung und Charakterisierung Piezoelektrischer Mikrospiegel*; BoD—Books on Demand: Hamburg, Germany, 2016.

12. Gu-Stoppel, S.; Quenzer, H.J.; Benecke, W. Design, fabrication and characterization of piezoelectrically actuated gimbal-mounted 2D micromirrors. In Proceedings of the 2015 Transducers—2015 18th International Conference on Solid-State Sensors, Actuators and Microsystems, Anchorage, AK, USA, 21–25 June 2015.

13. Hofmann, U.; Senger, F.; Janes, J.; Mallas, C.; Stenchly, V.; Wantoch, T.v.; Quenzer, H.-J.; Weiss, M. Wafer-level vacuum-packaged two-axis MEMS scanning mirror for pico-projector application. *Proc. SPIE* **2014**, *8977*. [CrossRef]

14. Specht, H. *MEMS-Laser-Display-System: Analyse, Implementierung und Testverfahrenentwicklung*; Technische Universität Chemnitz: Chemnitz, Germany, 2011.

15. Castelino, K.; Milanovic, V.; McCormick, D.T. MEMS-Based Low Power Portable Vector Display. Available online: http://www.adriaticresearch.org/Research/pdf/Vector_Display_MOEMS05.pdf (accessed on 20 January 2017).

16. Gu-Stoppel, S.; Janes, J.; Quenzer, H.J.; Hofmann, U.; Kaden, D.; Wagner, B.; Benecke, W. Design, fabrication and characterization of low-voltage piezoelectric two-axis gimbal-less microscanners. In Proceedings of the 2013 Transducers Eurosensors XXVII: The 17th International Conference on Solid-State Sensors, Actuators and Microsystems (TRANSDUCERS EUROSENSORS XXVII), Barcelona, Spain, 16–20 June 2013; pp. 2489–2492.

17. Rombach, S.; Marx, M.; Gu-Stoppel, S.; Manoli, Y. Low power and highly precise closed-loop driving circuits for piezoelectric micromirrors with embedded capacitive position sensors. *Proc. SPIE* **2016**, *9760*. [CrossRef]

18. Zhang, C.; Zhang, G.; You, Z. A two-dimensional micro scanner integrated with a piezoelectric actuator and piezoresistors. *Sensors* **2009**, *9*, 631–644. [CrossRef] [PubMed]

19. Gu-Stoppel, S.; Kaden, D.; Quenzer, H.J.; Hofmann, U.; Benecke, W. High speed piezoelectric microscanners with large deflection using mechanical leverage amplification. *Proced. Eng.* **2012**, *47*, 56–59. [CrossRef]

20. Yuan, C.C.; Xi, X.K. On the correlation of Young's modulus and the fracture strength of metallic glasses. *J. Appl. Phys.* **2011**, *109*. [CrossRef]

21. Sharpe, W.N.; Jackson, K.M.; Hemker, K.J.; Xie, Z. Effect of specimen size on Young's modulus and fracture strength of polysilicon. *J. Microelectromech. Syst.* **2001**, *10*, 317–326. [CrossRef]

22. Ritchie, R.O. Failure of Silicon: Crack Formation and Propagation. In Proceedings of the 13th Workshop on Crystalline Solar Cell Materials and Processes, Vail, CO, USA, 10–13 August 2003.

23. Greek, S.; Ericson, F.; Johansson, S.; Schweitz, J.-Å. In situ tensile strength measurement and Weibull analysis of thick film and thin film micromachined polysilicon structures. *Thin Solid Films* **1997**, *292*, 247–254. [CrossRef]

24. Gu-Stoppel, S.; Quenzer, H.J.; Heinrich, F.; Janes, J.; Benecke, W. A study of integrated position sensors for PZT resonant micromirrors. *Proc. SPIE* **2015**, *9375*. [CrossRef]

![micromachines logo] *micromachines*

MDPI

Article

Modelling and Experimental Verification of Step Response Overshoot Removal in Electrothermally-Actuated MEMS Mirrors

Mengyuan Li [1,2,*], Qiao Chen [3], Yabing Liu [3], Yingtao Ding [1] and Huikai Xie [2,*]

[1]   School of Information and Electronics, Beijing Institute of Technology, Beijing 100081, China; ytd@bit.edu.cn
[2]   Department of Electrical and Computer Engineering, University of Florida, Gainesville, FL 32611, USA
[3]   WiO Technology Co., Ltd., Wuxi 214035, China; wio@wiotek.com (Q.C.); ybliu@wiotek.com (Y.L.)
*   Correspondence: limengyuan@ufl.edu (M.L.); hkxie@ece.ufl.edu (H.X.);
     Tel.: +1-352-215-9990 (M.L.); +1-352-846-0441 (H.X.)

Received: 1 September 2017; Accepted: 20 September 2017; Published: 25 September 2017

Abstract: Micro-electro-mechanical system (MEMS) mirrors are widely used for optical modulation, attenuation, steering, switching and tracking. In most cases, MEMS mirrors are packaged in air, resulting in overshoot and ringing upon actuation. In this paper, an electrothermal bimorph MEMS mirror that does not generate overshoot in step response, even operating in air, is reported. This is achieved by properly designing the thermal response time and the mechanical resonance without using any open-loop or closed-loop control. Electrothermal and thermomechanical lumped-element models are established. According to the analysis, when setting the product of the thermal response time and the fundamental resonance frequency to be greater than $Q/2\pi$, the mechanical overshoot and oscillation caused by a step signal can be eliminated effectively. This method is verified experimentally with fabricated electrothermal bimorph MEMS mirrors.

Keywords: micro-electro-mechanical system (MEMS) mirror; bimorph; electro-thermal actuator; resonance frequency; thermal modelling; overshoot; ringing

---

1. Introduction

Micro-electro-mechanical system (MEMS) mirrors were reported to be in use as early as 1980 as an optical scanner [1]. Since then, MEMS mirrors have been used in a wide range of applications, such as optical switches or optical attenuators in telecommunications [2,3], object tracking [4], projection displays [5], and 3D sensing [6]. Different applications may have different requirements for MEMS mirrors, but stable switching or scanning is always needed. For example, in optical switching, MEMS mirrors are the optical engine for high-precision optical beam positioning, which requires fast and stable switching with minimal cross talk between channels [7]. In the application of object tracking, a MEMS mirror is used to steer a laser beam to a target, which requires the laser steering to be accurate, fast and smooth [4]. Mechanically, a typical MEMS mirror can be simply modelled as a spring-mass-damper system, where the mirror plate is the mass. Most MEMS mirrors operate in air or vacuums, which is typically an under-damped condition that will cause undesired oscillation and overshoot when applying a step input. The under-damped oscillation and overshoot will increase the settling time and may also seriously affect the performance of the whole optical system, such as introducing cross talk or missing the target [8,9].

Usually, to suppress or remove the under-damped oscillation, a control strategy such as open-loop control or closed-loop control may be employed [10]. Closed-loop control requires position sensing, which increases system complexity and cost [11]. Shaping input signals is an open-loop control method, where a pre-shaped input signal that corresponds to the reverse of the oscillation with proper time delay is constructed and then applied to control the system [12–14]. For example, Daqaq et al. employed an input shaping scheme to realize a desired scanning beam locus with an electromagnetic MEMS mirror [14], where the input signal was calculated based on mirror dynamic characteristics, and using the Laplace transform. Shi et al. reported a method to control a thermally-actuated MEMS mirror based on a high-order dynamic model, and the experiment shows the residual oscillation is greatly eliminated [15]. However, the accuracy of the model largely affects the performance of the open-loop control method, and it may not work well when it is a relatively complex system, such as a high-order system, or a time-varying system.

In this paper, we report a solution that can eliminate under-damped oscillation of electrothermal bimorph MEMS mirrors without using either the open-loop or closed-loop control strategy. This solution utilizes the low pass nature of thermal response to suppress the mechanical oscillation. This paper is organized as follows. In Section 2, the electrothermal bimorph actuation principle and the electrothermal bimorph based MEMS mirror design are introduced first, and then the static and dynamic characteristics of a fabricated electrothermal bimorph MEMS mirror with step response overshoot is presented. In Section 3, a thermomechanical model of the bimorph MEMS mirror is developed, from which the new method of suppressing the step response overshoot and oscillation is established. Section 4 presents the design and testing results of the improved electrothermal bimorph MEMS mirror based on the new method, which experimentally validates the effectiveness of the overshoot suppression.

2. Two-Axis Electrothermal MEMS Mirror

The two-axis electrothermal MEMS mirror used in this study is based on an inverted-series-connected (ISC) thermal bimorph actuator structure [16]. A thermal bimorph refers to a beam consisting of two layers of materials with different thermal expansion coefficients (TEC). When the temperature of the bimorph changes due to the Joule heating generated by a heater embedded in the bimorph, the bimorph bends because of the TEC difference. However, the tip of a single bimorph has both tangential tip-tilt and lateral shift upon actuation, as shown in Figure 1a. Thus, a unique ISC bimorph actuator design has been developed [17], as shown in Figure 1b, where each ISC bimorph consists of an inverted (IV) segment, a non-inverted (NI) segment, and an overlap (OL) segment, resulting in an "S" shape. Two such bimorphs are connected in a folded fashion, eliminating both tip-tilt and lateral shift at the end of the folded beam, as shown in Figure 1c. The two materials in the bimorph are typically aluminum (Al) and silicon dioxide (SiO_2) because of their large TEC difference. A titanium (Ti) layer is embedded in the SiO_2 along the ISC bimorph actuators to form a resistor, as shown in Figure 1d. When voltage is applied to the Ti resistor, Joule heat is generated, which changes the bimorph temperature.

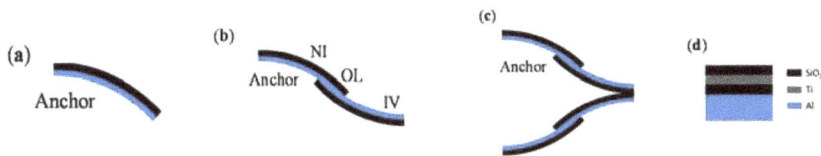

Figure 1. Electrothermal bimorph actuator design. (**a**) Simple bimorph beam; (**b**) inverted-series-connected (ISC) bimorph design; (**c**) Folded double ISC bimorph design; (**d**) A Ti heater embedded in the bimorph.

A fabricated two-axis ISC MEMS mirror is shown in Figure 2a, which consists of four ISC bimorph actuators and a mirror plate. The mirror plate is supported by the four ISC actuators on the four sides symmetrically. With four ISC actuators controlling the four sides of the mirror plate, the mirror plate can move vertically or generate angular scan in two axes. The device is fabricated using a hybrid bulk- and surface-micromachining process and SOI wafers are selected to ensure the flatness of the mirror plate. The process includes Ti heater lift-off, SiO$_2$ plasma-enhanced chemical vapor deposition (PECVD), Al sputter deposition, SiO$_2$ reactive ion etch (RIE), and silicon deep reactive-ion etching (DRIE), as described in [18]. A fabricated device is shown in Figure 2a, where the chip size is 2 mm × 2 mm, and the diameter and thickness of the mirror plate are 1 mm and 25 μm, respectively. A zoom-in view of an ISC actuator is shown in Figure 2b, which shows a clear "S" shape. There is a Ti resistor running along the entire bimorph loop. The mirror plate is 150 μm below the substrate surface.

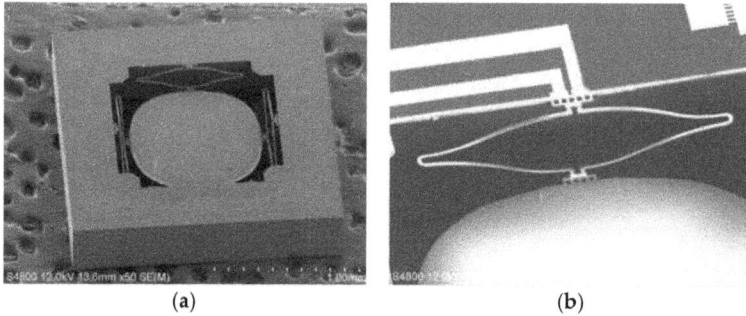

(a)   (b)

Figure 2. Two-axis ISC MEMS mirror. (a) Scanning electron microgram (SEM) of a fabricated device (2 mm × 2 mm); (b) SEM of an ISC actuator.

3. Step Response Modelling of the Electrothermal Bimorph MEMS Mirror

When a current is injected into the integrated Ti resistor, the bimorph actuator bends due to the different TEC of Al and SiO$_2$. The transfer function of the dynamic response of the MEMS mirror can be expressed as

$$H(s) = H_T(s) \cdot H_M(s) \tag{1}$$

where $H_T(s)$ and $H_M(s)$ are, respectively, the transfer functions of the electrothermal response, and mechanical response of the bimorph actuator.

Firstly, these two physical processes will be modelled separately. According to the study reported in [19], the temperature is quite uniform on bimorph actuators since the resistive heater is uniformly distributed along the bimorph, and the thermal isolations on both ends of the bimorph are good. Thus, the electrothermal response of the bimorph actuator can be approximately modelled as a first-order system. The corresponding heat transfer equation can be simplified as:

$$C_T \frac{\mathrm{d}\Delta T}{\mathrm{d}t} + \frac{\Delta T}{R_T} = P \tag{2}$$

where ΔT is the average temperature change on the bimorph, C_T is the heat capacitance of the bimorph actuator, R_T is the equivalent thermal resistance from the bimorph to the substrate and the ambient, and P is the input electrical power. From Equation (2), the transfer function of the thermal response of the system can be derived as:

$$H_T(s) = \frac{R_T}{sR_TC_T + 1} \tag{3}$$

Let us consider the case when only one actuator is activated. In this case, the opposing actuator will be stationary while the two neighboring actuators will be displaced as much as half of that of the activated actuator, as illustrated in Figure 3.

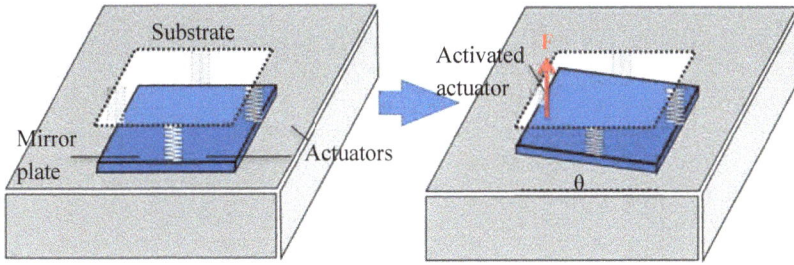

Figure 3. The mass-spring model of the MEMS mirror. Only the bimorph actuator on the left is activated.

Thus, the mirror can be modelled as a second-order mass-spring-damper system. The corresponding equation of motion is given by:

$$I\frac{d^2\theta}{dt^2} + D\frac{d\theta}{dt} + k_\theta\theta = FL \tag{4}$$

where I is the moment of inertia of the square mirror plate, θ is the rotation angle of the mirror plate, D is the air damping coefficient, k_θ is the equivalent torsional stiffness of all the bimorph actuators combined, L is length of the mirror plate, and F is the force generated by the activated actuator. Thus, the mechanical force-to-angle transfer function is readily obtained from Equation (4):

$$H_M(s) = \frac{L/I}{S^2 + (D/I)S + k_\theta/I} \tag{5}$$

Plugging Equations (3) and (5) into Equation (1) yields

$$H(s) = H_T(s) \cdot H_M(s) = \frac{R_T}{R_T C_T S + 1} \cdot \frac{L/I}{S^2 + (D/I)S + k_\theta/I} \tag{6}$$

$$\text{or } H(s) = \frac{1}{\tau S + 1} \cdot \frac{\omega_n^2}{S^2 + 2\omega_n\zeta S + \omega_n^2} = \frac{\frac{1}{\tau} \cdot \omega_n^2}{(S^2 + 2\omega_n\zeta S + \omega_n^2)\left(S + \frac{1}{\tau}\right)} \tag{7}$$

where $\tau = R_T C_T$, $\omega_n = \sqrt{k_\theta/I}$, and $\zeta = \frac{D}{2\sqrt{k_\theta I}}$ respectively represent the thermal time constant, natural resonant frequency of rotation, and damping ratio of the bimorph-mirror plate system.

For an under-damped system, the normalized step response in time domain can be obtained from Equation (7), i.e.,

$$\tilde{\theta}(t) = 1 - \frac{e^{-\beta\zeta\omega_n t}}{\zeta^2\beta(\beta-2)+1} - \frac{\zeta\beta e^{-\zeta\omega_n t}}{\sqrt{1-\zeta^2}\sqrt{\zeta^2\beta(\beta-2)+1}}\sin(\omega_d t + \alpha) \tag{8}$$

where $\beta = \frac{1}{\tau\omega_n\zeta}$, $\omega_d = \omega_n\sqrt{1-\zeta^2}$

$$\alpha = \arctan\frac{\zeta(\beta-2)\sqrt{1-\zeta^2}}{\zeta^2\beta(\beta-2)+1} \tag{9}$$

For an under-damping system, it is more convenient to use quality factor $Q = \frac{1}{2\zeta}$ to represent damping; also $\omega_n = 2\pi f_0$, where f_0 is the resonance frequency. The dynamical response of the whole system mainly depends on these parameters. As shown in (7), the dynamical response of the whole system includes a first-order low-pass filter sub-system and an under-damped second-order sub-system. Typically, the mirror plate of such a MEMS mirror is about 1 mm in size and surrounded by air, leading to a resonance frequency in the range of 0.3–3 kHz, a thermal time constant in the range of 1–50 ms, and a quality factor, or Q factor, of about 50 [17,18,20]. So, ζ is about 0.01.

Hypothetically, let us take $Q = 50$ and consider different τ and f_0. For example, take (a) $\tau = 1$ ms, and $f_0 = 300$ Hz, and (b) $\tau = 10$ ms, and $f_0 = 1.5$ kHz. Figure 4 shows the frequency response and step response of this system. Not surprisingly, there is overshoot and ringing in the step response as shown in Figure 4a. More interestingly, the step response in Figure 4b exhibits almost zero overshoot and ringing. This indicates that the step response's overshoot and ringing can be eliminated by properly designing the thermal response and mechanical response of the electrothermal bimorph actuator.

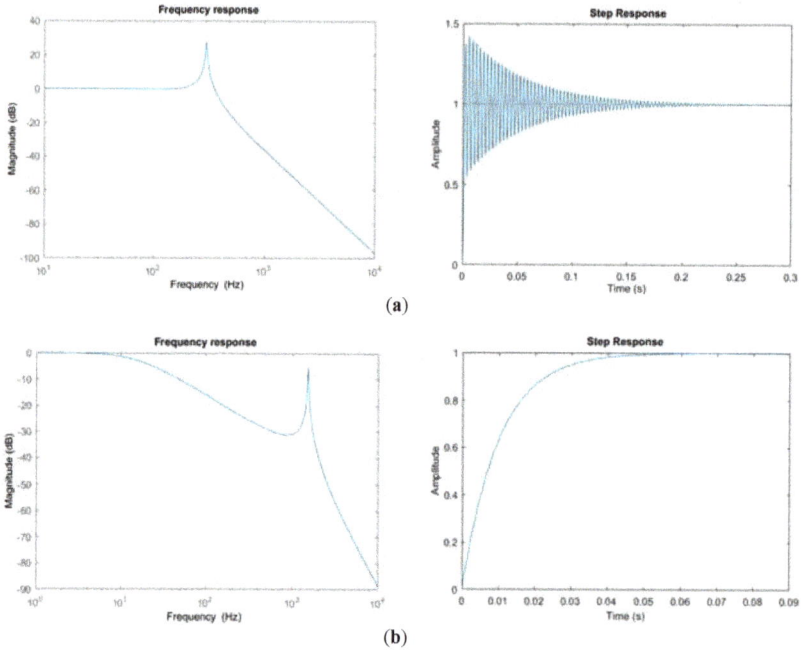

(a)

(b)

Figure 4. Frequency response and step responses of the whole system with (a) $Q = 50$, $\tau = 1$ ms and $f_0 = 300$ Hz; (b) $Q = 50$, $\tau = 10$ ms and $f_0 = 1,500$ Hz.

4. Elimination of Overshoot and Ringing

According to Equation (3), the cutoff frequency of the thermal response $f_{c,T}$ is given by

$$f_{c,T} = \frac{1}{2\pi\tau} \tag{10}$$

Since the thermal response functions as a low pass filter, the mechanical response will decrease rapidly with increasing frequency above $f_{c,T}$ by 20 dB/decade. Thereby, the peak of the mechanical resonance will be suppressed by a factor of $f_0/f_{c,T}$. As the mechanical gain at the resonance is equal to Q for an under-damped system, to completely remove the overshoot and ringing, we must have

$$f_0/f_{c,T} \geq Q \tag{11}$$

Plugging Equation (10) into (11) yields

$$\tau \cdot f_0 \geq \frac{Q}{2\pi} \tag{12}$$

Thus, according to Equation (12), for a given packaging environment (i.e., Q is fixed), an electrothermal bimorph actuator with fast thermal response must have high resonant frequency in order to suppress the overshoot and ringing of a step response. Numerical plotting Equation (8) will provide a better understanding. Let us still take $Q = 50$. Figure 5 shows the step response of a system with $f_0 = 1$ kHz and τ varying from 1 ms to 10 ms, while Figure 6 shows the step response of a system with $\tau = 5$ ms and f_0 varying from 200 Hz to 2 kHz. It can be observed that both the overshoot and ringing are effectively removed when $\tau \geq 10$ ms for $f_0 = 1$ kHz or when $f_0 \geq 2$ kHz for $\tau = 5$ ms. In both cases, $\tau \cdot f_0 > 50/(2\pi) \simeq 8$, which agrees very well with Equation (12).

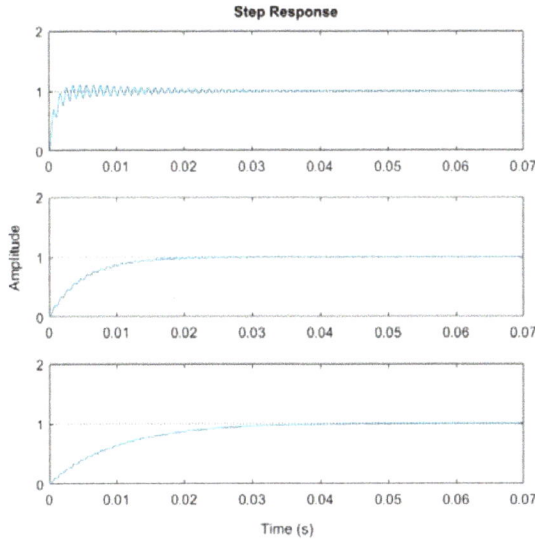

Figure 5. Step response with fixed resonant frequency of 1 kHz and different thermal time constants ($\tau = 1, 5, 10$ ms).

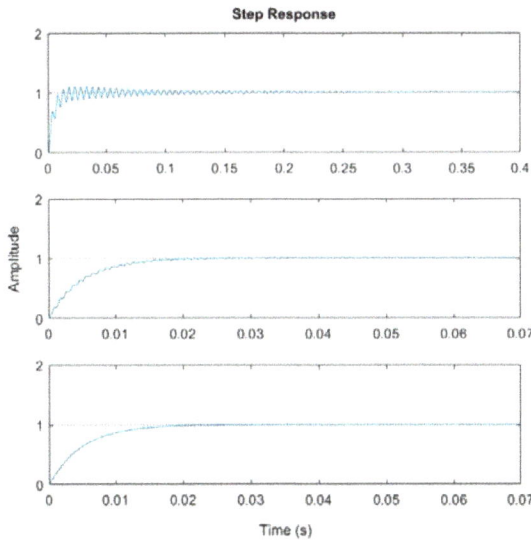

Figure 6. Step response with $\tau = 5$ ms and different resonant frequencies ($f_0 = 0.2, 1.0, 2.0$ kHz).

5. Experimental Verification

According to the analysis in Section 4, both the thermal response time and the resonant frequency determine the characteristics of the step response of an electrothermal MEMS mirror. Figure 7 shows a schematic diagram of a portion of an electrothermal MEMS mirror, including a complete thermal bimorph actuator, and part of the mirror plate. The bimorph actuator further consists of a folded double ISC bimorph, a thermal isolation A between the bimorph and the substrate, and a thermal isolation B between the bimorph and the mirror plate.

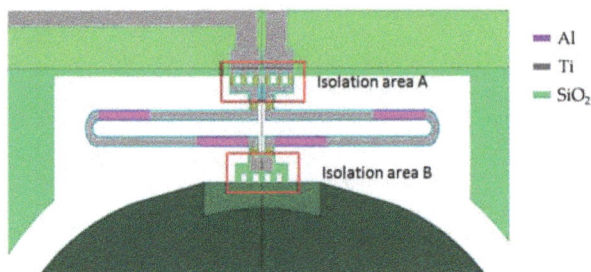

Figure 7. Schematic view of a single bimorph actuator design.

As we described above, the thermal response time and the resonant frequency are given by

$$\tau = R_T \, C_T \ \text{ and } \ f_0 = \frac{1}{2\pi}\sqrt{k_\theta / I} \tag{13}$$

where R_T is the inverse of the thermal conduction of the thermal isolation A plus the thermal conduction of the air around the bimorph, C_T is the thermal capacitance of the entire bimorph, k_θ is the torsional stiffness of the bimorph, and I is the moment of inertia which is proportional to the mass of the mirror plate. The stiffness of the bimorph k_θ can be changed largely by varying the length of the bimorph. Note that the thermal isolation region B blocks the heat flux from flowing into the mirror plate; it does not affect the thermal response time of the bimorph actuator and its thermal resistance is much greater than R_T in Equation (13). Table 1 lists two designs with different bimorph lengths leading to f_0. Figure 8 shows SEMs of both MEMS mirror designs.

Table 1. Parameters of two different MEMS mirror designs.

Design Type	Thermal Isolation A Length (µm)	Bimorph Length (µm)	Mirror Diameter (mm)
Design 1	110	845	1
Design 2	85	500	1

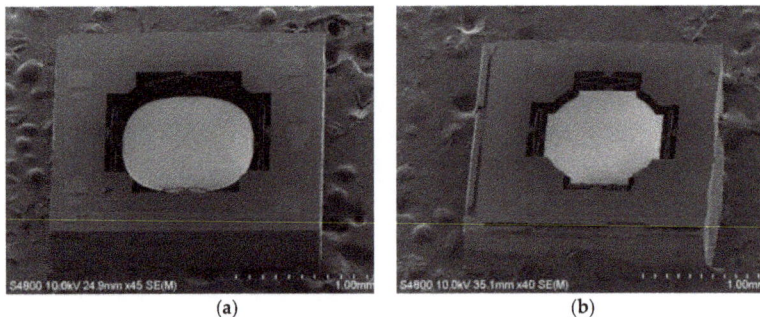

Figure 8. SEMs of two MEMS mirrors. (a) Design 1; (b) Design 2.

To obtain the step response of the MEMS mirrors, a setup as illustrated in Figure 9 was constructed, where a laser beam was directed to the center of a MEMS mirror through a beam splitter (BS), and then the laser beam was reflected by the MEMS mirror and incident on a position sensitive device (PSD) (OT-302D, On-Trak Photonics, Inc., California, USA). The output signal of the PSD was a direct measure of the lateral shift of the laser spot. The mirror tilt angle is readily calculated as follows:

$$\theta = \frac{1}{2}\arctan\frac{\Delta d}{d} \tag{14}$$

where θ is the mirror tilt angle, d is the distance between the mirror and the PSD, and Δd is the laser spot displacement on the PSD.

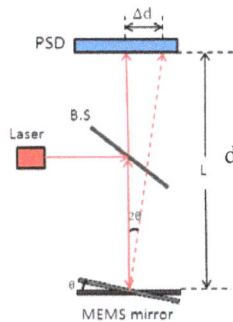

Figure 9. Experimental setup for measuring the mirror tilt angle.

First, the resonant frequencies of the two designs shown in Figure 8 were measured with simple frequency-sweeping using the setup shown in Figure 9, which were 0.592 kHz, and 1.89 kHz, respectively. Then a step voltage signal was applied to one of the actuators of a MEMS mirror, and the PSD output signal was recorded, which was the step response of the MEMS mirror. Figure 10 shows the step responses of the two designs. The experimental rise time t_r and f_0 for the two designs are listed in Table 2.

For a first order system, the step response is given by $\left(1 - e^{-\frac{t}{\tau}}\right)$. Thus, the 10% to 90% rise time can be readily derived as $t_r = (\ln 9)\tau \approx 2.2\tau$. Using this relation, the τ values are calculated and given in Table 2. Also plotted in Figure 8 are the simulated step responses with $Q = 50$. There is a small difference between the experiment and simulation, which is believed to be due to the fact that the assumed $Q = 50$ may not be accurate.

As shown in Figure 10 and Table 2, when the product of $\tau \cdot f_0$ is increased from 1 to about 4, the overshoot for Design 2 is reduced by a factor of 5. However, the $\tau \cdot f_0$ product for Design 2 is still less than 8, when the optimal value is calculated from Equation (12). Thus, just as predicted, a small overshoot remains in Design 2. In order to further reduce the overshoot, $\tau \cdot f_0$ must be increased. We may increase either τ or f_0 or both. Following this study, the bimorph length may be further reduced to increase f_0. According to Equation (13), other structural parameters, such as the thickness and width of each layer in the isolation and bimorph, the density of the holes on the isolation region, and the thickness and size of the mirror plate, can all be used to tune the product of $\tau \cdot f_0$.

Table 2. $\tau \cdot f_0$ of the two different MEMS mirror designs.

Design Type	Rise Time t_r (ms)	Thermal Response Time, τ (ms)	Resonant Frequency, f_0 (kHz)	$\tau \cdot f_0$	Overshoot (Test)	Overshoot (Simulation)
Design 1	4.3	1.95	0.592	1.16	10.2%	8.8%
Design 2	4.6	2.09	1.89	3.95	2.2%	0.66%

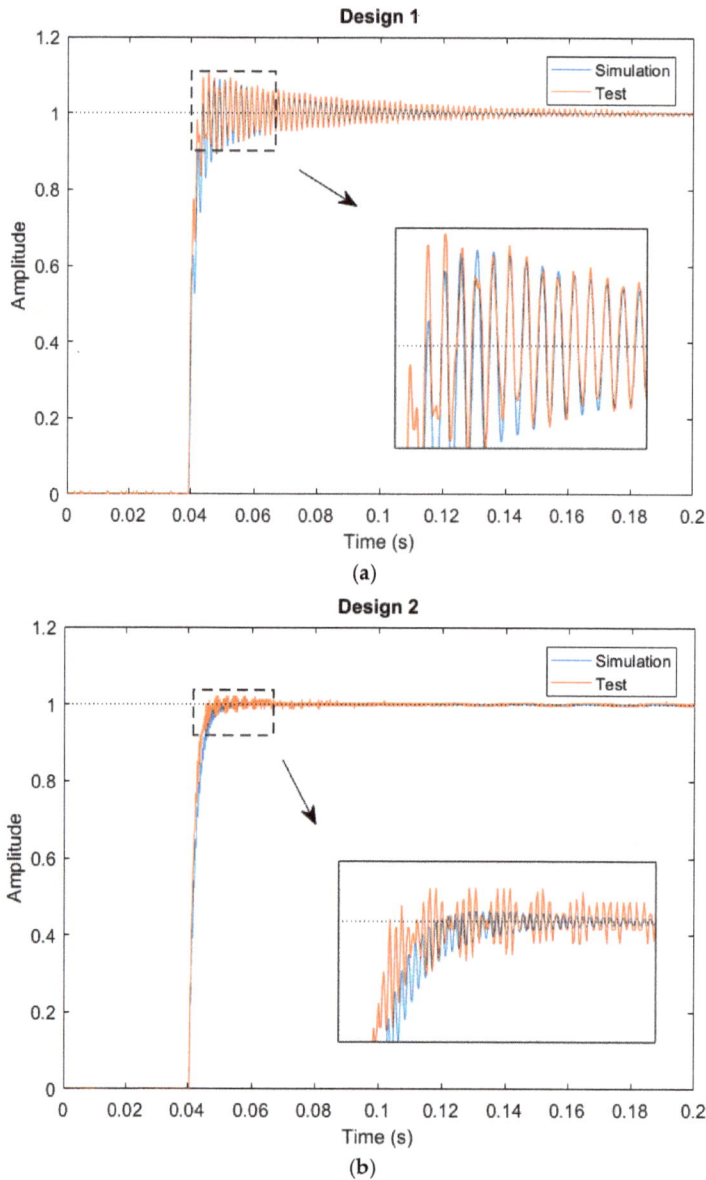

Figure 10. Step responses of the MEMS mirrors. (**a**) Design 1; (**b**) Design 2.

6. Conclusions

In this work, a solution that can suppress the overshoot and ringing of the step response of an under-damped electrothermal bimorph actuated MEMS mirrors is proposed and experimentally verified. A model based on the dynamical response of the electrothermal MEMS mirror is established. This model shows that the dynamical response of the electrothermal MEMS mirror can be considered as a first-order RC system connected with a second spring-mass system. Simply tuning the product to $\tau \cdot f_0$ can reduce, or even completely remove, the overshoot and ringing of the step response of the electrothermal MEMS mirror operating in air. This method provides a powerful venue for optimal use of electrothermal MEMS mirrors.

Acknowledgments: This work was supported in part by the National Natural Science Foundation of China under Grant 61574016 and 61404008, in part by 111 project of China under Grant B14010, and in part by the US National Science Foundation under award#1512531.

Author Contributions: Q.C. and H.X. conceived the concept; M.L. and Q.C. performed the theoretical analysis; Q.C. designed the devices; Y.L. performed the experiments; M.L. and Y.D. analyzed the data; M.L. and Q.C. wrote the paper; Y.D. and H.X. supervised the work.

Conflicts of Interest: The authors declare no conflict of interest.

References

1. Petersen, K.E. Silicon torsional scanning mirror. *IBM J. Res. Dev.* **1980**, *24*, 631–637. [CrossRef]
2. Tsai, J.C.; Fan, L.; Hah, D.; Wu, M.C. A High Fill-Factor, Large Scan-Angle, Two-Axis Analog Micromirror Array Driven by Leverage Mechanism. In Proceedings of the IEEE/LEOS International Conference on Optical MEMS and Their Applications, Takamatsu, Japan, 22–26 August 2004.
3. Fan, K.C.; Lin, W.L.; Chiang, L.H.; Chen, S.H.; Chung, T.T. A 2 × 2 Mechanical Optical Switch with a Thin MEMS Mirror. *J. Lightwave Technol.* **2009**, *27*, 1155–1161. [CrossRef]
4. Milanovic, V.; Lo, W.K. Fast and high-precision 3D tracking and position measurement with MEMS micromirrors. In Proceedings of the 2008 IEEE/LEOS International Conference on Optical MEMs and Nanophotonics, Freiburg, Germany, 11–14 August 2008; pp. 72–73.
5. Van Kessel, P.F.; Hornbeck, L.J.; Meier, R.E.; Douglass, M.R. A MEMS-based projection display. *Proc. IEEE* **1998**, *86*, 1687–1704. [CrossRef]
6. Ito, K.; Niclass, C.; Aoyagi, I.; Matsubara, H.; Soga, M.; Kato, S.; Maeda, M.; Kagami, M. System Design and Performance Characterization of a MEMS-Based Laser Scanning Time-of-Flight Sensor Based on a 256 × 64-pixel Single-Photon Imager. *IEEE Photonics J.* **2003**, *5*, 6800114. [CrossRef]
7. Lee, C.D.; Huang, L.S.; Kim, C.J.; Wu, M.C. Free-space fiber-optic switches based on MEMS vertical torsion mirrors. *J. Lightwave Technol.* **1999**, *17*, 7–13.
8. Chu, P.B.; Lee, S.S.; Park, S. MEMS: The path to large optical crossconnects. *IEEE Commun. Mag.* **2002**, *40*, 80–87. [CrossRef]
9. Kim, J.; Nuzman, C.J.; Kumar, B.; Lieuwen, D.F. 1100 × 1100 port MEMS-based optical cross-connect with 4-dB maximum loss. *IEEE Photonics Technol. Lett.* **2003**, *15*, 1537–1539. [CrossRef]
10. Borovic, B.; Liu, A.Q.; Popa, D.O. Open-loop versus closed-loop control of MEMS devices: Choices and issues. *J. Micromech. Microeng.* **2005**, *15*, 1917–1924. [CrossRef]
11. Lani, S.; Bayat, D.Z.; Despont, M. 2D tilting MEMS micro mirror integrating a piezoresistive sensor position feedback. In Proceedings of the SPIE OPTO 2015, San Francisco, CA, USA, 7–12 February 2015.
12. Pal, S.; Xie, H. Pre-Shaped Open Loop Drive of Electrothermal Micromirror by Continuous and Pulse Width Modulated Waveforms. *IEEE J. Quantum Electron.* **2010**, *46*, 1254–1260. [CrossRef]
13. Popa, D.O.; Kang, B.H.; Wen, J.T.; Stephanou, H.E.; Skidmore, G.; Geisberger, A. Dynamic modeling and input shaping of thermal bimorph MEMS actuators. In Proceedings of the 2003 IEEE International Conference on Robotics and Automation (Cat. No. 03CH37422), Taipei, Taiwan, 14–19 September 2003; Volume 1, pp. 1470–1475.
14. Daqaq, M.F.; Reddy, C.K.; Nayfeh, A.H. Input-shaping control of nonlinear MEMS. *Nonlinear Dyn.* **2008**, *54*, 167–179. [CrossRef]
15. Shi, M.; Zhang, H.; Chen, Q. The input shaping control of electro-thermal MEMS micromirror. In Proceedings of the 2014 IEEE International Conference on Mechatronics and Automation, Tianjin, China, 3–6 August 2014; pp. 583–587.
16. Todd, S.T.; Xie, H. An Electrothermomechanical Lumped Element Model of an Electrothermal Bimorph Actuator. *J. Microelectromech. Syst.* **2008**, *17*, 213–225. [CrossRef]
17. Jia, K.; Pal, S.; Xie, H. An Electrothermal Tip–Tilt–Piston Micro-mirror Based on Folded Dual S-Shaped Bimorphs. *J. Microelectromech. Syst.* **2009**, *5*, 1004–1014.

18. Chen, Q.; Zhang, H.; Zhang, X.; Xu, D.; Xie, H. Repeatability Study of 2D MEMS Mirrors Based on S-shaped Al/SiO$_2$ bimorphs. In Proceedings of the 8th Annual IEEE International Conference on Nano/Micro Engineered and Molecular Systems, Suzhou, China, 7–10 April 2013; pp. 817–820.
19. Pal, S.; Xie, H. A parametric dynamic compact thermal model of an electrothermally actuated micromirror. *J. Micromech. Microeng.* **2009**, *19*, 065007. [CrossRef]
20. Liu, L.; Pal, S.; Xie, H. MEMS mirrors based on a curved concentric electrothermal actuator. *Sens. Actuators A Phys.* **2012**, *188*, 349–358. [CrossRef]

micromachines

MDPI

Article

Design and Modeling of Polysilicon Electrothermal Actuators for a MEMS Mirror with Low Power Consumption

Miguel Lara-Castro [1], Adrian Herrera-Amaya [2], Marco A. Escarola-Rosas [1],
Moisés Vázquez-Toledo [3], Francisco López-Huerta [4,*], Luz A. Aguilera-Cortés [2] and
Agustín L. Herrera-May [1]

[1]   Micro and Nanotechnology Research Center, Universidad Veracruzana, Calzada Ruiz Cortines 455,
      Boca del Río, VER 94294, Mexico; septmig@gmail.com (M.L.-C.); maerescarola@gmail.com (M.A.E.-R.);
      leherrera@uv.mx (A.L.H.-M.)
[2]   Depto, Ingeniería Mecánica, Campus Irapuato-Salamanca, Universidad de Guanajuato/Carretera
      Salamanca-Valle de Santiago Km. 3.5 + 1.8 km, Salamanca, GTO 36885, Mexico;
      herreraugto@gmail.com (A.H.-A.); aguilera@ugto.mx (L.A.A.-C.)
[3]   Sistemas Automatizados, Centro de Ingeniería y Desarrollo Industrial/Av. Pie de la Cuesta No. 702,
      Desarrollo San Pablo, Querétaro 76125 México; moises.vazquez@cidesi.edu.mx
[4]   Engineering Faculty, Universidad Veracruzana, Calzada Ruiz Cortines 455, Boca del Río,
      Veracruz 94294, Mexico
*   Correspondence: frlopez@uv.mx; Tel.: +52-229-775-2000

Received: 14 January 2017; Accepted: 20 June 2017; Published: 25 June 2017

Abstract: Endoscopic optical-coherence tomography (OCT) systems require low cost mirrors with small footprint size, out-of-plane deflections and low bias voltage. These requirements can be achieved with electrothermal actuators based on microelectromechanical systems (MEMS). We present the design and modeling of polysilicon electrothermal actuators for a MEMS mirror (100 µm × 100 µm × 2.25 µm). These actuators are composed by two beam types (2.25 µm thickness) with different cross-section area, which are separated by 2 µm gap. The mirror and actuators are designed through the Sandia Ultra-planar Multi-level MEMS Technology V (SUMMiT V®) process, obtaining a small footprint size (1028 µm × 1028 µm) for actuators of 550 µm length. The actuators have out-of-plane displacements caused by low dc voltages and without use material layers with distinct thermal expansion coefficients. The temperature behavior along the actuators is calculated through analytical models that include terms of heat energy generation, heat conduction and heat energy loss. The force method is used to predict the maximum out-of-plane displacements in the actuator tip as function of supplied voltage. Both analytical models, under steady-state conditions, employ the polysilicon resistivity as function of the temperature. The electrothermal-and structural behavior of the actuators is studied considering different beams dimensions (length and width) and dc bias voltages from 0.5 to 2.5 V. For 2.5 V, the actuator of 550 µm length reaches a maximum temperature, displacement and electrical power of 115 °C, 10.3 µm and 6.3 mW, respectively. The designed actuation mechanism can be useful for MEMS mirrors of different sizes with potential application in endoscopic OCT systems that require low power consumption.

Keywords: electrothermal actuators; endoscopic optical-coherence tomography; microelectromechanical systems (MEMS) mirror; polysilicon; SUMMiT V

1. Introduction

Microelectromechanical systems (MEMS) have allowed the develop of devices with advantages such as low cost, small size, high reliability, fast response and easy integration with electronic circuits [1–3]. Among these devices, MEMS mirrors have potential applications such as projection displays [4], tunable optical filter [5], tunable laser [6], Fourier transform spectrometer system [7], confocal scanning microendoscope [8], optical bio-imaging [9] and optical coherence tomography [10]. For 3D endoscopic optical-coherence tomography (OCT) systems are necessary low cost MEMS mirrors composed by compact structures that have large out-of-plane deflections, minimum bias voltage and orthogonal scanning capacity [11,12]. These systems are minimally invasive and can have high resolution and reliability [12]. For this, the mirrors need high precision actuators that allow the variation of their tilting angles with low power consumption [13]. To adjust and control the mirror motion can use different actuators types, including the electromagnetic [14,15], electrostatic [16], electrothermal [17,18] or piezoelectric [19,20] actuators.

Mirrors with electrostatic actuators have a fast speed, a small mechanical scanning range at non-resonance (generally 2°–3°) and a large actuator footprint, which can be increased at resonance [21,22]. This actuation mechanism requires complex fabrication and high drive voltages about 100 V [23], which constraints its application in endoscopic OCT systems. Other actuators are the electromagnetics that generate large displacements with small driving voltage and have fast response time as well as high resonance frequency [24–26]. Although electromagnetic mirrors register problems with electromagnetic interference (EMI) and need precise assembly techniques of magnetic materials and metallic coils, limiting they use in endoscopic imaging [26]. On the other hand, piezoelectric actuators offer a large motion range combined with high speed and low electric energy [27]. Nevertheless, there are several challenges of the MEMS mirrors to develop endoscopic imaging such as charge leakage, coupling nonuniformity and hysteresis [28]. Other option is a MEMS mirror with an electrothermal actuation mechanism, which has large deflections caused by low bias voltage and does not present EMI and electrostatic discharging problems [28–32]. However, these mirrors require to decrease their footprint size, operation temperature and bias voltage as well as simplify their mechanical structure and performance. To overcome several of these challenges, we propone the design of polysilicon electrothermal actuators for MEMS mirrors based on the Sandia Ultra-planar Multi-level MEMS Technology V (SUMMiT V®) process from Sandia National Laboratories. This electrothermal actuation mechanism has a simple structural configuration composed by an array of four polysilicon actuators, which can achieve out-of-plane displacements with low dc voltages. These actuators do not require materials with different thermal expansion coefficients due to that employ polysilicon layers with distinct wide, which are separated by 2 µm gap. This device has a small footprint size (1028 µm × 1028 µm), compact structure and simple performance with reduced temperatures. The proposed design includes the modeling of temperature behavior and maximum displacements of the actuators under steady-state conditions. Our actuation mechanism can be used for the rotation of MEMS mirrors of different sizes. The rotation orientation of the mirror can be adjusted through the selective biasing of the four actuators. Thus, the proposed design could be considered for potential applications in endoscopic OCT systems.

This paper is organized as follows. Section 2 contains the design and modeling of the proposed actuation mechanism, which includes its electrothermal and structural behavior. Section 3 shows the results and discussions of temperature and out-of-plane displacements of the actuators using analytical models. Finally, the paper ends with the conclusion and future researches.

2. Design and Modeling

This section presents the design and modeling of the electrothermal actuators for a MEMS mirror. It considers the temperature distribution and out-of-plane displacements of the actuators generated by different dc biasing voltages under steady-state conditions.

2.1. Structural Configuration

Figure 1 shows the design of a MEMS mirror with an array of four polysilicon electrothermal actuators and springs, which are based on the SUMMiT V process [33]. The surface of the silicon substrate below of the actuators and mirror must be etched to allow the free motion of the actuators and mirror, as shown in Figure 2. Each actuator has two polysilicon structural layers (i.e., poly3 and poly4 of the SUMMiT V process) of 2.25 μm thickness with different cross-section area, separated by 2 μm gap. Thus, the electrical resistances of these layers are not equal, which allow a temperature change along the actuator when an electrical current is applied. It generates out-of-plate displacements of the actuator due to Joule effect, whose amplitudes can be controlled varying the current values. Thus, this actuator does not need materials layers with different thermal expansion coefficients that simplify its fabrication process. This design includes actuators with inverted structural layers to achieve out-of-plane motions with opposite directions, as shown in Figure 3a,b. Thereby, the mirror is connected to two pair actuators with inverted layers that can have displacements in opposite directions, increasing the tilting angle of the mirror. In addition, four polysilicon springs (508 μm length, 5 μm width and 2.25 μm thickness) with low stiffness are employed to connect the actuators with the mirror. Due to the small cross-section area and large length of each spring, the four springs have high electrical resistance that constraint the current flow through them. In this work, the effect of the thermal energy through the springs and mirror is not considered.

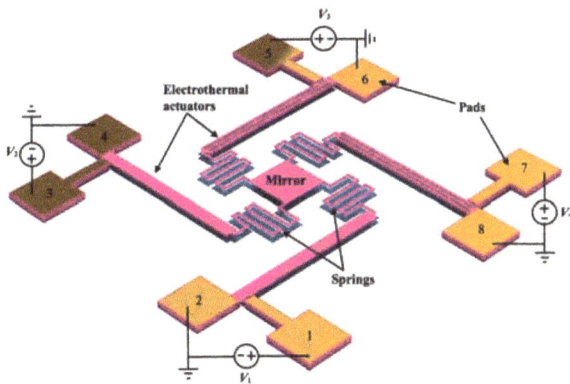

Figure 1. Design of an electrothermal actuation mechanism for the rotation of a microelectromechanical systems (MEMS) mirror.

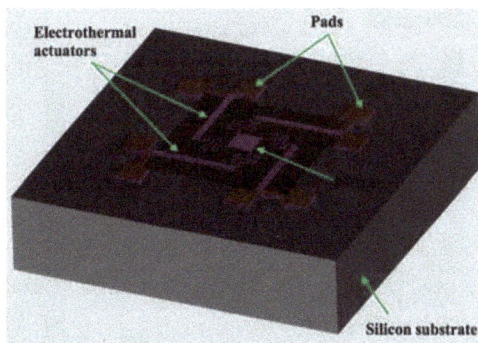

Figure 2. View of the MEMS mirror design in a silicon die.

In the design stage, the temperature and out-of-plane displacements of the actuators considering different dimensions of length (L_i) and width (w_h and w_c) of the upper (hot) and bottom (cold) beams are studied. The first structural layer is formed by a polysilicon beam (w_c) and the second layer is composed by three polysilicon beams of width w_h each one, in which $w_h \ll w_c$. Figure 3a,b depicts views of the hot and cold beams in two electrothermal actuators with deflections in opposite directions. In addition, the mirror and springs are designed using the poly4 layer of SUMMiT V process. In this fabrication process, on the mirror surface can be deposited an aluminum layer (96 μm × 96 μm × 0.7 μm). The springs have a connection with low stiffness between the actuators and mirror, which lets higher mirror tilting.

The operating principle of the electrothermal actuator with bending motion is caused by the asymmetrical thermal expansion of the two structural layers with different cross section area and electrical resistance. The resistance of the narrower layer is higher than that of the wider layer. If a dc bias voltage is applied at the end of the two layers (see Figure 3a,b) then a current flows through them, generating an increase of temperature in both layers. Due to the difference in the electrical resistance of the two layers, the temperature and dissipated energy in the narrower layer (high electrical resistance) is larger than the wider layer (low electrical resistance). This allows more thermal deformation of the narrower layer, which forces the actuator tip to an out-of-plane motion towards the wider layer. Therefore, the difference of the thermal deformation between the two actuator layers generates an out-of-plane motion. Figure 3c depicts the main geometrical parameters of an electrothermal actuator. In this work, we consider actuators with three different lengths (350 μm, 450 μm and 550 μm), constant thickness ($t_h = t_c = 2.25$ μm) and variable width (i.e., w_h of 2 μm to 5 μm and w_c of 20 μm to 30 μm).

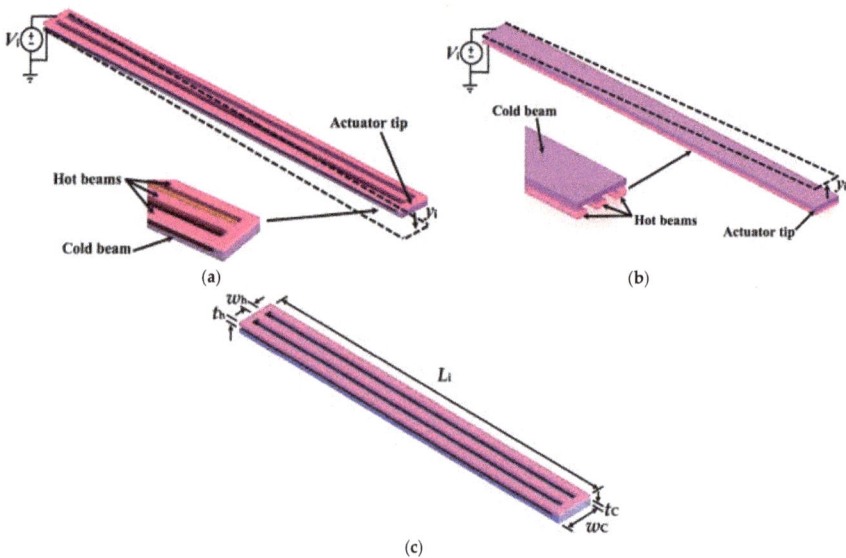

Figure 3. View of out-of-plane displacements, y_i, with directions (**a**) downward and (**b**) upward of two electrothermal actuators with inverted structural layers due to Joule effect; (**c**) geometrical parameters of the hot and cold beams of an electrothermal actuator.

2.2. Electrical Model of Electrothermal Actuators

An equivalent electric circuit of the electrothermal actuator is developed to predict the voltage drop along its hot and cold beams, as shown in Figure 4. For this case, R_1, R_2 and R_3 are the electrical resistance values obtained for each hot beam (w_h), cold beam (w_c) and connection between both beams, respectively. These resistances are calculated including the dimensions of the beams and the resistivity of the polysilicon layers. For instance, Table 1 shows the values of the electrical resistances for an electrothermal actuator with the following dimensions: $L_h = L_c = 450$ μm, $w_h = 5$ μm, $w_c = 30$ μm and $t_h = t_c = 2.25$ μm.

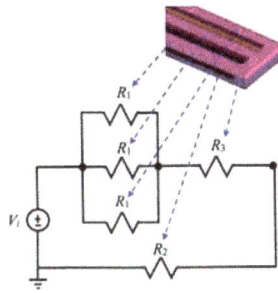

Figure 4. Schematic of equivalent electrical circuit of an electrothermal actuator.

Table 1. Resistance values of the equivalent electrical circuit of an electrothermal actuator considering the following dimensions: $w_h = 2$ μm, $w_c = 30$ μm and $t_h = t_c = 2.25$ μm.

Parameter	Electrical Resistance (Ω)		
	$L_h = 350$ μm	$L_h = 450$ μm	$L_h = 550$ μm
R_1	1576.6	2027	2077.4
R_2	105.1	135.1	165.2
R_3	1.7	1.7	1.7

2.3. Analytical Modeling of the Electrothermal and Structural Behavior

The electrothermal behavior of a polysilicon beam with length larger than its thickness and width can be simplified using an analysis in one dimension [31]. The electrothermal actuator (see Figure 3a) can be decomposed into three line-shape beams connected in series. For this, the first line-shape beam is obtained combining the three upper beams (hot beams) in a wider beam. Thus, the first line-shape beam has an equivalent electrical resistance equal to a third of the resistance of an upper beam. The second line-shape beam is formed by the connection between the upper and bottom beams, which has a 2.5 μm gap. In addition, the bottom (cold) beam forms the third line-shape beam. For this case, we assumed that the length of the upper (hot) beam (L_h) is equal to the length of the bottom beam (L_c): $L_h = L_c = L$. Figure 5 shows a differential element for the thermal analysis of the actuator.

In Figure 5b, heat flow equation is obtained by examining a differential element of polysilicon beam of width w, thickness t and length Δs. Assuming steady-state conditions, resistive heating power in the differential element is equal to heat conduction out of the element. Therefore, the energy balance of the differential element of the beam with heat losses can be expressed as [31]:

$$- k_p w t \left[\frac{dT}{ds}\right]_s + J^2 \rho w t \Delta s - Q \Delta s w \frac{T - T_0}{R_t} = -k_p w t \left[\frac{dT}{ds}\right]_{s+\Delta s} \tag{1}$$

where J is the current density, k_p is the thermal conductivity and ρ is the resistivity of the polysilicon, T is the operation temperature, T_0 is the substrate temperature, Q is the shape factor that includes the impact of the element shape on heat conduction to the substrate and R_t is the thermal resistance generated by the substrate and actuator that are considered wide enough [31]:

$$R_t = \frac{t_a}{k_a} + \frac{t_n}{k_n} + \frac{t_s}{k_s} \tag{2}$$

where t_a is the distance between both the bottom beam of the actuator and Si$_3$N$_4$ surface, t_n is the thickness of the Si$_3$N$_4$ film, t_s is the thickness of the SiO$_2$ film and k_a, k_n and k_s are the thermal conductivity of air, Si$_3$N$_4$ and SiO$_2$ films, respectively.

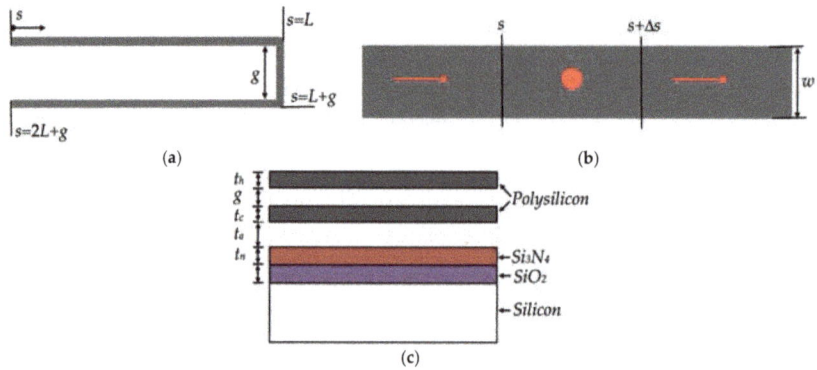

Figure 5. (a) Schematic of the one-dimensional model for an electrothermal actuator; (b) its differential element; and (c) cross-section of the different layers for the thermal analysis.

The shape factor Q for the heat conduction is given by [34]:

$$Q = \frac{t}{w}\left(\frac{2t_a}{t} + 1\right) + 1 \tag{3}$$

To apply Equation (3) in the electrothermal actuator, we approximated $t = t_h = t_c$ and $w = w_c$. The resistivity of polysilicon, $\rho(T)$, depends of the temperature and its value is determined by:

$$\rho(T) = \rho_0[1 + \xi(T - T_0)] \tag{4}$$

where ρ_0 is the initial resistivity at the substrate temperature and ξ is the linear temperature coefficient.
Considering the limit as $\Delta s \to 0$ for Equation (1), the following second-order differential equation is obtained:

$$k_p \frac{d^2T}{ds^2} + J^2\rho = \frac{Q}{t}\frac{(T - T_0)}{R_t} \tag{5}$$

The first term on the left of Equation (5) indicates the net rate of heat conduction into the element per unit volume. The rate of heat energy generation inside the element per unit volume is represented by the second term on the left. Finally, the rate of heat energy loss of the element per unit volume is considered in the term of the right side. Substituting Equation (4) into Equation (5), we obtain:

$$\frac{d^2T}{ds^2} - m^2T = -m^2T_0 - \frac{J^2\rho_0}{k_p} \tag{6}$$

with

$$m^2 = \frac{Q}{k_p R_t t} - \frac{J^2\rho_0\xi}{k_p} \tag{7}$$

Solving Equation (6) and applying the solution to the upper (hot) and bottom (cold) beams, we get the following temperature distribution:

$$T_h(s) = C_1 e^{m_h s} + C_2 e^{-m_h s} + T_o + \frac{J_h^2\rho_0}{k_p m_h^2} \tag{8}$$

$$T_c(s) = C_3 e^{m_c s} + C_4 e^{-m_c s} + T_o + \frac{J_c^2\rho_0}{k_p m_c^2} \tag{9}$$

with

$$m_h^2 = \frac{Q}{k_p R_t t} - \frac{J_h^2 \rho_0 \varsigma}{k_p} \tag{10}$$

$$m_c^2 = \frac{Q}{k_p R_t t} - \frac{J_c^2 \rho_0 \varsigma}{k_p} \tag{11}$$

where $T_h(s)$ and $T_c(s)$ are the temperature distribution along the upper (hot) and bottom (cold) beams, respectively, and J_h and J_c are the current density through the upper and bottom beams, respectively. To determine the constants C_i, we assume a temperature on the anchor pads equal to the substrate temperature (i.e., $T_h(0) = T_0$ and $T_c(2L + g) = T_0$), a continuity of both temperature (i.e., $T_h(L) = T_c(L)$) and rate of heat conduction (i.e., $3w_h dT_h(L)/ds = w_c dT_c(L)/ds$) across the join point of the upper and bottom beams. By assuming these boundary conditions, the following matrix equation is determined as:

$$\begin{bmatrix} 1 & 1 & 0 & 0 \\ e^{m_h L} & e^{-m_h L} & -e^{m_c L} & -e^{-m_c L} \\ 3w_h m_h e^{m_h L} & -3w_h m_h e^{-m_h L} & -w_c m_c e^{m_c L} & w_c m_c e^{-m_c L} \\ 0 & 0 & e^{m_c(2L+g)} & e^{-m_c(2L+g)} \end{bmatrix} \begin{bmatrix} C_1 \\ C_2 \\ C_3 \\ C_4 \end{bmatrix} = \begin{bmatrix} -\frac{J_h^2 \rho_0}{k_p m_h^2} \\ \frac{J_c^2 \rho_0}{k_p m_c^2} - \frac{J_h^2 \rho_0}{k_p m_h^2} \\ 0 \\ -\frac{J_c^2 \rho_0}{k_p m_c^2} \end{bmatrix} \tag{12}$$

The coefficients C_i of Equation (12) are determined using operations on matrices. Next, these coefficients are employed into Equations (8) and (9) to calculate the temperature increase along the upper and bottom beams due to bias voltages. These coefficients are calculated as:

$$C_1 = \frac{A(3+d) - e^{m_h L}\left[Bd\left(1 + e^{2m_c(L+g)}\right) + 2Dd e^{m_c(L+g)}\right] - A(3-d)e^{2m_c(L+g)}}{(d-3)e^{2m_c(L+g)} - (d+3)e^{2(m_c(L+g)+m_h L)} + e^{m_h L}(3-d)\left(e^{m_h L} + e^{-m_h L}\right)} \tag{13}$$

$$C_2 = \frac{\left(Bd + A(3-d)e^{m_h L} + \left(d(B+2D) - A(3+d)e^{m_h L}\right)e^{2m_c(L+g)}\right)e^{m_h L}}{(d-3 - (d+3)e^{2m_h L})e^{2m_c(L+g)} + 2(3\cosh(m_h L) - d\sinh(m_h L))e^{m_h L}} \tag{14}$$

$$C_3 = \frac{D\left(2(9+d^2) + (9-d^2)\left(e^{-2m_h L} + e^{2m_h L}\right)\right) + 18Be^{-m_c(L+g)} + F + 24A(G+H)}{4(G+H)\left((3-d)e^{(m_c+m_h)L} + (3+d)e^{(m_c-m_h)L}\right)} \tag{15}$$

$$C_4 = \frac{\left(-6Ae^{m_c(2L+g)} + D\left((3+d)e^{-m_h L} + (3-d)e^{m_h L}\right)e^{m_c L} + 3Be^{m_c(2L+g)}\left(e^{m_h L} + e^{-m_h L}\right)\right)e^{2m_c(L+g)}}{(3-d)e^{m_c(L+g)+m_h L} + (3+d)e^{m_c(L+g)-m_h L} - (3+d)e^{3m_c(L+g)+m_h L} - (3-d)e^{3m_c(L+g)-m_h L}} \tag{16}$$

with

$$A = -\frac{J_h^2 \rho_0}{k_p m_h^2} \tag{17}$$

$$B = \frac{J_c^2 \rho_0}{k_p m_c^2} - \frac{J_h^2 \rho_0}{k_p m_h^2} \tag{18}$$

$$D = -\frac{J_c^2 \rho_0}{k_p m_c^2} \tag{19}$$

$$F = -6A\left((3+d)e^{m_c g} + (3-d)e^{-m_h L}\right)e^{m_c(L+g)} + 3B\left((3+d)e^{-2m_h L} + (3-d)e^{2m_h L}\right)e^{-m_c(L+g)} \tag{20}$$

$$G = (3\cosh(m_h L)\sinh(m_c g) + d\sinh(m_h L)\cosh(m_c g))\cosh(m_c L) \tag{21}$$

$$H = (3\cosh(m_h L)\cosh(m_c g) + d\sinh(m_h L)\sinh(m_c g))\sinh(m_c L) \tag{22}$$

For the deflection analysis of the actuators, the linear thermal expansion for both upper (ΔL_h) and bottom (ΔL_c) beams can be determined as:

$$\Delta L_h = \alpha \int_0^L (T_h(s) - T_0)ds \tag{23}$$

$$\Delta L_c = \alpha \int_L^{2L+g} (T_c(s) - T_0)ds \tag{24}$$

where α is the thermal expansion coefficient of polysilicon.

By substituting Equations (8), (9) and (13)–(16) into Equations (23) and (24), the thermal expansions of the upper and bottom beams are given by:

$$\Delta L_h = \alpha \left\{ \frac{C_1}{m_h} \left(e^{m_h L} - 1 \right) - \frac{C_2}{m_h} \left(e^{-m_h L} - 1 \right) + \frac{J_h^2 \rho_0 L}{k_p m_h^2} \right\} \tag{25}$$

$$\Delta L_c = \alpha \left\{ \frac{C_3}{m_c} \left[e^{m_c L} \left(e^{m_c (L+g)} - 1 \right) \right] - \frac{C_4}{m_c} \left[e^{-m_c L} \left(e^{-m_c (L+g)} - 1 \right) \right] + \frac{J_c^2 \rho_0 (L+g)}{k_p m_c^2} \right\} \tag{26}$$

The structure of the electrothermal actuator can be considered as a plane rigid frame with two fixed ends. This actuator (see Figure 6) has a statically indeterminate structure with the degree of indeterminacy of 3 [35,36]. The bending moment of the actuator structure due to three unknowns (X_1, X_2 and X_3) is studied using the force method [35]. These unknowns are internal forces (horizontal force X_1, vertical force X_2 and bending moment X_3). The force method will be used to find the redundant unknowns followed by the virtual work method to obtain the deflection at the tip of the frame.

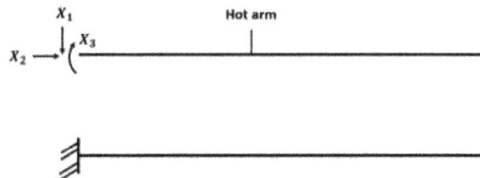

Figure 6. Rigid structure simplified for the electrothermal actuator regarding three redundant forces and moments (X_1, X_2 and X_3).

The three redundants (X_1, X_2 and X_3) are calculated through the canonical equations of the force method, which satisfy the compatibility conditions of the deformations [36]. For this case, the canonical equations are given by the following matrix form:

$$\begin{bmatrix} \delta_{11} & \delta_{12} & \delta_{13} \\ \delta_{21} & \delta_{22} & \delta_{23} \\ \delta_{31} & \delta_{32} & \delta_{33} \end{bmatrix} \begin{bmatrix} X_1 \\ X_2 \\ X_3 \end{bmatrix} = \begin{bmatrix} 0 \\ \Delta L_h - \Delta L_c \\ 0 \end{bmatrix} \tag{27}$$

where the coefficients δ_{ij} are called unit displacements that represent the displacements along the direction of unknown X_i caused by action of unit unknown X_j. δ_{ij} can be determined by the diagram product of the bending moments related with the unit unknowns X_i and X_j. These coefficients are obtained as:

$$\delta_{11} = \frac{L^2}{3EI_c}(L + 3g) + \frac{L^3}{3EI_h} \tag{28}$$

$$\delta_{12} = \delta_{21} = -\frac{Lg}{2EI_c}(L+g) \tag{29}$$

$$\delta_{13} = \delta_{31} = -\frac{L^2}{2EI_h} - \frac{L}{2EI_c}(L+2g) \tag{30}$$

$$\delta_{22} = \frac{g^2}{3EI_c}(g+3L) \tag{31}$$

$$\delta_{23} = \delta_{32} = \frac{g}{2EI_c}(g+2L) \tag{32}$$

$$\delta_{33} = \frac{L}{EI_h} + \frac{1}{EI_c}(g+L) \tag{33}$$

where E is the Young's modulus of polysilicon, I_h and I_c are the moment of inertia of the hot and cold beams, respectively.

Taking at account the method of virtual work, a unit force F is applied to the free end of actuator to calculate the maximum out-of-plane displacement:

$$\delta_{max} = \int \frac{M_F M}{EI_h} ds = \frac{L^2}{6EI_h}(X_1 L - 3X_3) \tag{34}$$

where M_F is the bending moment due to the virtual unit force and M the bending moment related with the thermal expansion. The physical and mechanical properties of the polysilicon used in the above analysis are listed in Table 2.

Table 2. Physical and mechanical properties of the polysilicon beams.

Property	Value
Young's Modulus, E	169 GPa
Thermal expansion, α	2.5×10^{-6} K^{-1}
Thermal conductivity, k_p	125 W·m^{-1}·K^{-1}
Substrate Temperature, T_0	300 K
Linear temperature coefficient, ξ	1.25×10^{-3} K^{-1}
Resistivity at T_0, ρ_0	20.27×10^{-6} Ω·m
Density	2330 kg·m^{-3}
Poisson ratio	0.23

Solving Equation (27), the unknowns X_1, X_2 and X_3 are the follows:

$$X_1 = \frac{18EI_c I_h(\Delta L_h - \Delta L_c)(I_h g + I_c L + I_h L)}{L(6I_c^2 L^3 + 2I_c^2 L^2 g + 6I_c I_h L^3 + 40I_c I_h L^2 g + 8I_c I_h L g^2 + 2I_h^2 L^2 g + 17I_h^2 L g^2 + 3I_h^2 g^3)} \tag{35}$$

$$X_2 = \frac{6EI_c(\Delta L_h - \Delta L_c)(I_c^2 L^2 + 2I_c I_h L^2 + 7I_c I_h L g + I_h^2 L^2 + 7I_h^2 L g + 6I_h^2 g^2)}{g^2[6I_c^2 L^3 + 2I_c^2 L^2 g + 6I_c I_h L^3 + 40I_c I_h L^2 g + 8I_c I_h L g^2 + 2I_h^2 L^2 g + 17I_h^2 L g^2 + 3I_h^2 g^3]} \tag{36}$$

$$X_3 = \frac{6EI_c I_h L(\Delta L_c - \Delta L_h)(5I_h g - I_c g + I_c L + I_h L)}{g[6I_c^2 L^3 + 2I_c^2 L^2 g + 6I_c I_h L^3 + 40I_c I_h L^2 g + 8I_c I_h L g^2 + 2I_h^2 L^2 g + 17I_h^2 L g^2 + 3I_h^2 g^3]} \tag{37}$$

Substituting Equations (35) and (36) into Equation (34), we determine the maximum out-of-plane displacement (δ_{max}) of the electrothermal actuator:

$$\delta_{max} = \frac{3I_c L^2(\Delta L_h - \Delta L_c)(I_c L^2 + I_h L^2 + I_h g^2 + 6I_h L g)}{g(6I_c^2 L^3 + 2I_c^2 L^2 g + 6I_c I_h L^3 + 40I_c I_h L^2 g + 8I_c I_h L g^2 + 2I_h^2 L^2 g + 17I_h^2 L g^2 + 3I_h^2 g^3)} \tag{38}$$

3. Results and Discussions

This section presents the results of the temperature shift and displacements of the actuator caused by different bias voltages. For this, we considered several variations in the dimensions (width and length) of the actuator.

By using Equations (8), (9) and (38), we determine the temperature and maximum out-of-plane displacement of the electrothermal actuator generated by low dc bias voltages. In this analysis, the initial temperature of the actuator is 20 °C and the length of each actuator is modified between 350 and 550 µm. In addition, we regard a variable width (i.e., w_h of 2 µm to 5 µm and w_c of 20 µm to 30 µm) for the upper and bottom beams and a constant thickness (i.e., $t_h = t_c = 2.25$ µm). We compared the results of our models with respect to analytical models of temperature and displacements of electrothermal actuators reported by reference [31]. For this, we use Equations (7), (8) and (23) of reference [31] and assume negligible the flexure beam length (i.e., $L_f = 0$). However, these models are applied for electrothermal actuators of variable cross-section area with in-plane deflections. In order to employ these models to our actuators with out-of-plane deflections, we considered that the variables of width and thickness of their hot and cold beams are equals to the thickness and width of our hot and cold beams. Figure 7a,b shows the results of the temperature along of the surface of the upper (hot) and bottom (cold) beams, which are generated by a bias voltage of 2.5 V. This distribution considers different lengths (350 and 550 µm) and two values of width for each upper beam (2 and 5 µm). For all the cases, the maximum temperature is achieved close to the half of the length of the upper beam. The shorter beams present higher temperatures than the larger beams due to their less electrical resistance, which produce higher currents for a bias voltage. For the upper beams of 5 µm width, the temperature decays more slowly along of the electrothermal actuator, as shown in Figure 7b. In the actuator tip, we observed a significant variation in the behavior of the temperature distribution along of the hot and cold beams. The results of our analytical models have a similar behavior respect to those of reference [31]; although, our results register the highest temperature values in all the cases. Next, we calculate the temperature distribution regarding two actuators of different lengths (450 and 550 µm), which are supplied by different dc bias voltages, as shown in Figure 8a,b. The maximum voltage of 2.5 V generates the higher temperature magnitudes (147.3 °C and 114.9 °C) for both actuators, considering our models. For the actuator of 450 µm length, the bias voltages of 1.0 V, 1.5 V and 2.0 V increase the temperature up 38.2 °C, 62.1 °C and 97.6 °C, respectively. For the same voltages, the actuator of 550 µm length has an increment of temperature of 34.0 °C, 52.0 °C and 78.5 °C, respectively. Also, the temperature distribution along the actuator of 450 µm length was determined varying the width of the upper and bottom beams, as shown in Figure 9a,b. For upper beams of 2 µm width and bias voltage of 2.5 V, the temperature has a low increment of 16.6 °C when the width of the bottom beam increases from 20 to 30 µm. Instead, the temperature distribution decays more slowly for upper beams of 5 µm width, keeping 30 µm width for the bottom beam. For these cases, the results of our models have good agreement respect to those of reference [31].

Figure 10a,b depicts the maximum out-of-plane displacements of the actuator tip as a function of bias voltage and assuming different length and width values. For these cases and considering 2.5 V, the beam of 550 µm length has the larger displacements (10.3 µm and 6.8 µm) when $w_h = 2$ µm and $w_c = 30$ µm, respectively. These displacements have direction down due to the higher temperature of the upper beams. However, if the position of the beams is inverted then the motion of the actuator will be upward. If the length of the actuator is 450 µm and the bias voltage is 2.5 V then the maximum displacements are 8.9 µm and 5.6 µm, respectively. The response of our models has good agreement respect to results of reference [31]. Although the displacements obtained with our analytical models have higher values than those of the reference [31]. Figure 11a,b shows the maximum out-of-plane displacements of the actuator (450 µm length and 2.5 V voltage) considering different dimensions in the width of its upper and bottom beams. For these cases, the larger displacement (8.9 µm) is obtained with 2.5 V voltage for beams with $w_h = 2$ µm and $w_c = 30$ µm, respectively. In addition, the displacement of the actuator tip decreases when the width of the upper beams increases. Moreover, if the width of the bottom beam increases then the actuator tip will have larger displacements. The electrical power of each actuator is determined using the equivalent electrical circuit of Figure 5. For an actuator with w_h = 2 µm, w_c = 30 µm and three different lengths L_h: 350 µm, 450 µm and 550, we obtain the following electrical power: 9.9 mW, 7.7 mW and 6.3 mW. Finally, the displacements of the actuator tip can be increased with bias voltages higher than 2.5 V, which also will increment the electrical power. For instance, if the actuator of L_h = 550 µm and w_h = 2 µm is biased with 5 V then its maximum displacement, temperature and power are increased up 59.2 µm, 570.3 °C and 25.2 mW, respectively. Furthermore, the mirror surface area can be scalable to achieve larger values than 10000 µm². On the other hand, the surface of the silicon substrate below of the actuators array and mirror must be etched using DRIE process to allow the free motion of the actuators and mirror under different bias voltages. Nevertheless, the maximum displacement of the actuators must generate stress less than the rupture stress of the polysilicon.

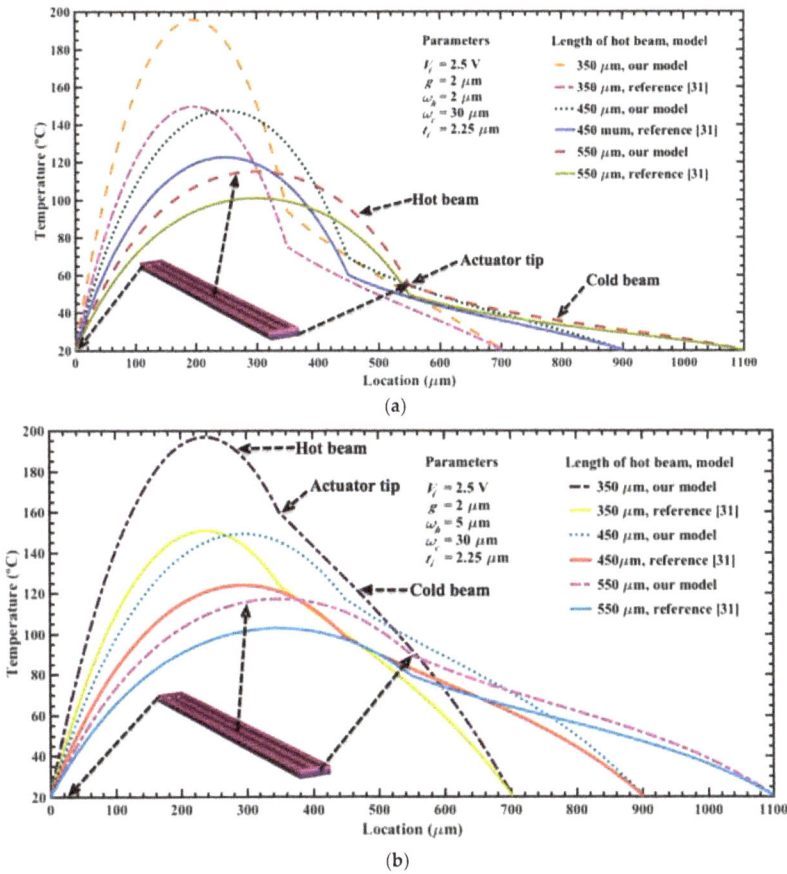

Figure 7. Distribution of the temperature along of the upper (hot) and bottom (cold) beams of an electrothermal actuator, which considers different lengths (350 µm to 550 µm) and two width values for the upper beams: (**a**) 2 µm; and (**b**) 5 µm.

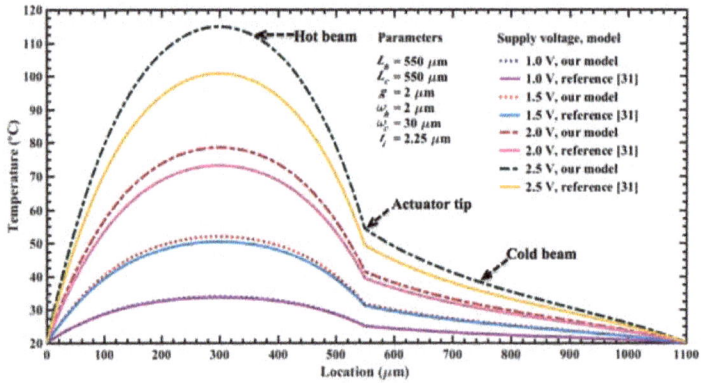

(a)

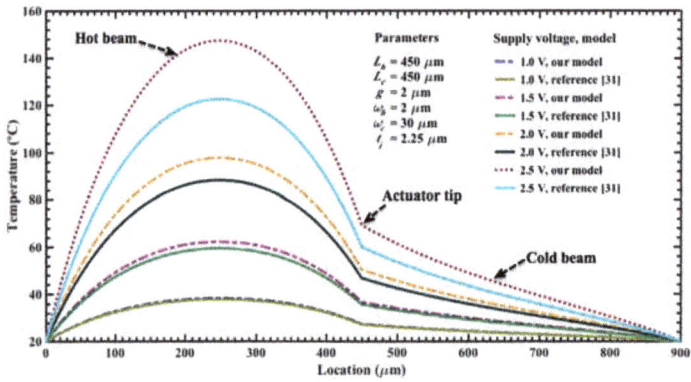

(b)

Figure 8. Distribution of the temperature along of the upper (hot) and bottom (cold) beams of two electrothermal actuators with lengths of (**a**) 550 μm and (**b**) 450 μm. This temperature is due to different bias voltages, whose values change from 0.5 to 2.5 V.

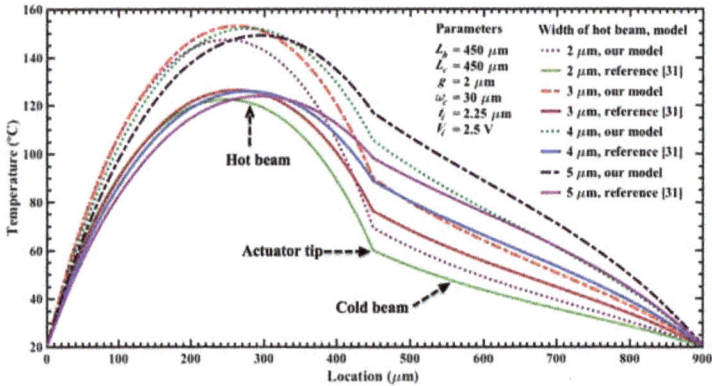

(a)

Figure 9. *Cont.*

(b)

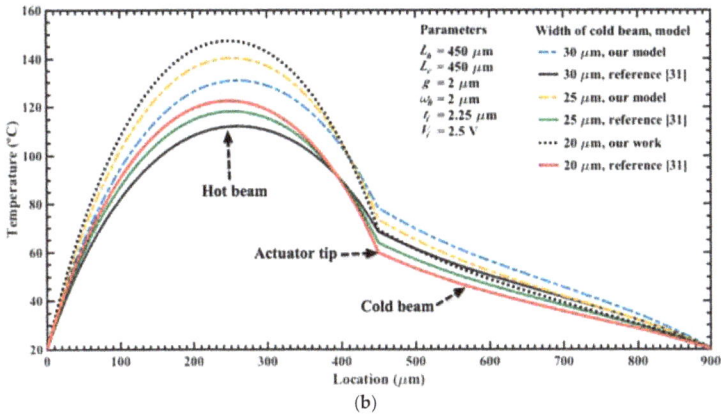

Figure 9. Distribution of the temperature along of the upper (hot) and bottom (cold) beams of an electrothermal actuator, modifying the width of the (**a**) upper and (**b**) bottom beams. For both cases, the length of the actuator is 450 μm and bias voltage is 2.5 V, respectively.

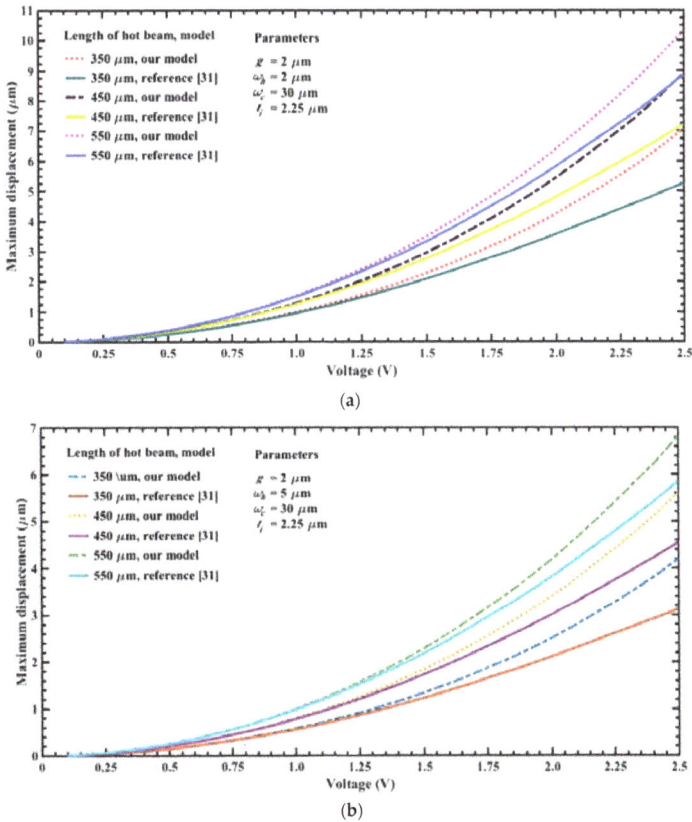

(a)

(b)

Figure 10. Maximum out-of-plane displacements of the electrothermal actuator tip as a function of bias voltage, regarding different lengths and two width values for the upper beams: (**a**) 2 μm and (**b**) 5 μm.

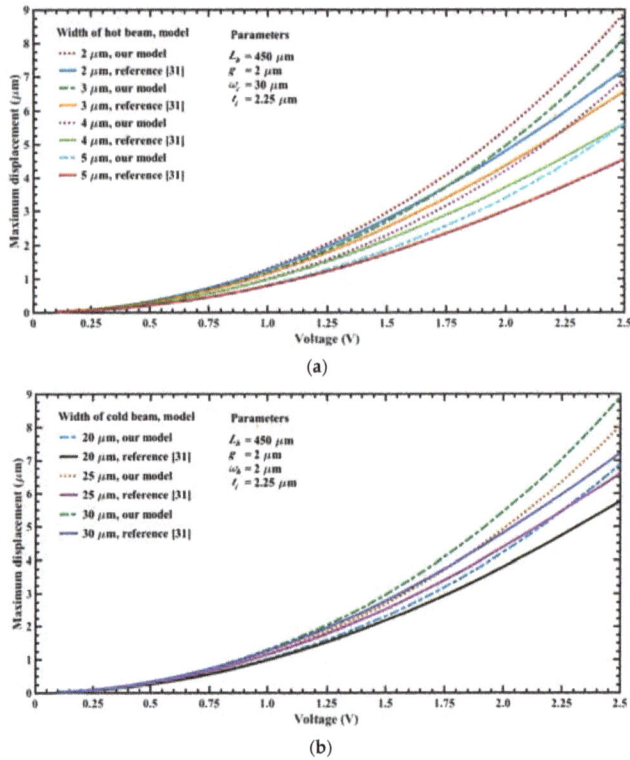

Figure 11. Maximum out-of-plane displacements of the electrothermal actuator tip as a function of bias voltage, varying the width of the (a) upper and (b) bottom beams. For both cases, the length of the actuator is 450 μm and bias voltage is 2.5 V, respectively.

Finally, we developed finite elements method (FEM) models using the ANSYS® software (version 15.0, ANSYS, Berkeley, CA, USA) to predict the out-of-plane displacements of the proposed actuation mechanism. For this, the pads were negligible and the initial end of each actuator was considered as fixed support. For these supports were applied a bias voltage of 2.5 V and initial temperature of 20 °C. The FEM models regard polysilicon actuators with the following dimensions: $L_h = L_c = 550$ μm, $\omega_h = 2$ μm, $\omega_c = 30$ μm, $t_i = 2.25$ μm and $g = 2$ μm. Our FEM models include elements solid226 type with a hexahedral mesh. First, we use a FEM model of a single electrothermal actuator under 2.5 V bias voltage. Figure 12 depicts the out-of-plane displacements of this actuator, achieving a maximum downward deflection of 10.3 μm that well agree with the results (10.3 μm and 8.8 μm) of both our analytical model and that of the reference [31], as shown in Figure 10a. Next, we used a FEM model composed by four polysilicon electrothermal actuators, four springs (508 μm length, 5 μm width and 2.25 μm thickness) and a mirror. Each one of these actuators has the same dimension respect to the previous actuator. The initial ends of the four actuators have boundary conditions of clamped support and temperature of 20 °C. For this FEM model, we studied four different cases modifying the bias voltage values of the four actuators. For the first case, one actuator was only supplied with a voltage of 2.5 V, keeping the other three actuators without bias voltage (see Figure 13). Thus, the actuator and mirror have maximum out-of-plane deflections of 7.4 μm and 4.8 μm, respectively. For this case, the displacement of the actuator decreases (3.9 μm) respect the response of a single actuator without connection with springs and mirror. This displacement reduction is due to an increment of the model stiffness when the four actuators are joined to the mirror. In the second case two actuators are biased with 2.5 V, obtaining out-of-plane displacements with opposite directions (downward and upward) that allow the mirror rotation with respect to two of its vertices, as shown in Figure 14. The absolute value of the maximum displacement of the two biased actuators is 6.7 μm, which is 3.5 μm less than that obtained with a single actuator. Two mirror vertices reach maximum displacements of 3.7 μm and −3.7 μm, respectively. For the third case, a 2.5 V bias voltage is applied for three actuators, achieving maximum displacements of 9.2 μm, 7.7 μm and −4.5 μm (see Figure 15). Indeed, two mirror vertices have displacements of 6.2 μm and −1.4 μm that enable the mirror tilting. In the last case all the actuators are biased with 2.5 V, obtaining the downward and upward deflection of two actuator pairs as well as the mirror rotation along the *x*-axis (see Figure 16). The larger displacements of the actuators and mirror are 7.1 μm, −7.1 μm, 3.9 and −3.9 μm, respectively. In order to reach larger deflection and tilting of the actuators and mirror, the bias voltage can be increased. Moreover, the rotation orientation of the mirror can be regulated through the selective biasing of the four actuators. Also, the proposed actuation mechanism can be employed for MEMS mirrors of larger surface area and their rotation angles can be controlled using different bias voltages.

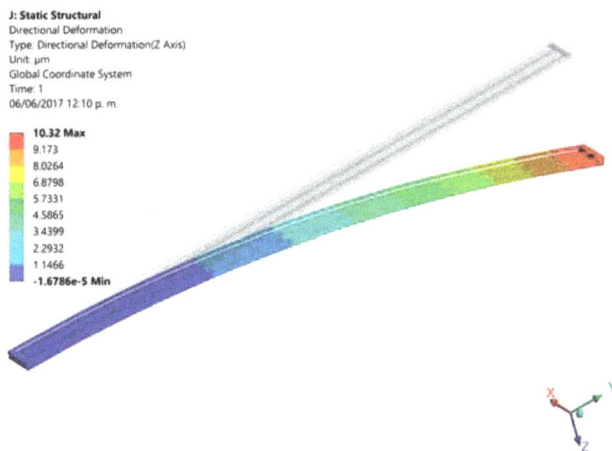

Figure 12. Out-of-plane displacements of one polysilicon electrothermal actuator ($L_h = L_c = 550$ μm) caused by a 2.5 V bias voltage.

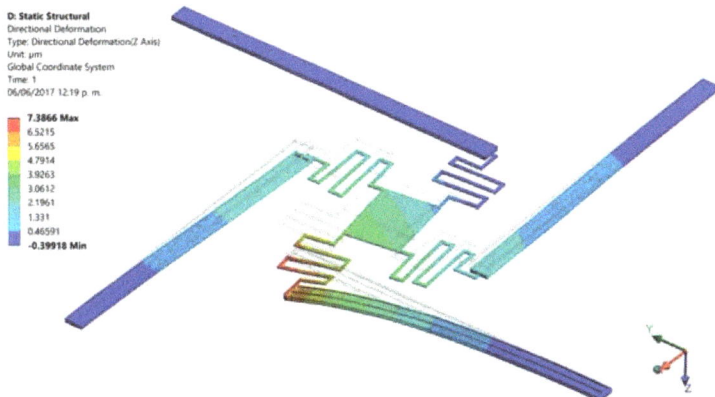

Figure 13. Out-of-plane deflections of the MEMS mirror when one polysilicon electrothermal actuator ($L_h = L_c = 550$ μm) is biased with 2.5 V.

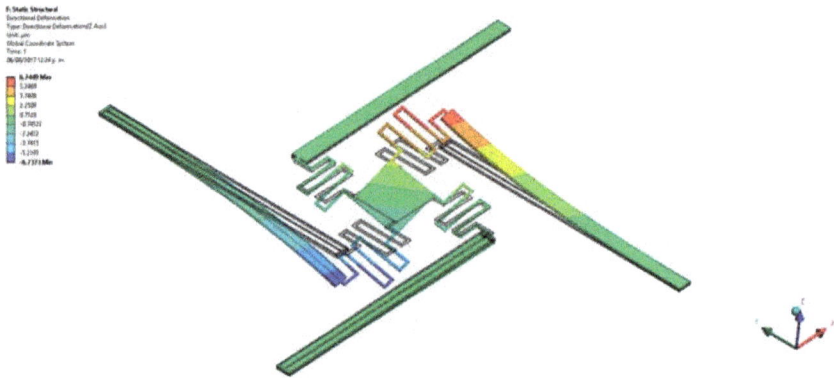

Figure 14. Out-of-plane displacements of the MEMS mirror when two polysilicon electrothermal actuators ($L_h = L_c = 550$ µm) are biased with 2.5 V.

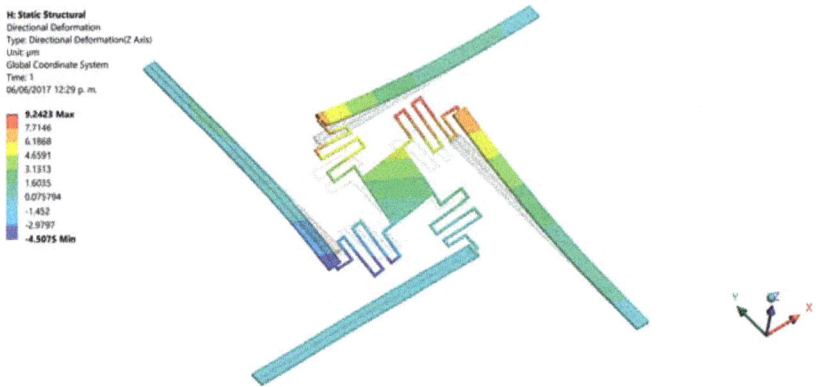

Figure 15. Out-of-plane displacements of the MEMS mirror when three polysilicon electrothermal actuators ($L_h = L_c = 550$ µm) are biased with 2.5 V.

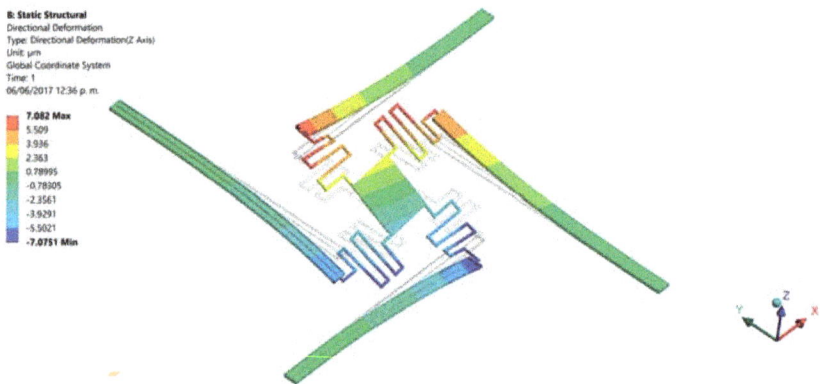

Figure 16. Out-of-plane displacements of the MEMS mirror when four polysilicon electrothermal actuators ($L_h = L_c = 550$ µm) are biased with 2.5 V.

Table 3 depicts the characteristics of several MEMS mirrors that use electrothermal actuators. Based on these devices, our design provides an easy actuation mechanism that does not require materials layers with different thermal expansion coefficients. It can simplify the actuators fabrication process and reduce the thermal residual stresses due to the fabrication. The proposed design is based on SUMMiT V process, which improves the flatness of the structures and minimize thermal residual strains. Indeed, our design has a minimum footprint size (1028 × 1028) and mirror surface area (100 μm × 100 μm), achieving different rotation orientations of the mirror that are well controlled using reduced bias voltages. Most of the other designs need different metallic films (e.g., Al, Cu, W or Pt) deposited on the actuators by sputtering process, which can generate initial thermal strains (i.e., initial displacement offset) that can affect the actuators performance. Indeed, our actuation mechanism can be adjusted for MEMS mirrors with larger surface area than 10,000 μm^2, which can be suitable for potential applications in endoscopic OCT systems.

Table 3. Characteristics of several MEMS mirrors based on electrothermal actuators.

Authors	Mirror Size	Device Footprint (μm × μm)	Maximum Displacement (μm)	Bias Voltage (V)
Zhang et al. [18]	900 μm × 900 μm	2500 × 2500	312	3
Kawai et al. [37]	3000 μm diameter	5000 × 5000	*-	20
Zhang et al. [38]	1000 μm × 1000 μm	1500 × 1500	70	2
Li et al. [39]	1000 μm diameter	2000 × 2000	227	0.8
Espinosa et al. [40]	1000 μm × 1000 μm	1500 × 1500	174	3.5
Koh et al. [41]	1500 μm × 1000 μm	6000 × 6000	*-	5
Our work	100 μm × 100 μm	1028 × 1028	59.2	5

*- Data not available in literature.

4. Conclusions

The design and modeling of an electrothermal actuation mechanism for a polysilicon mirror (100 μm × 100 μm × 2.25 μm) was developed. These actuators were designed based on the SUMMiT V surface micromachining process from Sandia National Laboratories. The actuators are composed by two polysilicon structural layers, which are vertically separated by 2 μm. The temperature and out-of-plane displacements of the actuators were determined using electrothermal and structural models and assuming the polysilicon resistivity as a function of temperature. The electrothermal models included the rate of heat energy generation, heat conduction and heat energy loss. On the other hand, the structural model was obtained with the force method and assuming low dc voltages (0.5 V to 2.5 V). For actuators with lengths of 450 and 550 μm, the higher temperatures and out-of-plane displacements generated by 2.5 V are: 147.3 °C, 115 °C, 8.9 μm and 10.3 μm, respectively. These actuators can have upward and downward motion if their structural layers are inverted. Thus, the mirror tilting can be controlled modifying the position of the structural layers and altering the actuators dimensions and magnitudes of the dc bias voltages. In addition, the device footprint size is 1028 μm × 1028 μm considering electrothermal actuators of 550 μm length. With a bias voltage of 2.5 V, the electrical power for an actuator of 550 μm length was 6.3 mW. The proposed actuation mechanism could be used to obtain the rotation of MEMS mirrors with different surface area. The rotation orientation of the mirrors can be modified through the selective biasing of the actuators. This actuation mechanism for MEMS mirrors could be considered for potential applications in endoscopic OCT systems.

Future researches will include the fabrication and characterization of several electrothermal actuators array for MEMS mirrors with different surface area using the SUMMiT V process.

Acknowledgments: The work was partially supported by the MEMS University Alliance Program of Sandia National Laboratories, CONACYT and FORDECYT-CONACYT through grants 48757 and 115976, project PROINNOVA "Ecoplataforma biomimética para agricultura de precision aplicando micro/nanotecnología" through grant 231500 and projects PRODEP "Estudio de Dispositivos Electrónicos y Electromecánicos con Potencial Aplicación en Fisiología y Optoelectrónica" and PFCE "2016–2017 DES Técnica Veracruz P/PFCE-2017-30MSU0940B-22".

Author Contributions: Miguel Lara-Castro, Adrian Herrera-Amaya, Francisco López-Huerta and Agustín L. Herrera-May develop the design and modeling of the electrothermal actuators for a MEMS mirror. Moisés Vazquez-Toledo and Marco A. Escarola-Rosas made the layout of the actuators using the SUMMiT V fabrication process. Adrian Herrera-Amaya, Francisco López-Huerta, Luz A. Aguilera-Cortés and Agustín L. Herrera-May wrote all the sections of the paper.

Conflicts of Interest: The authors declare no conflict of interest.

References

1. Chang, C.I.; Tsai, M.H.; Liu, Y.C.; Sun, C.M.; Fang, W. Pick-and-place process for sensitivity improvement of the capacitive type CMOS MEMS 2-axis tilt sensor. *J. Micromech. Microeng.* **2013**, *23*, 095029. [CrossRef]
2. Juárez-Aguirre, R.; Domínguez-Nicolás, S.M.; Manjarrez, E.; Tapia, J.A.; Figueras, E.; Vázquez-Leal, H.; Aguilera-Cortés, L.A.; Herrera-May, A.L. Digital Signal Processing by Virtual Instrumentation of a MEMS Magnetic Field Sensor for Biomedical Applications. *Sensors* **2013**, *13*, 15068–15084. [CrossRef] [PubMed]
3. Huang, J.Q.; Li, F.; Zhao, M.; Wang, K.A. Surface Micromachined CMOS MEMS Humidity Sensor. *Micromachines* **2015**, *6*, 1569–1576. [CrossRef]
4. Hung, A.C.-L.; Lai, H.Y.-H.; Lin, T.-W.; Fu, S.-G.; Lu, M.S.-C. An electrostatically driven 2D micro-scanning mirror with capacitive sensing for projection display. *Sens. Actuators A* **2015**, *222*, 122–129. [CrossRef]
5. Liu, Y.; Xu, J.; Zhong, S.; Wu, Y. Large size MEMS scanning mirror with vertical comb drive for tunable optical filter. *Opt. Lasers Eng.* **2013**, *51*, 54–60. [CrossRef]
6. Holmström, S.T.S.; Baran, U.; Urey, H. MEMS laser scanners: A review. *J. Microelectromech. Syst.* **2014**, *23*, 259–275. [CrossRef]
7. Wang, W.; Chen, J.; Zivkovic, A.S.; Tanguy, Q.A.A.; Xie, H. A compact Fourier transform spectrometer on a silicon optical bench with an electrothermal MEMS mirror. *J. Microelectromech. Syst.* **2016**, *25*, 347–355. [CrossRef]
8. Liu, L.; Wang, E.; Zhang, X.; Liang, W.; Li, X.; Xie, H. MEMS-based 3D confocal scanning microendoscope using MEMS scanner for both lateral and axial scan. *Sens. Actuators A* **2014**, *215*, 89–95. [CrossRef] [PubMed]
9. Liu, L.Y.; Keeler, E.G. Progress of MEMS scanning micromirrors for optical bio-imaging. *Micromachines* **2015**, *6*, 1675–1689.
10. Haindl, R.; Trasischker, W.; Baumann, B.; Pircher, M.; Hitzenberger, C.K. Three-beam Doppler optical coherence tomography using a facet prism telescope and MEMS mirror for improved transversal solution. *J. Mod. Opt.* **2015**, *62*, 1781–1788. [CrossRef] [PubMed]
11. Jung, W.; McCormick, D.; Ahn, Y.; Sepehr, A.; Brenner, M.; Wong, B.; Tien, N.; Chen, Z. In vivo tree-dimensional spectral domain endoscopic optical coherence tomography using a microelectromechanical system mirror. *Opt. Lett.* **2007**, *32*, 3239–3241. [CrossRef] [PubMed]
12. Jung, W.; McCormick, D.T.; Zhang, J.; Wang, L.; Tien, N.C.; Chen, Z. Three-dimensional endoscopic optical coherence tomography by use of a two-axis microelectromechanical scanning mirror. *Appl. Phys. Lett.* **2006**, *88*, 163901. [CrossRef]
13. Solgaard, O.; Godil, A.A.; Howe, R.T.; Lee, L.P.; Peter, Y.-A.; Zappe, H. Optical MEMS: From micromirrors to complex systems. *J. Microelectromech. Syst.* **2014**, *23*, 517–535. [CrossRef]
14. Kim, J.-H.; Jeong, H.; Lee, S.-K.; Ji, C.-H.; Park, J.-H. Electromagnetically actuated biaxial scanning micromirror fabricated with silicon on glass wafer. *Microsyst. Technol.* **2017**, *23*, 2075–2085. [CrossRef]
15. Cho, A.R.; Han, A.; Ju, S.; Jeong, H.; Park, J.-H.; Kim, I.; Bu, J.-U.; Ji, C.-J. Electromagnetic biaxial microscanner with mechanical amplification at resonance. *Opt. Express* **2015**, *23*, 16792–16802. [CrossRef] [PubMed]
16. Fan, C.; He, S. A microelectrostatic repulsive-torque rotation actuator with-width finger. *J. Micromech. Microeng.* **2015**, *25*, 095006. [CrossRef]
17. Zhang, X.; Duan, C.; Liu, L.; Li, X.; Xie, H. A non-resonant fiber scanner based on a electrothermally-actuated MEMS stage. *Sens. Actuators A* **2015**, *233*, 239–245. [CrossRef] [PubMed]
18. Zhang, X.; Zhou, L.; Xie, H. A fast, large-stroke electrothermal MEMS mirror based on Cu/W bimorph. *Micromachines* **2015**, *6*, 1876–1889. [CrossRef]
19. Naono, T.; Fujii, T.; Esashi, M.; Tanaka, S. Non-resonant 2-D piezoelectric MEMS optical scanner actuated by Nb doped PZT thin film. *Sens. Actuators A* **2015**, *233*, 147–157. [CrossRef]

20. Chen, C.D.; Lee, Y.H.; Yeh, C.S. Design and vibration analysis of a piezoelectric-actuated MEMS scanning mirror and its application to laser projection. *Smart Mater. Struct.* **2014**, *23*, 125007. [CrossRef]

21. Jung, W.; Tang, S.; McCormick, D.T.; Xie, T.; Anh, Y.-C.; Su, J.; Tomov, I.V.; Krasieva, T.B.; Tromberg, B.J.; Chen, Z. Miniaturized probe based on a microelectromechanical system mirror for multiphoton microscopy. *Opt. Lett.* **2008**, *33*, 1324–1326. [CrossRef] [PubMed]

22. Sun, J.; Guo, S.; Wu, L.; Liu, L.; Choe, S.-W.; Sorg, B.S.; Xie, H. 3D in vivo optical coherence tomography based on a low-voltage, large-scan-range 2D MEMS mirror. *Opt. Express* **2010**, *18*, 12065. [CrossRef] [PubMed]

23. Bauer, R.; Li, L.; Uttamchandani, D. Dynamic properties of angular vertical comb-drive scanning micromirros with electrothermally controlled variable offset. *J. Microelectromech. Syst.* **2014**, *23*, 999–1008. [CrossRef]

24. Li, F.; Zhou, P.; Wang, T.; He, J.; Yu, H.; Shen, W. A large-size MEMS scanning mirror for speckle reduction application. *Micromachines* **2017**, *8*, 140. [CrossRef]

25. Ataman, C.; Lani, S.; Noell, W.; de Rooij, N. A dual-axis pointing mirror with moving-magnet actuation. *J. Micromech. Microeng.* **2013**, *23*, 025002. [CrossRef]

26. Choi, Y.-M.; Gorman, J.J.; Dagalakis, N.G.; Yang, S.H.; Kim, Y.; Yoo, J.M. A high-bandwidth electromagnetic MEMS motion stage for scanning applications. *J. Micromech. Microeng.* **2012**, *22*, 105012. [CrossRef]

27. Koh, K.H.; Kobayashi, T.; Lee, C. Investigation of piezoelectric driven MEMS mirrors based on single and double S-shaped PZT actuator for 2-D scanning applications. *Sens. Actuators A* **2012**, *184*, 149–159. [CrossRef]

28. Duan, C.; Wang, D.; Zhou, Z.; Liang, P.; Samuelson, S.; Pozzi, A.; Xie, H. Swept-source common-path optical coherence tomography with a MEMS endoscopic imaging probe. In Proceedings of the SPIE Optical Coherence Tomography and Coherence Domain Optical Methods Biomedicine XVIII, San Francisco, CA, USA, 3–5 February 2014; Volume 8934. [CrossRef]

29. Samuelson, S.R.; Xie, H. A large piston displacement MEMS mirror with electrothermal ladder actuator arrays for ultra-low tilt applications. *J. Microelectromech. Syst.* **2014**, *23*, 39–49. [CrossRef]

30. Koh, K.H.; Lee, C. A two-dimensional MEMS scanning mirror using hybrid actuation mechanisms with low operation voltage. *J. Microelectromech. Syst.* **2012**, *21*, 1124–1135. [CrossRef]

31. Huang, Q.A.; Lee, N.K.S. Analysis and design of polysilicon thermal flexure actuator. *J. Micromech. Microeng.* **1999**, *9*, 64–70. [CrossRef]

32. Torres, D.; Wang, T.; Zhang, J.; Zhang, X.; Dooley, S.; Tan, X.; Xie, H.; Sepúlveda, N. VO2-based MEMS mirrors. *J. Microelectromech. Syst.* **2016**, 780–787. [CrossRef]

33. Sandia National Laboratories. Available online: http://www.sandia.gov/mstc/_assets/documents/design_documents/SUMMiT_V_Dmanual.pdf (accessed on 3 June 2016).

34. Lin, L.; Chiao, M. Electrothermal responses of lineshape microstructures. *Sens. Actuators A* **1996**, *55*, 35–41. [CrossRef]

35. Karnovsky, I.A.; Lebed, O. *Advanced Methods of Structural Analysis*; Springer: New York, NY, USA, 2010; pp. 211–270.

36. Megson, T.H.G. *Structural and Stress Analysis*, 3rd ed.; Elsevier Ltd.: Amsterdam, The Netherlands, 2014.

37. Kawai, Y.; Kim, J.H.; Inomata, N.; Ono, T. Parametrically actuated resonant micromirror using stiffness tunable torsional springs. *Sens. Mater.* **2016**, *28*, 131–139.

38. Zhang, H.; Xu, D.; Zhang, X.; Chen, Q.; Xie, H.; Li, S. Model-based angular scan error correction of an electrothermally-actuated MEMS mirror. *Sensors* **2015**, *15*, 30991–31004. [CrossRef] [PubMed]

39. Liu, L.; Pal, S.; Xie, H. MEMS mirrors based on a curved concentric electrothermal actuator. *Sens. Actuators A* **2012**, *188*, 349–358. [CrossRef]

40. Espinosa, A.; Rabenorosoa, K.; Clevy, C.; Komati, B.; Lutz, P.; Zhang, X.; Samuelson, S.R.; Xie, H. Piston motion performance analysis of a 3DOF electrothermal MEMS scanner for medical applications. *Int. J. Optomech.* **2014**, *8*, 179–194. [CrossRef]

41. Koh, K.H.; Qian, Y.; Lee, C. Design and characterization of a 3D MEMS VOA driven by hybrid electromagnetic and electrothermal actuation mechanics. *J. Micromech. Microeng.* **2012**, *22*, 105031. [CrossRef]

micromachines

MDPI

Article

Design and Fabrication of a 2-Axis Electrothermal MEMS Micro-Scanner for Optical Coherence Tomography [†]

Quentin A. A. Tanguy [1,2,*], Sylwester Bargiel [1], Huikai Xie [2], Nicolas Passilly [1], Magali Barthès [1], Olivier Gaiffe [1], Jaroslaw Rutkowski [1], Philippe Lutz [1] and Christophe Gorecki [1]

[1] FEMTO-ST Institute, CNRS UMR6174, University of Bourgogne Franche-Comté, 25000 Besançon, France; sylwester.bargiel@femto-st.fr (S.B.); nicolas.passilly@femto-st.fr (N.P.); magali.barthes@femto-st.fr (M.B.); olivier.gaiffe@femto-st.fr (O.G.); jaroslaw.rutkowski@hotmail.com (J.R.); philippe.lutz@femto-st.fr (P.L.); christophe.gorecki@femto-st.fr (C.G.)

[2] Department of Electrical & Computer Engineering, University of Florida, Gainesville, FL 32611, USA; hkx@ufl.edu

* Correspondence: quentin.tanguy@femto-st.fr; Tel.: +33-769-388-345

† This paper is an extended version of our paper published in 2016 IEEE International Conference on Optical MEMS and Nanophotonics (OMN), Tanguy, Q.A.A., Duan, C., Wang, W., Xie, H., Bargiel, S., Struk, P., Lutz, P. & Gorecki, C., A 2-axis electrothermal MEMS micro-scanner with torsional beam.

Academic Editor: Kazunori Hoshino
Received: 31 March 2017; Accepted: 1 May 2017; Published: 5 May 2017

Abstract: This paper introduces an optical 2-axis Micro Electro-Mechanical System (MEMS) micromirror actuated by a pair of electrothermal actuators and a set of passive torsion bars. The actuated element is a dual-reflective circular mirror plate of 1 mm in diameter. This inner mirror plate is connected to a rigid frame via a pair of torsion bars in two diametrically opposite ends located on the rotation axis. A pair of electrothermal bimorphs generates a force onto the perpendicular free ends of the mirror plate in the same angular direction. An array of electrothermal bimorph cantilevers deflects the rigid frame around a working angle of 45° for side-view scan. The performed scans reach large mechanical angles of 32° for the frame and 22° for the in-frame mirror. We denote three resonant main modes, pure flexion of the frame at 205 Hz, a pure torsion of the mirror plate at 1.286 kHz and coupled mode of combined flexion and torsion at 1.588 kHz. The micro device was fabricated through successive stacks of materials onto a silicon-on-insulator wafer and the patterned deposition on the back-side of the dual-reflective mirror is achieved through a dry film photoresist photolithography process.

Keywords: optical Micro Electro-Mechanical System (MEMS); Micro Optical Electro-Mechanical System (MOEMS); electrothermal actuation; torsion bar; dry photoresist; dual-reflective mirror; optical coherence tomography

1. Introduction

Optical Micro Electro-Mechanical System (MEMS) micro-scanners are exploited by a large variety of applications that usually require large displacement range, high operating frequencies, miniaturization, simplicity of packaging and integration. Various methods, such as piezoelectric, electrostatic, electromagnetic and electrothermal technologies [1] have been used to develop devices able to measure each application's requirements. Among them, electrothermal actuation clearly stands out in terms of high performance, real time diagnosis, miniaturization of devices and endoscopy-based imaging. Although its working frequency is usually lower than for other actuation techniques, it still adequately reaches paces compatible with real time imaging [2]. MEMS electrothermal micro-scanners have a small size, high fill factor, high displacement range, low-voltage actuation and are relatively linear which makes them particularly adapted for in vivo endoscopic Optical Coherence Tomography (OCT) imaging applications [3].

The micro-scanner proposed in this paper (shown in Figure 1a) was designed and fabricated in order to be, in a future perspective, embedded into a Swept-Source OCT (SS-OCT) endomicroscopic probe (Figure 1b) based on a Mirau micro-interferometer [4].

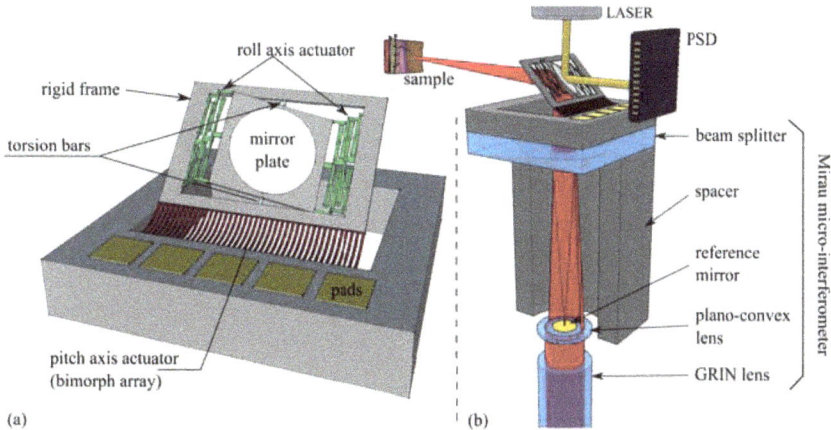

Figure 1. (**a**) Survey of the 2-axis Micro Electro-Mechanical System (MEMS) micro-scanning device. (**b**) Section plane of the different elements constituting the future endoscopic probe with the MEMS micro-scanner on top of the Mirau micro-interferometer for Optical Coherence Tomography (OCT) imaging process along with dynamical feedback control of the mirror position.

Many MEMS micromirrors use a set of four electrothermal bimorph actuators located on the four sides of the central mirror plate [5–7]. During actuation or scanning, the center of these mirrors' plate has to be partially maintained into a fixed position; first, by applying an offset voltage and second, by driving each pair of opposite actuators with a differential drive scheme [7]. However, the mirror plate is still subject to fluctuation with surrounding temperature and to uncontrolled changes due to vibrations or disturbances. In addition to these flaws, angular sensing mechanisms are usually unavailable, so that they are left uncontrolled [8] or with mere open-loop controls [9]. Concerning the few systems that demonstrate a close loop control, a single surface is used for both target operation and position sensing as in [10–12]. Conversely, for applications where one reflective side is to be exclusively dedicated to the main task as for OCT, phosphorescence or two-photon microscopy, exploiting the other side of the mirror is a reliable trade-off for direct position sensing compared to intermediate sensing methods [13,14], easy to be carried out at a macro scale in a preliminary stage. Our MEMS device is a 2-axis electrothermal scanning system characterized by a large scanning range, a torsion bar (Figure 2), a novel actuation mechanism (Figures 3 and 4a) and a dual-reflective aluminum-coated mirror plate (Figure 4a,c). The Mirau micro-interferometer associated with the swept source performs an axial scan (A-scan). Once the micro-scanning device is embedded on top of it, two additional B-scan axes can be realized so that a 3D image can be obtained.

2. Design of the Device

This micromirror was designed to increase the stability of the in-frame mirror and to provide large scanning ranges over a large bandwidth at low driving voltage in order to allow in vivo operation and remedy to the lack of possibility of feedback control of the micromirrors. It shows off two reflective surfaces on both sides of the plate, appreciated for multi-use applications where the dynamics of the mirror plate need to be accurately controlled. Indeed, an optical position detector can sense the real-time angle on one of the two reflective sides. For actuation, a pair of meshed electrothermal actuators is associated to a set of torsion bars that helps keeping the central axis of the mirror steady. These structures are represented in Figure 1 in green and blue colors, respectively. The mirror plate is consequently tilted inside the frame using the pair of Meshed Inverted-Series-Connected (MISC) electrothermal actuators located on both sides of the plate. The actuators are inverted one from another and apply a force in the opposite direction on the mirror plate generating the rotation around the axis of roll. Meanwhile, the pair of torsional bars, that are collinear to the virtual axis of roll, maintains the axis of the mirror in the plane of the outer frame, thus bringing stability to the system over a wide frequency range. A Silicon On Insulator (SOI) substrate ensures mechanical and electrical bonding support to the outer frame which also bends out of plane. This rotative motion around an axis of pitch is made possible by a bimorph cantilever array (sketched in red color in Figure 1). Although it is actuated, the frame acts as a support for the in-frame mirror plate. The main frame and mirror plate are made of a 30 µm-thick SOI device layer.

2.1. Torsion Bar

The torsion bars are used to prevent the mirror plate from oscillating around the roll axis, thus restricting the motion to a pure rotation. The materials used for the torsion beams are limited to those used in the bimorph to simplify the fabrication process. They are made of a "sandwich" structure, composed of layers of $SiO_2/Pt/Al/SiO_2$ respectively. The torsion bars were purposely dimensioned so that the expression of the bending mode of the torsion rods is minimized and does not impact the torsional motion. The stiffness of the bending mode is reported in [5,15,16] and is related to the resonance frequency through Equations (1)–(6):

$$k_b = \sqrt{\frac{Ewt^3}{4L^3}}, \tag{1}$$

where L, w, t refer respectively to the length, width, and thickness of the torsion bar and m to the mass of the mirror plate. The frequency of the bending mode is given by:

$$f_b = \frac{1}{2\pi}\sqrt{\frac{k_b}{m}} \tag{2}$$

The torsion mode stiffness of the system can be estimated from:

$$k_\phi = 2k_t + 2\frac{L_m^2}{4}k_b \tag{3}$$

$$\text{with } k_t = \frac{\mu wt^3}{3L}\cdot\left(1 - \frac{192}{\pi^5}\left(\frac{t}{w}\right)\sum_{n=1,3,5,...}^{\infty}\frac{1}{n^5}\tanh\left(\frac{n\pi t}{2w}\right)\right) \tag{4}$$

the free torsion stiffness as reported in [17,18]. Finally, the frequency of the torsion mode is given by [1]:

$$f_t = \frac{1}{2\pi}\sqrt{\frac{k_t}{J_t}}, \tag{5}$$

with $k_t = \frac{2\mu I_t}{L}$ where I_t is the second moment of area of the torsion shaft, J_t the moment of inertia of the mirror plate, $\mu = \frac{E}{2(1+v)}$ the shear modulus of elasticity, E the average Young's modulus and v the Poisson's ratio. The torsion frequency can also be calculated via the second moment of area for a rectangular-sectioned bar given by [18]:

$$I_t = wt^3\left(\frac{1}{3} - 0.21\frac{t}{w}(1 - \frac{t^4}{12w^4})\right), \tag{6}$$

where w and t are respectively the width and thickness of the torsion bar.

The bending mode frequency of the torsion bar is chosen to be twice as high as its torsion mode frequency. To do so, the bar is 3.3 µm thick, 180 µm long and 28 µm wide. Figure 2 shows the torsion bar before and after release for different conditions and a schematic cross section of the torsion bar can be found in Figure 5j. The Si layer from the device layer located underneath the torsion bar (Figure 2a) does not remain in the released structure. Otherwise, it would hold the whole structure and eventually culminate in the breaking of the MISC electrothermal actuators. The layer of Al is sandwiched between the main layers of SiO_2 and brings ductility to the torsion. The aluminum somewhat pushes away the yield stress breaking point of the structure making it more reliable regarding dynamical torsion and fatigue resistance. If not, the high residual stress initially induced in the actuators during fabrication would lead to fatal damages as pointed out in Figure 2d.

Figure 2. SEM pictures of torsion bar: (**a**) Sandwich bar from the backside before complete release. A narrow bridge of Si still holds the structure. (**b**) Sandwich bar from the backside after release. (**c**) Sandwich bar from the front side after release. (**d**) Example of Al-free torsion bar after release, broken under excessive torsion stress.

2.2. Electrothermal Actuation & MISC Actuators

The actuators are often cumbersome and are responsible for a much larger footprint of the final device than the size of the mirror plate. This issue has been tackled in some cases by modifying the shape of the actuators as in [19]. We present here an actuator based on Inverted-Series-Connected (ISC) electrothermal actuators as demonstrated in [7] but providing more flexibility and a higher displacement. It is a mesh of ISC actuators in series and in parallel that optimizes the space around the mirror plate to increase the displacement and the force of the actuators without degrading the fill factor. The principle of the meshed ISC (MISC) actuator is shown in Figure 3.

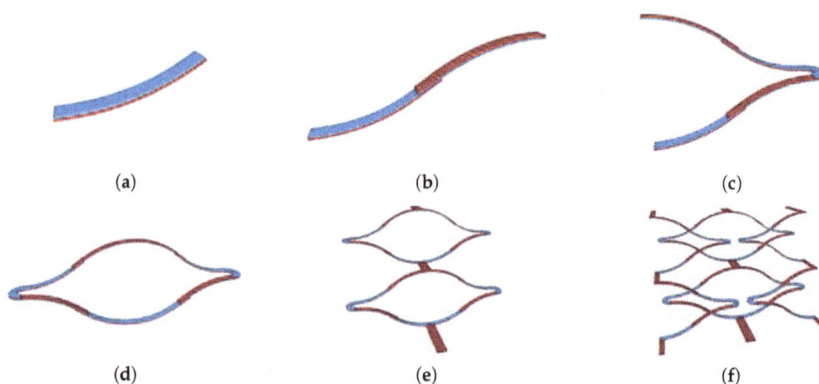

Figure 3. Schematic build up of the Meshed Inverted Series Connected (MISC) actuator. (a) Fundamental bimorph cantilever beam (tip-tilt and lateral displacement). (b) ISC actuator (Lateral shift). (c) Double S-shaped configuration (piston motion). (d) Pair of double S-shaped actuators in parallel (Stiffness and stability increased). (e) Cumbersome double actuator (increased displacement). (f) MISC actuator.

The MISC actuator is the latest evolution of four generations of shapes of electrothermal actuators: the single bimorph cantilever is the core element shown in Figure 3a and reported in [20,21]. In [6,7], bimorphs were connected in series as in Figure 3b,c including inverted and non-inverted bimorphs (whose cross sections are shown respectively in Figure 5k,l) to get rid of the tip-tilt effect, bypass the lateral shift and end up into a pure vertical translative motion called piston motion. These latter structures were then interconnected in parallel as in Figure 3d to increase the overall motion stability. Figure 3e shows an intermediate structure and was reported by [22]. The MISC actuator shown in Figure 3f is the structure actuating the micromirror and can be seen, as fabricated in Figure 4b. The torsion bars generate a counter momentum in the opposite direction of the momentum created by the two actuators. Hence, the actuators need to be able to provide a higher force and a larger displacement than that which can be provided by conventional ISC actuators. For a comparable space occupied, the MISC actuators provide a higher force, a larger displacement and a higher flexibility. This latter advantage is also highly appreciated during the release process and brings more suppleness for industrial fabrication where the dispersion of parameters on a single wafer can be significant. The bimorph is a sandwich of 1.1 μm of Al and 1 μm of SiO_2. A thin heater layer of 1500 Å of Pt insulated in a sheath of thin SiO_2 is wrapped between the Al and the SiO_2 as shown in Figure 5c.

2.3. Dual-Reflective Mirror Plate

The mirror is coated with aluminum on both sides of the plate using E-beam evaporation. The deposition on the upper side is 1.1 μm thick and is performed during the same Al metalization as for the bimorphs. The Al layers of the front side mirror and of the electrothermal bimorph cantilevers are realized in one step using the same photomask to simplify the complete fabrication process. Therefore, the Al of the mirror plate has the same thickness as the layer of the bimorphs. The backside of the mirror is the side used to scan the focused laser beam and its smoothness is critical for the OCT image quality. Hence, the deposition is done at very low deposition rate (1.2 Å s^{-1}) while the substrate is being rotated at a speed of 10 rpm. SEM pictures of the reflective front and back side are shown respectively in Figure 4a,c.

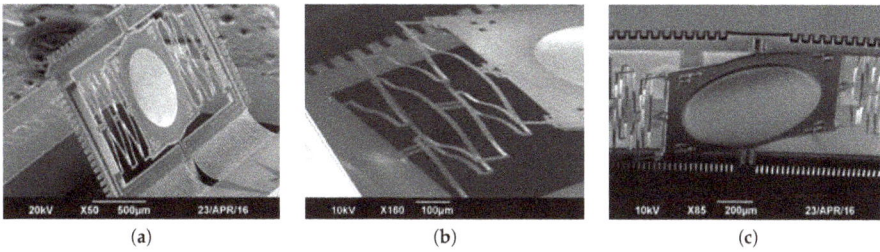

Figure 4. Scanning Electron Microscopy (SEM) pictures of the micro-scanner. (**a**) Overview of the front side of the mirror plate and the frame. (**b**) Detail of the MISC actuators after final release of the device. (**c**) Close-up view of the backward reflective side of the mirror plate.

3. Fabrication

The complete fabrication process is described in Figure 5. The devices are fabricated on an SOI wafer of 500 µm of handle layer, 30 µm of device layer and 1 µm of BOX. After a thorough clean up of the wafer, the first step (Figure 5a) consists of a deposition of 1 µm of Plasma-Enhanced Chemical Vapor Deposition (PECVD) SiO_2 on the device layer which is subsequently wet etched to form the bottom layer of the non-inverted bimorphs, the hard frame, the torsion bars and the thermal bridges. It is then followed by another PECVD deposition of a thin layer of SiO_2 as an insulator and a lift-off of platinum (Figure 5b) to pattern the heater throughout the actuators, the electrical paths and the pads.

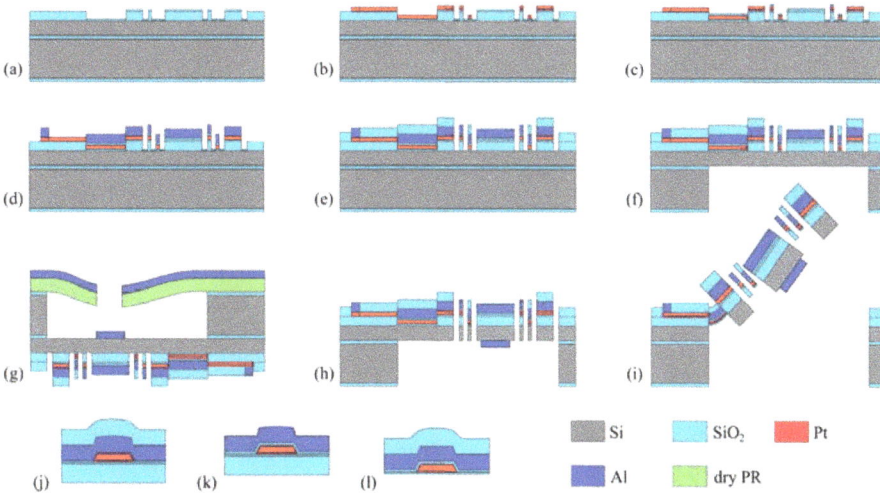

Figure 5. Fabrication steps (**a**) PECVD first layer of SiO_2. (**b**) Sputter of the Pt heater. (**c**) PECVD of an insulation layer of SiO_2 followed by via opening. (**d**) Evaporation of Al. (**e**) PECVD second layer of SiO_2. (**f**) Anisotropic dry etch of the handle layer & BOX dry etch. (**g**) Lamination of dry PR & evaporation of Al. (**h**) Anisotropic etching of device layer from the front side. (**i**) Si isotropic etching to release. (**j**) Cross section of torsion bar. (**k**) Cross section of inverted bimorph. (**l**) Cross section of non-inverted bimorph.

The platinum is also used in the sandwich of the torsion bars. The three central pads control the inner actuators of the roll axis and are connected to the Al path of the bimorph array and isolated from its Pt layer via the SiO_2 insulation film (Figure 5c). By doing so, the heat transfer generated by the current driven through the bimorph array is minimized. As shown in Figure 5d, a thick layer of

1.1 µm of Al is deposited by evaporation (to facilitate the lift-off process) following a photolithography of 3.5 µm of AZ nLOF2035 for the bimorphs, the mirror plate on the front side, the torsion bars, the electrical paths and the pads. We used pure Al, which was then protected by a thin coat of Cr of 150 Å to prevent oxidation.

A second layer of 1 µm of SiO_2 is deposited by PECVD and patterned through RIE/ICP dry etch to form the top layer of the inverted bimorphs. This step is represented in Figure 5e.

The handle layer is anisotropically etched through DRIE to form the device's backside cavity (Figure 5f). The exposed BOX is also etched with RIE/ICP until the buried face of the device layer is reached. Then Al is deposited onto the mirror plate's backside by evaporation. A dry film photoresist DuPont ^{TM} WBR2050 was laminated at 85 °C on the backside of the SOI wafer held by a carrier wafer before exposition.

The final release stage of the device divides into two substages respectively shown in Figure 5h,i. The first one consists of an anisotropic dry etch all the way through the device layer followed by an isotropic etch to release the actuators. The isotropic etch should not be performed longer than necessary to avoid ablating SiO_2 from the deformable elements which could eventually damage or break them. The isotropic process time is interrupted when the frame and the inner actuators pop out of the plane. At that step, a plasma O_2 can be used to get rid of the impurities remaining on the chip. An SEM picture of the micro scanner after release can be found in Figure 4a.

Finally, several released chips are packaged onto a generic PCB support customized for handling and characterization of the micro-devices (Figure 6a). The micro-scanners are bonded onto the central Au pad with silver epoxy glue and electrically connected to the PCB pads by wire-bonding (Figure 6b).

(a)   (b)

Figure 6. PCB support for the MEMS micro-scanner handling and testing. (**a**) Overview of the multi-use PCB with the micro device in the center, and connectors on both sides. (**b**) Zoomed-in picture of the micro device bonded onto the central gold pad with silver epoxy glue and wire bonding on Cu/Au pads for electrical routing.

4. Characterization

After release, the electrical resistances of the roll axis actuators in parallel and the pitch axis actuator are 1.07 kΩ and 1.34 kΩ, respectively. The optical setup is shown in Figure 7.

A laser beam is directed onto the MEMS micromirror which reflects it towards a diffusing screen. The latter is observed from its backside by an ultra fast Phantom ^{TM}Miro M120 camera. The frame declines by 32° from an initial angle of 70° to a final angle of 38° reached at a voltage of 17 V (178 mW) while the mirror plate achieves a mechanical sweep range of 22° deflecting from an initial angle of 18° to −4° for a voltage of 16.5 V (188 mW) (Figure 8a). The characteristics of power consumption and angular displacement as a function of the voltage applied are also shown in Figure 8.

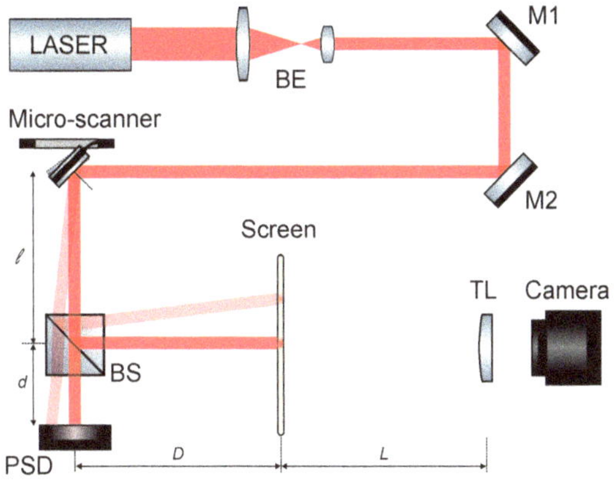

Figure 7. Optical setup implemented for the MEMS micro-scanner statical and dynamical characterization. BE: beam expander, M1, M2: mirrors, BS: beam splitter, PSD: Photo Sensing Detector, TL: tube lens.

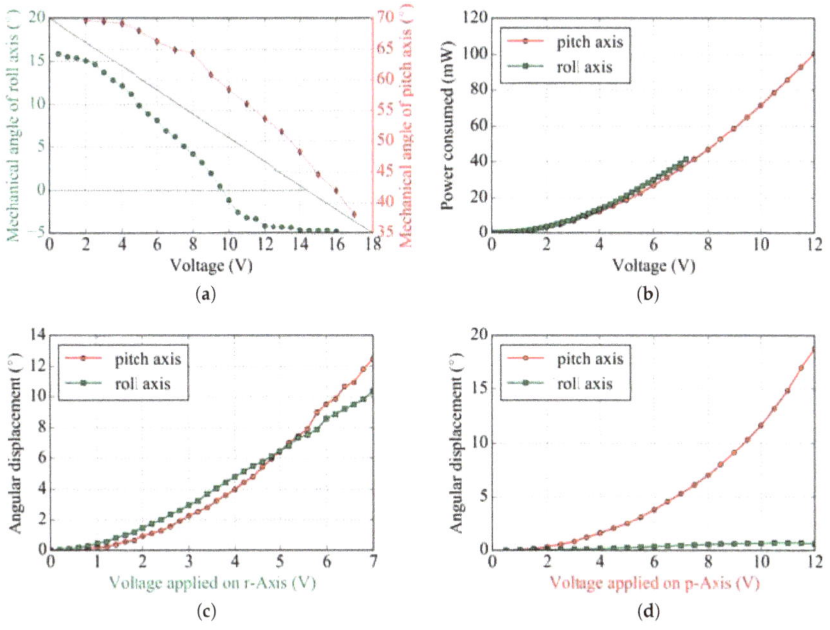

Figure 8. (**a**) Statical angular displacement of the mirror on the roll axis (in green) and of the frame on the pitch axis (in red). (**b**) Power consumption vs. voltage applied for both axes. (**c**) Statical relative angular displacement of both axes when only the roll axis is driven. (**d**) Statical relative angular displacement of both axes when only the pitch axis is driven.

A Polytec ^{TM}MEMS Analyser was used to establish the frequency response of the micromirror. A white noise with an amplitude 1.5 V and an offset of 3 V was applied on the actuators one by one, and the magnitude of the deflection of the frame and the mirror plate was measured in dB. The Bode diagrams are shown in Figure 9. The coupling between the roll axis and the pitch axis is unilateral: when the inner actuator is driven, the heat is dissipated into the mirror plate, through the frame and through the bimorph array whose temperature increases at the same time, contributing into the cross-coupling of the two axes. In this situation, we observe four resonant modes: pure pitch motion at 205 Hz, pure roll motion of the torsional mirror plate at 1.286 kHz, a mode with both components at 1.588 kHz and a fourth mode that is less influential because of its high damping. Conversely, when the bimorph array is actuated, only the first pitch mode is observed at 205 Hz.

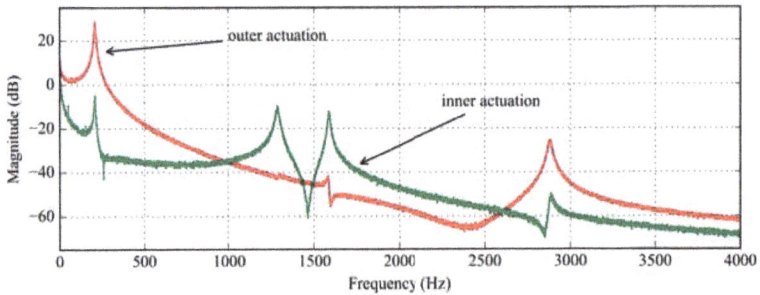

Figure 9. Superimposed frequency responses of the system when the voltage is applied on the pitch axis (outer actuator) in red and on the roll actuator (inner actuator) in green.

Finally, Lissajous laser scans have been recorded by the high-speed camera (at 30 kfps) when the micromirror is actuated at its resonance frequencies. Corresponding time elapsed scans are shown in Figure 10 (after 4 ms, 17 ms and 45 ms). In these conditions, if, on the one hand, a resolution of 10 μm is sought at a working distance of 5 mm from the mirror as in [4], and on the other hand, a 90 kHz A-scan rate swept-source is employed (requiring to interpolate the 30 kHz experimental scans), it would then require 45 ms, corresponding to an imaging frequency of 22 Hz, to cover 99 % of a scanned area of 770 μm × 270 μm. At this frequency, and because of the Lissajous type of scanning, a significant number of pixels is averaged. Larger averaging, e.g., when 95 % of the scanned area is illuminated more than 9 times, can be reached at a frequency of 5 Hz.

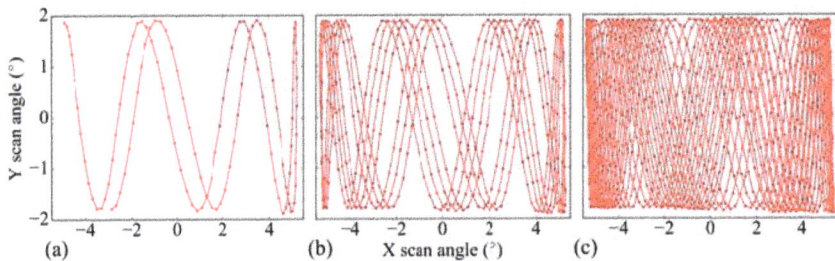

Figure 10. Time elapsed Lissajous laser scanning patterns recorded by the high speed camera at 30 kfps, after (**a**) 4 ms. (**b**) 17 ms. (**c**) 45 ms.

Acknowledgments: This work was supported by the LabEx Action program (contract ANR-11-LABX-0001-01), by the French RENATECH network and its FEMTO-ST technological facility as well as the US National Science Foundation under award #1512531. I would also like to give thanks to Vincent Maurice and Jean-Marc Cote for their support.

Author Contributions: S.B., H.X., P.L. and C.G. supervised the work; Q.T. and S.B. designed the devices; Q.T. fabricated the devices; J.R., Q.T. and N.P. designed the instrumentation and the experimental setup; Q.T., N.P., O.G. and M.B. modeled and characterized the devices; N.P. and O.G. carried out calculations and data treatment for the results interpretation; H.X. and S.B wrote a draft of the manuscript and Q.T., S.B., P.L., H.X., N.P., O.G., J.R. and M.B. contributed to the scientific interpretation of the resutls and the edition of the manuscript.

Conflicts of Interest: The authors declare no conflict of interest.

Abbreviations

The following abbreviations are used in this manuscript:

MEMS	Micro Electro-Mechanical System
MOEMS	Micro Optical Electro-Mechanical System
SCS	Single Crystal Silicon
OCT	Optical Coherence Tomography
SS	Swept Source
ISC	Inverted Series Connected
MISC	Meshed ISC
SOI	Silicon On Insulator
BOX	Buried Oxide
BOE	Buffered Oxide Etch
PECVD	Plasma-Enhanced Chemical Vapor Deposition
CTE	Coefficient of Thermal Expansion
RIE	Reactive-Ion Etching
ICP	Inductive Coupled Plasma
GRIN	GRadient INdex
PSD	Position Sensing Detector

References

1. Petersen, K.E. Silicon torsional scanning mirror. *IBM J. Res. Dev.* **1980**, *24*, 631–637.
2. Sun, J.; Xie, H. MEMS-based endoscopic optical coherence tomography. *Int. J. Opt.* **2011**, *2011*, 825629.
3. Sun, J.; Guo, S.; Wu, L.; Liu, L.; Choe, S.W.; Sorg, B.S.; Xie, H. 3D In Vivo optical coherence tomography based on a low-voltage, large-scan-range 2D MEMS mirror. *Opt. Express* **2010**, *18*, 12065–12075.
4. Struk, P.; Bargiel, S.; Froehly, L.; Baranski, M.; Passilly, N.; Albero, J.; Gorecki, C. Swept source optical coherence tomography endomicroscope based on vertically integrated mirau micro interferometer: Concept and technology. *IEEE Sens. J.* **2015**, *15*, 7061–7070.
5. Wu, L.; Xie, H. A large vertical displacement electrothermal bimorph microactuator with very small lateral shift. *Sens. Actuators A Phys.* **2008**, *145–146*, 371–379.
6. Todd, S.T.; Jain, A.; Qu, H.; Xie, H. A multi-degree-of-freedom micromirror utilizing inverted-series-connected bimorph actuators. *J. Opt. A Pure Appl. Opt.* **2006**, *8*, S352.
7. Jia, K.; Pal, S.; Xie, H. An electrothermal tip-tilt-piston micromirror based on folded dual S-shaped bimorphs. *J. Microelectromech. Syst.* **2009**, *18*, 1004–1015.
8. Kobayashi, T.; Maeda, R. Piezoelectric optical micro scanner with built-in torsion sensors. *Jpn. J. Appl. Phys. Part 1 Regul. Pap. Short Notes Rev. Pap.* **2007**, *46*, 2781–2784.
9. Wang, W.; Chen, J.; Zivkovic, A.S.; Tanguy, Q.A.; Xie, H. A compact Fourier transform spectrometer on a silicon optical bench with an electrothermal MEMS mirror. *J. Microelectromech. Syst.* **2016**, *25*, 347–355.
10. Han, F.; Wang, W.; Zhang, X.; Xie, H. Modeling and control of a large-stroke electrothermal MEMS mirror for fourier transform microspectrometers. *J. Microelectromech. Syst.* **2016**, *25*, 750–760.
11. Wang, W.; Chen, J.; Zivkovic, A.S.; Xie, H. A Fourier Transform Spectrometer based on an electrothermal MEMS mirror with improved linear scan range. *Sensors* **2016**, *16*, 1611.
12. Zhao, Y.; Tay, F.E.H.; Zhou, G.; Chau, F.S. Fast and precise positioning of electrostatically actuated dual-axis micromirror by multi-loop digital control. *Sens. Actuators A Phys.* **2006**, *132*, 421–428.
13. Fujita, T.; Maenaka, K.; Takayama, Y. Dual-axis MEMS mirror for large deflection-angle using SU-8 soft torsion beam. *Sens. Actuators A Phys.* **2005**, *121*, 16–21.
14. Tseng, V.F.G.; Xie, H. Simultaneous piston position and tilt angle sensing for large vertical displacement micromirrors by frequency detection inductive sensing. *Appl. Phys. Lett.* **2015**, *107*, 214102.
15. Lowet, G.; Audekercke, R.V.; der Perre, G.V.; Geusens, P.; Dequeker, J.; Lammens, J. The relation between resonant frequencies and torsional stiffness of long bones in vitro. Validation of a simple beam model. *J. Biomech.* **1993**, *26*, 689–696.

16. Timoshenko, S. Analysis of Bi-metal thermostats. *J. Opt. Soc. Am.* **1925**, *11*, 233–255.
17. Ji, C.H.; Kim, Y.K. Electromagnetic micromirror array with single-crystal silicon mirror plate and aluminum spring. *J. Lightw. Technol.* **2003**, *21*, 584–590.
18. Young, W.; Roark, R.; Budynas, R. *Roark's Formulas for Stress and Strain*. McGraw-Hill: New York, NY, USA, 2002; Volume 7.
19. Liu, L.; Pal, S.; Xie, H. MEMS mirrors based on a curved concentric electrothermal actuator. *Sens. Actuators A Phys.* **2012**, *188*, 349–358.
20. Todd, S.T.; Xie, H. Steady-state 1D electrothermal modeling of an electrothermal transducer. *J. Micromech. Microeng.* **2006**, *16*, 665–665.
21. Jain, A.; Kopa, A.; Pan, Y.; Feeder, G.K.; Xie, H. A two-axis electrothermal micromirror for endoscopic optical coherence tomography. *IEEE J. Sel. Top. Quantum Electron.* **2004**, *10*, 636–642.
22. Samuelson, S.R.; Xie, H. A large piston displacement MEMS mirror with electrothermal ladder actuator arrays for ultra-low tilt applications. *J. Microelectromech. Syst.* **2014**, *23*, 39–49.

micromachines

MDPI

Article

Modeling of MEMS Mirrors Actuated by Phase-Change Mechanism

David Torres [1], Jun Zhang [2], Sarah Dooley [3], Xiaobo Tan [1] and Nelson Sepúlveda [1,*]

[1] Department of Electrical & Computer Engineering, Michigan State University, East Lansing, MI 48840, USA; torresd5@egr.msu.edu (D.T.); xbtan@egr.msu.edu (X.T.)
[2] Department of Electrical & Computer Engineering, University of California, San Diego, La Jolla, CA 92093, USA; j5zhang@ucsd.edu
[3] Air Force Research Laboratory, Sensors Directorate, WP-AFB, Dayton, OH 45433, USA; sarah.dooley@us.af.mil
* Correspondence: nelsons@egr.msu.edu; Tel.: +1-517-432-2130

Academic Editor: Huikai Xie
Received: 7 March 2017; Accepted: 19 April 2017; Published: 26 April 2017

Abstract: Given the multiple applications for micro-electro-mechanical system (MEMS) mirror devices, most of the research efforts are focused on improving device performance in terms of tilting angles, speed, and their integration into larger arrays or systems. The modeling of these devices is crucial for enabling a platform, in particular, by allowing for the future control of such devices. In this paper, we present the modeling of a MEMS mirror structure with four actuators driven by the phase-change of a thin film. The complexity of the device structure and the nonlinear behavior of the actuation mechanism allow for a comprehensive study that encompasses simpler electrothermal designs, thus presenting a general approach that can be adapted to most MEMS mirror designs based on this operation principle. The MEMS mirrors presented in this work are actuated by Joule heating and tested using optical techniques. Mechanical and thermal models including both pitch and roll displacements are developed by combining theoretical analysis (using both numerical and analytical tools) with experimental data and subsequently verifying with quasi-static and dynamic experiments.

Keywords: MEMS mirrors; vanadium dioxide; phase-change materials; hysteresis; dynamic model

1. Introduction

Microeletromechanical system (MEMS) mirrors are microstructures capable of redirecting an incident beam of light to a desired position. MEMS mirror devices can be characterized by their dynamic performance, degrees of freedom, size, and power consumption. The size and power consumption parameters are determined by the actuation mechanism that is implemented in the design, while the movement capability (degrees of freedom) and tilt angle amplitude are dependent on the mechanical design of the device. The speed will depend on the time response of the mechanical structure and actuation processes. The four main mechanisms implemented in MEMS mirrors are: electrostatic (ES), piezoelectric (PE) material, electromagnetic (EM) and electrothermal (ET). ES and PE use electrostatic fields for actuation—the ES commonly uses repelling/attracting forces between two plates to move the mirror platform from a resting state [1], while the PE method uses piezoelectric materials such as lead zirconate titanate (PZT) [2], where small unorganized dipoles generate material expansion and contraction upon an applied electric field. In both cases (ES and PE), mechanical forces are generated by an electric potential signal of relative large amplitude (for example: 115 V and 40 V for ES and PE, respectively), but the total power consumed by these devices is low due to the low current consumption [3,4]. The EM mechanism generates movement as the result of the force between interacting magnetic fields (Lorentz force). A possible configuration for EM consists of the interaction of a static magnetic field (created by a magnetic material) with a dynamic magnetic field (created by applying a current through a metal-trace loop inside a mirror device) [5]. The current amplitude of this mechanism can be as large as 515.17 mA for optimum performance [6]. Although the electrical actuation signals for these devices can be made very fast, their speed is ultimately determined by the dynamics of the mechanical structure. Finally, the ET mechanism uses a current to generate heat (Joule heating) on a structure, which can reach temperatures of $\approx 300\,°C$ [7]. The main advantage of the ET mechanism (over ES and PE) is the much lower voltage signals required for operation. Perhaps the most common configuration for this mechanism is a bimorph structure formed by two materials (thin films) with different thermal expansion coefficients (TEC). As the temperature increases, one material will expand more than the other, generating a bending in the structure [8], concave towards the film of lower TEC. In this case, the speed of actuation will depend on the thermal dynamics of the system, making it the slowest mechanism of all for devices of similar size and thermal mass. The power consumed in this mechanism can be lower than the EM, but higher than the ES and PE, since the temperature increase depends on the amplitude of the applied current, which can be as high as 252 mA for maximum displacement [9].

A new method of actuation for MEMS mirror was presented in [10], where the conventional ET actuation mechanism (i.e., using two materials with different TEC) was replaced with a smart material, vanadium dioxide (VO_2), which goes through a phase transition that can be induced by a gradient of temperature. VO_2 has a reversible solid-to-solid phase transition that comes with drastic changes in the mechanical [11], electrical [12], and optical properties [13] of the material. When induced thermally, the transition of VO_2 occurs at $\approx 68\,°C$, but this transition temperature can be reduced by doping [14] or adding extrinsic stress to the material [15,16]. The integration of VO_2 with the MEMS mirror technology decreases the temperature required in the conventional TE mechanism, from $300\,°C$ to $90\,°C$ for full actuation, which lowers the total power consumed by the device. Another advantage of using VO_2 as the actuation mechanism is the large strain energy density generated during the transition, with values higher than conventional actuation mechanisms such as thermal expansion, electrostatic, electromagnetic, and piezoelectric [17]. Furthermore, the intrinsic hysteretic behavior of VO_2 properties (including the mechanical stress that generates deflection in VO_2-based MEMS [18,19]) across the phase transition has been exploited to design programmable MEMS actuators [20] and resonators [21], and can be used as well to program tilting angles in MEMS mirrors. However, all of these advantages come at the cost of added nonlinear effects that make the modeling and control more complicated than other actuation mechanisms.

The modeling of MEMS mirror is a necessity to better understand the behavior of the devices. A general dynamic equation (second-order differential equation) in terms of summation of torques has been used to describe the dynamic behavior of a MEMS mirror [22–24]. The parameters of the equation are dependent on the mechanical structure and the actuation mechanism of the device. Different modeling and control methods have been proposed for the nonlinear hysteresis in VO_2-based MEMS devices. Nonlinear mathematical models such as the Prandtl–Ishlinskii model [25] and the Preisach model [26] have been adopted to capture and estimate the hysteresis behaviors. Unlike the identification of the Prandtl–Ishlinskii model, which requires solving a nonlinear optimization problem, the Preisach model identification problem can be reformulated as a linear least-squares problem and solved efficiently [26]. The Preisach model is thus adopted in this work. In order to control the systems with hysteresis, feedforward control can be realized by inverting the hysteresis nonlinearity [26], and feedback control can also be implemented, where the feedback signal can be obtained based on external sensors or with self-sensing methods [27]. In self-sensing, the correlation between the electrical and mechanical properties across the transition is utilized [28].

In this work, we present a mathematical model that describes the movement of a VO_2-based MEMS mirror. The modeling is focused on one of the four actuators of the device. First, the mechanical model of the system is derived, where the nonlinear behavior of the VO_2 is incorporated in the model as an external force applied to the system. A Preisach model is used to capture the hysteresis behavior of the VO$_2$. The parameters for the whole model are identified using simulation and experimental results. Finally, the hysteresis model is validated with a different set of experimental results including quasi-static and dynamic responses. The proposed model can be translated to other actuators of the MEMS mirror, and this work facilitates the control of the device.

2. Experimental Procedures

The VO$_2$-based MEMS mirror used in this paper is shown in Figure 1 and the design has been reported in [7,29,30]. The device consists of four mechanical actuators (legs) coupled with a reflective platform (mirror). Tilting of the mirror platform is achieved by individual actuation of the legs, which is independently controlled, or by actuation of all the legs using the same input signal simultaneously, which generates a piston-like movement. There are two actuation mechanisms: stress due to the thermal expansion difference of the materials forming the bimorph regions of the device, and stress generated during the phase-change transition of the VO$_2$. Outside the phase transition region of VO$_2$, the only active mechanism is thermal expansion, while, during the phase transition, both mechanisms exist, but the phase transition of VO$_2$ dominates [26]. The generated stress is capable of bending a thin bimorph rectangular structure composed of continuous SiO$_2$ ($\approx$1.4 μm) and VO$_2$ (250 nm) layers and a thin patterned metal (130 nm) layer. To increase the vertical displacement of the leg, a rigid structured (frame), composed of a thick (50 μm) layer of Si, connects the bimorphs. The transition of the VO$_2$ film is induced using Joule heating, where an input current is applied through the monolithically integrated resistive heater of the leg. The metal traces are designed to have a smaller width in the bimorph parts of the leg. This is done to create a higher resistance in these regions of the heater, which increases the dissipated power and localizes the generated heat in the bimorph regions.

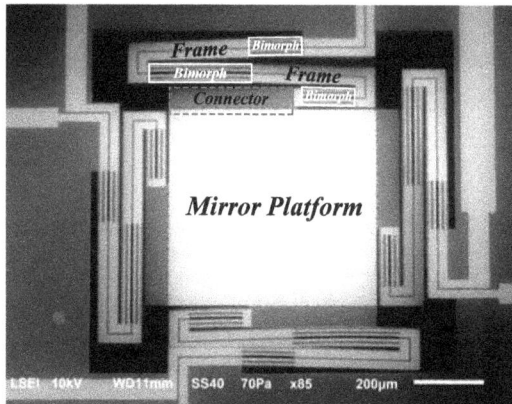

Figure 1. SEM image of a VO$_2$-based micro-electro-mechanical system (MEMS) mirror (top view), where the different parts of an actuator leg are labeled: frame, bimorph and the connector between the mirror platform and the actuator leg.

2.1. Design and Fabrication of VO$_2$-Based MEMS Mirrors

The VO$_2$-based MEMS mirror presented here follows the same fabrication process as in [10], but different metal layers are used to increased the yield per wafer—this is discussed in more detail in Section 2.2. Device fabrication starts with a two–inch double–sided p-type <111> polished Si wafer as the substrate with a thickness of 300 μm. A thin SiO$_2$ layer (1 μm) is deposited on both sides of the wafer by plasma enhance chemical vapor deposition (PECVD) at a temperature of 300 °C. One of the SiO$_2$ layers is used as a mask for the Si back-side etch, while the other (top side) forms the first layer of the bimorphs over which the VO$_2$ films are deposited. A thin film from any material that can survive most chemicals used in standard MEMS processing (i.e., Si$_3$N$_4$) would have been an acceptable choice for the backside. However, the selection of SiO$_2$ material for the top side is based on the larger mechanical actuation across the VO$_2$ phase transition, which is due to the higher orientation of the VO$_2$ grains with monoclinic (011)$_M$ planes parallel to the substrate [20]. Although higher VO$_2$ orientations are expected from crystalline substrates (i.e., quartz, sapphire), their processing represents major fabrication processing hurdles. The VO$_2$ is deposited by pulse laser deposition (PLD) and patterned with dry etching using reactive ion etch (RIE), following the procedure shown in [10].

After the deposition of the VO$_2$, the remaining processes are performed at temperatures lower than 250 °C to avoid any degradation in the VO$_2$ due to over-oxidation of the film. A 200 nm SiO$_2$ layer is grown by PECVD using a temperature of 250 °C on top of the VO$_2$, for electrical isolation from the metal traces that will be deposited next. The electrical connections and resistive heaters are fabricated by depositing and patterning via lift-off layers of Cr (20 nm)/Au (110 nm), where the Cr is used as an adhesion layer between the SiO$_2$ and the Au. Another 200 nm of SiO$_2$ is deposited to insulate the metal traces from the ambient (air), reducing the thermal losses. This is followed by a sequence of SiO$_2$ dry etch steps by RIE in order to expose the metal contacts, pattern the legs and platform of the device, and expose the Si substrate. The same SiO$_2$ etching is repeated on the back-side SiO$_2$ layer to expose the Si substrate. During the processing of the back-side, the top side was protected by spinning PMGA (polymethylglutarimide) resist. After processing the backside, the PMGA is removed by submerging the sample in photoresist stripper (Microposit Remover 1165). Using the SiO$_2$ as a hard mask, the exposed Si layer on the backside was etched with deep reactive ion etch (DRIE). The DRIE etching is timed to remove 250 µm of the Si layer, reducing the Si substrate from 300 µm to 50 µm. The mirror structure is released by etching the remaining 50 µm of the Si substrate from the top by DRIE. Finally, to remove the Si from certain parts of the legs and create the bimorph sections, a Si isotropic etch is performed using XeF$_2$ gas. This process is timed to only etch the desired parts avoiding any undesired over etch that would affect the frame regions of the legs.

2.2. Increasing Yield by Reducing Intrinsic Stress

In our previous work [10], we used Ti/Pt for the metallization. The rationale for the combination of these metals was to have a high-temperature metal in case it was necessary to change the order of the deposition of materials in the fabrication process flow that would require the deposition of VO$_2$ after the metallization. However, the use of Ti/Pt created a low yield ($\approx$12.5%) in the final devices, due to peeling of the metal. We thought this was due to the intrinsic stress of the evaporated Ti/Pt metal layer on SiO$_2$, which could be as high as 340 MPa (compressive stress) [31,32]. In order to address the issue, in the present work, we have substituted the Ti/Pt layer with evaporated Cr/Au, which has lower intrinsic stress (250 MPa tensile stress) [33,34]. This has increased the yield to $\approx$75%.

2.3. Experimental Setup

The device is characterized by actuating only one of the four actuators and measuring the tilting angle produced by the movement. Due to symmetry of the device, the characterization of one leg can then be used to describe all the other actuators and derive the model for the entire device.

A schematic of the experimental setup is presented in Figure 2. The movement of the mirror is tracked using a laser scattering technique, where an infrared laser (λ = 985 nm) with a low power is focused on the platform of the mirror and aligned with a two-dimensional (2D) position sensing detector (PSD). The 2D PSD is used to facilitate the alignment of the setup and capture any 2D movement of the device. A digital camera (Nikon 1 J1, Nikon Co., Tokyo, Japan) with an external objective lens (10× Mitutoyo Plan Apo Infinity Corrected Long WD Objective lens, Mitutoyo Corp., Kawasaki, Japan) is used for calibration purposes, by enabling the conversion of the voltage output from the PSD sensor to the tilt angle of the mirror platform. The camera is placed at the side of the sample to monitor the movement of the device. The resolution of the optical setup is 0.577 µm/pixel with a speed of 30 frames/s. The video is analyzed using Tracker Video Analysis and Modeling software tool (Version 4.94). All the electrical signals to/from the experimental setup were controlled through a virtual instrument in LabVIEW and a National Instrument data acquisition system (DAQ). The electrical current applied to the legs is controlled through the base voltage applied to a BJT NPN transistor (2n3904) in an emitter follower configuration. A resistance in series between the transistor and the actuator was used to measure the voltage and calculate the current. The transistor device was used in forward active mode.

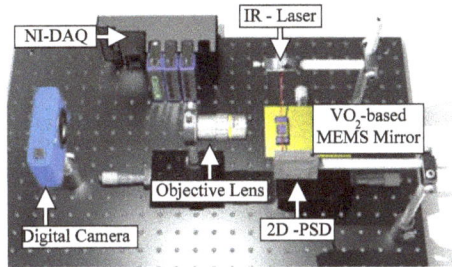

Figure 2. Measurement setup used for characterization of VO_2-based MEMS mirror.

3. Modeling

The connection of each mirror leg to the mirror (as shown in Figure 1) is not symmetrical along a perpendicular axis crossing the platform's center. This offset causes the VO_2-based MEMS mirror to have a 2D movement upon actuation. Therefore, the description and modeling of the system will involve movements along two perpendicular axes: pitch and roll. Figure 3 shows a schematic of the platform with the two axes used to describe the tilting movement of the platform. The force ($\vec{F}$) represents the actuation generated by the legs. A set of two equations (one per degree of freedom: pitch and roll) is used to model the movement of the mirror. The inclusion of the VO_2 in the device adds a nonlinear term to the equations due to the hysteretic behavior of the material. A non-monotonic Preisach model is developed to capture the hysteresis term. The effect of the VO_2 is included in the external force that generates actuation. The parameters for the linear part of the equation that describe the system's mechanical response are obtained from a combination of experimental measurements and finite element method (FEM) simulations (details in Section 3.1). The coefficients of the nonlinear part of the equation are calculated from a set of experiments (details in Section 3.2).

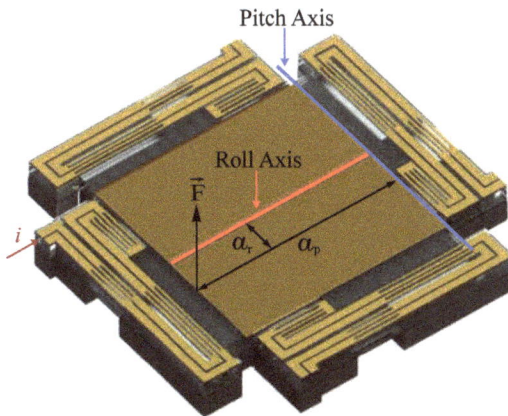

Figure 3. Schematic of the mirror platform showing the force ($\vec{F}$) applied by the actuated leg, and the axes of rotation for pitch and roll angles, where a_r (115 μm) and a_p (300 μm) are the distance between the force and each axis. In this case, the current i is applied to the bottom-left leg.

3.1. Linear Model

The linear model characterizes the dynamic behavior of the system, which is approximated with a second-order differential equation

$$J\ddot{\theta} + G\dot{\theta} + k\theta = \vec{T},$$ (1)

where J is the moment of inertia, G is the rotational damping coefficient, k is the rotational spring constant of the actuated leg, $\vec{T}$ is the external torque produced by the force $\vec{F}$, and θ is the angle of the mirror's platform generated during actuation. The characteristic equation of the system can be derived by applying the Laplace transformation to Equation (1) and rearranging the expression to get Equation (2):

$$s^2 + s\frac{G}{J} + \frac{k}{J} = 0,$$ (2)

which is in the format of a second-order characteristic equation:

$$s^2 + 2\zeta\omega_n s + \omega_n{}^2 = 0,$$ (3)

where ω_n is the resonant frequency and ζ is the damping ratio of the actuated leg. By combining Equations (2) and (3), the moment of inertia and the rotational damping coefficient can be expressed in terms of resonant frequency and rotational spring constant:

$$J = \frac{k}{\omega_n{}^2},$$ (4)

$$G = 2\zeta\frac{k}{\omega_n}.$$ (5)

The rotational spring constant (k) is calculated using FEM simulation of the mechanical structure, while ω_n and ζ are obtained from experimental results. The actuation force ($\vec{F}$) from VO$_2$ is represented in $\vec{T}$ via

$$\vec{T} = a \times \vec{F},$$ (6)

where a is the distance between the force and the axis of rotation. The generated force (F) from the VO$_2$ can be expressed as:

$$F = \Gamma[T - T_0],$$ (7)

where T is the temperature of the leg, T_0 is the ambient temperature, and Γ is the hysteresis operator describing the relationship between the generated force and the temperature of the mirror leg. The thermal process (i.e., Joule heating) can be represented as follows [35]:

$$\frac{dT(t)}{dt} = -d_1[T(t) - T_0] + d_2 i^2(t),$$

$$T' = T(t) - T_0,$$

where d_1 and d_2 are positive coefficients related to the properties of the materials, and i is the input current. Applying the Laplace transform to the previous equation results in

$$sT' = -d_1T' + d_2i^2(s),$$

$$T' = \frac{d_2}{s + d_1}i^2,$$

$$T' = \frac{\frac{d_2}{d_1}}{\frac{s}{d_1} + 1}i^2 \Rightarrow \frac{A_T}{\tau_{th}s + 1}i^2,$$

where τ_{th} is the thermal time constant, and A_T is the gain of the thermal transfer function. The external force can now be expressed as:

$$F = \Gamma[T'] = \Gamma\left[\frac{A_T}{\tau_{th}s+1}i^2\right]. \tag{8}$$

The time constant and the gain A_T are found from experimental results. Finally, the torque generated by this force is expressed as:

$$\vec{T} = a\cos(\theta) \times \Gamma\left[\frac{A_T}{\tau_{th}s+1}i^2\right], \tag{9}$$

which, since θ is close to zero and thus $\cos(\theta) \approx 1$, is simplified to

$$\vec{T} = a \times \Gamma\left[\frac{A_T}{\tau_{th}s+1}i^2\right], \tag{10}$$

where the value a for the roll axis (a_r) is 115 μm and for the pitch axis (a_p) is 600 μm, as shown in Figure 3.

3.2. Nonlinear Model

The actuation force (F) is generated by two actuation mechanisms: one is the stress due to the thermal expansion difference of the materials forming the bimorph regions, and the other is the stress generated during the phase-change transition of the VO$_2$. Similar to [26], a non-monotonic hysteresis model is developed:

$$\Gamma[T'] = \Gamma_C[T'] + \Gamma_E(T'), \tag{11}$$

where $\Gamma_C[T']$ is the phase transition-induced force captured by a Preisach model, and $\Gamma_E(T')$ is the differential thermal expansion-induced force modeled as a linear term.

3.2.1. Phase Transition-Induced Force

The relationship between the phase transition-induced force and the temperature is monotonically hysteretic, and a Preisach model [36] is employed:

$$\Gamma_C[T'](t) = \int_{\mathcal{P}_0} \mu(\beta, \alpha)\gamma_{\beta,\alpha}[T'(\cdot); \zeta_0](t)\mathrm{d}\beta\,\mathrm{d}\alpha + c_0, \tag{12}$$

where $\mathcal{P}_0$ is the Preisach plane $\mathcal{P}_0 \triangleq \{(\beta, \alpha) : T_{min} \leq \beta \leq \alpha \leq T_{max}\}$, $[T_{min}, T_{max}]$ define the phase transition range, μ is the density function, $\gamma_{\beta,\alpha}$ denotes the basic hysteretic unit (hysteron), $T'(\cdot)$ is the temperature history, $T'(\eta)$, $0 \leq \eta \leq t$, and c_0 is a constant bias.

The hysteron is a memory-dependent operator. With the initial condition, $\zeta_0 \in \{-1, 1\}$, the output of the hysteron can be expressed as

$$\gamma_{\beta,\alpha}[T'(\cdot); \zeta_0] = \begin{cases} +1 & \text{if } T'(t) > \alpha, \\ -1 & \text{if } T'(t) < \beta, \\ \zeta_0 & \text{if } \beta \leq T'(t) \leq \alpha. \end{cases} \tag{13}$$

In practical usage, the integral expression of the Preisach model is typically approximated by discretizing the weight function μ to a finite number of parameters [36,37]. The weight function is approximated as a piecewise constant function—the weight w_{ij} is constant within cell (i, j), $i = 1, 2, \cdots, N$; $j = 1, 2, \cdots, N - i + 1$, where N is called the discretization level and w_{ij} is the model parameter. At time n, the output of the discretized Preisach model is expressed as

$$\Gamma_C[T'(n)] = \sum_{i=1}^{N} \sum_{j=1}^{N+1-i} w_{ij} s_{ij}(n) + c_0, \tag{14}$$

where w_{ij} is the weight for the cell (i, j) that is non-negative, and $s_{ij}(n)$ is the signed area of the cell (i, j), which is determined by the history of the temperature values up to time n.

The model parameters consist of the weights $\{w_{ij}\}$ and the constant bias c_0. The model identification can be reformulated as a constrained linear least-squares problem and solved efficiently with the MATLAB R2010b (Mathworks, Natick, MA, USA) command *lsqnonneg* [26,37].

3.2.2. Differential Thermal Expansion-Induced Force

The differential thermal expansion-induced force is resulted from the thermal expansion difference of the VO$_2$ and SiO$_2$ layers. This component was modeled as a linear term and a quadratic term in previous studies [26,27]. The following linear model is adopted in this work:

$$\Gamma_E(T') = -k_0 T', \tag{15}$$

where k_0 is a constant term related to thickness, modulus of elasticity, and thermal expansion coefficients of VO$_2$ layer and SiO$_2$ layer, and the negative term is introduced due to the fact that the thermal expansion-induced force has an opposite direction as the phase transition-induced force.

It is noted that the nonlinear model (Equations (11), (14), and (15)) can be conveniently identified with the linear least-squares method [26]. It is shown in Section 4.3 that the proposed model can accurately capture and estimate the non-monotonic hysteresis behavior of the MEMS mirror.

Finally, combining all the terms, the equations describing the movement of the mirror can be expressed as:

$$J_p \ddot{\theta}_p + G_p \dot{\theta}_p + k_p \theta_p = \vec{T}_p = a \times \Gamma_p \left[\frac{A_T}{\tau_{th_p} s + 1} i^2 \right], \tag{16}$$

$$J_r \ddot{\theta}_r + G_r \dot{\theta}_r + k_r \theta_r = \vec{T}_r = a \times \Gamma_r \left[\frac{A_T}{\tau_{th_r} s + 1} i^2 \right], \tag{17}$$

where the subscript p and r are references for the pitch and roll motions, the values for the linear parameters are presented in Table 3, Γ_p and Γ_r are the nonlinear models (Equation (11)).

4. Results and Discussion

4.1. Simulation Results

An FEM model is created in COMSOL (Version 5.2a, COMSOL Inc., Burlington, MA, USA) to calculate the rotational spring constant of the leg. The parameters used for the materials on the simulation are taken from the COMSOL library and from [19] shown in Table 1. The FEM model consisted of the entire MEMS mirror structure, including the four legs, the mirror platform, and all the material layers. A force sequence of increasing magnitude is applied as a point load at the top of the leg that connects to the mirror's platform. Four different simulations are run where the force is applied at different locations, as shown in Figure 4. The rotational spring constant at each location is extracted from the simulation results. To obtain the value of k, the displacement caused by the force is converted to an angle (θ_{sim}) with respect to each axis by using Equation (18):

$$\theta_{sim} = \sin^{-1} \frac{(h_1 - h_2)}{d}, \tag{18}$$

where h_1 and h_2 are the displacements at the point load (caused by the force) and at the axis, and d is the distance between the point load and the axis. The torque is then calculated by using the distance between the point load location and the corresponding axis. Finally, the torque is divided by the angle resulting in the rotational spring constant of the leg. This is done for both angles (pitch and roll) and the results are shown in Table 2.

Table 1. Parameters of the materials used in finite element method (FEM) simulations, where the Si, SiO_2, and Au are obtained from the COMSOL library, while the VO_2 properties are reported in [19].

Properties	Materials			
	Si	SiO_2	Au	VO_2
Density [Kg/m^3]	2320	2200	19300	4670
Young's Modulus [GPa]	187 [38,39]	70	70	140
Poisson Ratio	0.22	0.17	0.44	0.33

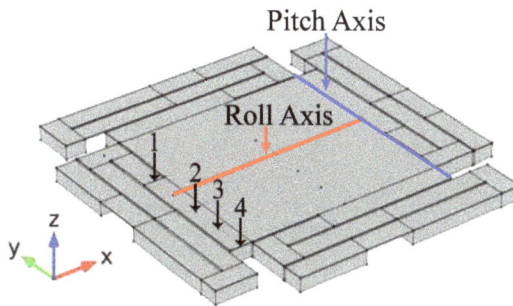

Figure 4. Finite element method (FEM) model schematic of the VO_2-based MEMS mirror used to find the rotational spring constant by applying a sequence of increasing force as a point load. The force is applied at different locations (1, 2, 3 & 4) for each simulation.

Table 2. Rotational spring constant from FEM simulation.

Point Load Location	Rotational Spring Constant	
	Pitch ($\times 10^{-9} \frac{N \cdot m}{deg}$)	Roll ($\times 10^{-9} \frac{N \cdot m}{deg}$)
1	2.29	1.219
2	2.29	1.217
3	2.29	1.217
4	2.29	1.217

4.2. MEMS Mirror Mechanical Model

A set of experiments are used to characterize the mechanical response of the structure and the nonlinear behavior of the VO_2 when actuating only one leg. Before each experiment, a pre-heating stage is performed to improve the stability and repeatability of the measurements, caused by the use of gold as the metal trace [40,41]. A similar process was performed in [7], where a sine wave was applied as the input voltage to anneal the metal layers. For the VO_2-based MEMS mirrors in this work, the pre-heating stage consisted of applying a 12 mA to all of the actuators for a total of 10 min. An input sequence of increasing voltage steps is used to measure the thermal time response of the actuated leg. The input is applied to the base of the transistor and had increasing amplitude steps of 0.5 V, which corresponds to ≈0.7 mA (once the transistor is on). Each step is held for 1 s before the next step started. The thermal time response (τ_{th}) within steps is calculated from the rise time using the following equation:

$$\tau_{th} = \frac{t_{rise}}{2.2},$$

(19)

where t_{rise} is the time taken for the structure to go from 10% to 90% of the output signal for one step. The results are shown in Figure 5. The thermal time response is calculated where the main dominant actuation mechanism is the thermal expansion difference of the materials forming the bimorph and not the transition of the VO_2, the values for the pitch and roll movements are 14 ms and 14.79 ms. During the transition of the VO_2, the system showed a pseudo-creep effect where each step took longer to reach steady state compared to outside the transition. This effect can be caused by the added stress from the legs that are not actuated. The added stress can move the transition temperature of the VO_2, which has been observed previously in VO_2 thin films [42,43]. Even more relevant to the present case, this effect was also observed in VO_2-based MEMS mirrors [10], where it was found that individual leg actuation and piston-like actuation required different actuation voltages—note that, during individual actuation, the remaining mirror legs add a stress that is not present during piston-like movement. The pseudo-creep is not included in the modeling of the device, in order to focus on the fundamental thermal and mechanical dynamics in the general case, and, as verified in later experiments, the presented model (ignoring the creep effect) shows adequate capability in predicting the mirror dynamics.

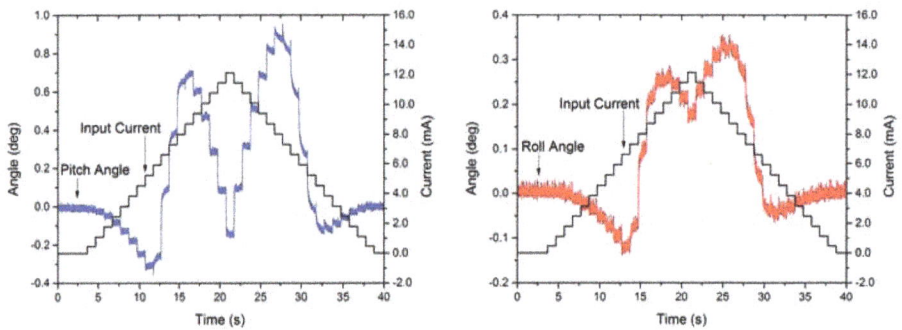

Figure 5. Time response measurements from actuating one leg for both variables: pitch (**left**) and roll (**right**) angles.

A frequency response measurement is performed to observe the mechanical response of the system. A sine wave signal ($i = 1.4\sin(2\pi ft) + 0.00714$ mA) is applied as the input of one of the legs while the frequency is swept from 0.1 Hz to 2000 Hz. The magnitude of the displacement is measured across the whole range of frequency, and then it is divided by the magnitude of the input current. Using the software Origin Pro9.0 (OriginLab Corporation, Northampton, MA, USA), the data is fitted using the magnitude of Equation (20), with a R^2 of 0.856 and 0.7797 for roll and pitch, respectively. Equation (20) is a linear approximation of the system, including the thermal and mechanical dynamics. Although the thermal response of the system in Equation (20) cannot capture the nonlinear behavior of the VO_2, it does capture the mechanical response of the system. The values for the resonance frequency (w_n) and the damping ratio (ζ) for each degree of freedom are found by a curve fit, and fitting parameters are shown in Figure 6. It is worth noted that the presented curve fitting method works well at low frequencies and produces larger errors at higher frequencies above 150 Hz. This is likely due to the fact that the fitting uses linear approximation and that the mechanical couplings between each leg and between the legs and the mirror are not fully captured. The highest frequency considered in the mechanical response of the system is 10 Hz, and the model follows the experimental results fairly well in this frequency range. Analysis and modeling at higher frequencies are potential extensions to this study:

$$\frac{\theta}{i^2} = \frac{A_T}{\tau s + 1} \frac{w_n^2}{s^2 + 2(w_n)\zeta + w_n^2}. \tag{20}$$

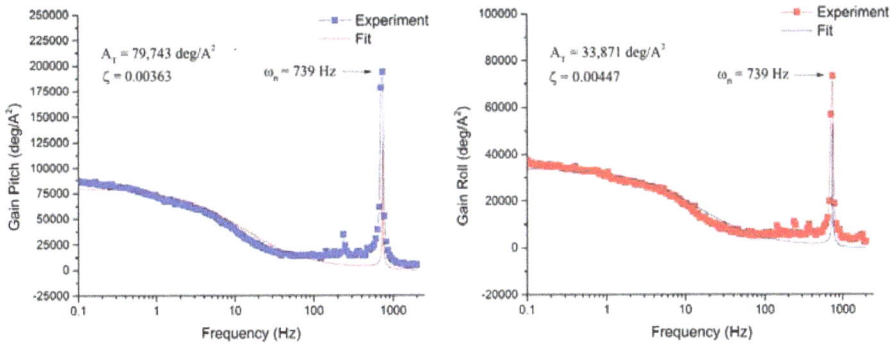

Figure 6. Frequency response for the actuation of one leg. A fitted curve is used to find the damping ratio (ζ) and the gain A_T. Both pitch (**left**) and roll (**right**) angles have the same resonant frequency with the value of 739 Hz.

4.3. Identification and Verification

4.3.1. Identification

A quasi-static measurement is performed to observe the static behavior of the leg across the phase transition. A series of current steps (each held for 550 ms) are applied to one of the legs with intervals of 0.1 V. The steady-state values are obtained by averaging the last 50 ms of the pitch and roll angles. This measurement will also be used to identify the unknown variables of the hysteresis model, since it contains the minor loops of the hysteresis. The plots are shown in Figure 7. It is shown that both of the hysteresis curves exhibit non-monotonic behavior.

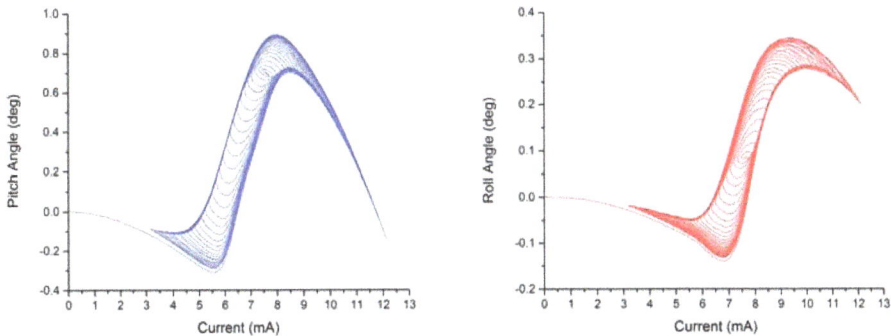

Figure 7. Identification plots of the pitch (**left**) and roll (**right**) angles, used to find the coefficients of the hysteresis model.

In order to identify the proposed model, the discretization level (N) of the Preisach model (Γ_C) is chosen to be 20. Further increasing the discretization level would increase the model complexity, but does not generate significant improvement in modeling accuracy. The root-mean-square error (RMSE) is chosen to quantify the accuracy of the model identification and verification results.

Figure 8a,b shows the identified weights of the Preisach models for the pitch and roll motions, respectively. Figure 8c shows the modeling performance for the hysteresis between the pitch angle and the current input, and Figure 8d shows the corresponding modeling error. The RMSE is 0.007 degrees. Similarly, Figure 8e–f shows the modeling performance for the hysteresis between the roll angle and the current. The RMSE is 0.003 degrees. It is shown that the proposed model can accurately capture the non-monotonic hysteresis of the MEMS mirror.

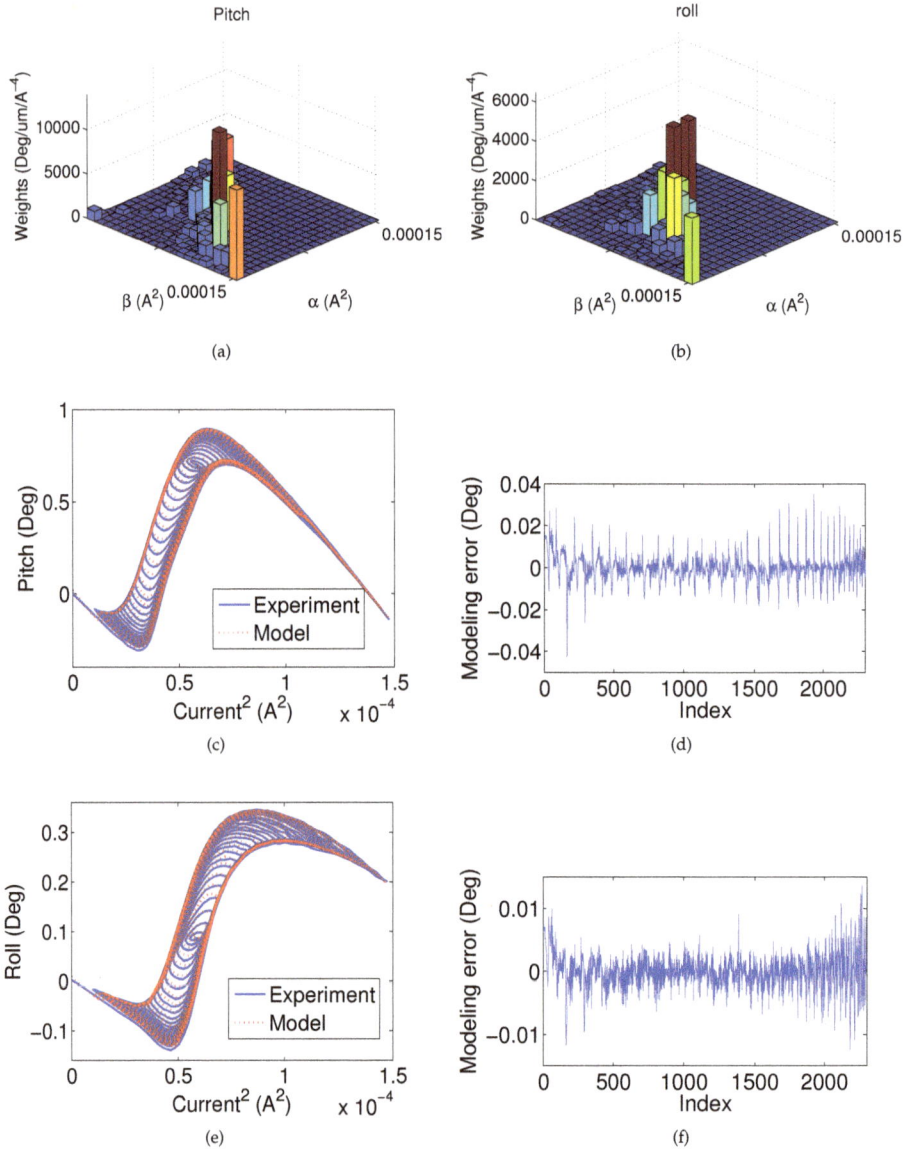

Figure 8. Parameters values (weights) used in the Preisach model for the (**a**) pitch and (**b**) roll; (**c**) the modeling performance; and (**d**) modeling error for the hysteresis between pitch angle and the current input; (**e**) the modeling performance; and (**f**) modeling error for the hysteresis between roll angle and the current input.

4.3.2. Quasi-Static Verification

The model identification results show that the proposed model can effectively capture the hysteresis under the chosen current step input. To confirm that the model can reliably and robustly predict the pitch and roll angles under any reasonable step input, additional experiments utilizing random step inputs are conducted. A randomly-chosen current step input, as shown in Figure 9a, is applied to the MEMS mirror. Each step is held for 1 s and the corresponding steady-state pitch angle and roll angle are recorded. With the chosen current input, the pitch angle and the roll angle estimations are calculated based on identified model shown in Equations (16) and (17) with parameters provided in Figure 8a,b and Table 3. Since the quasi-static condition is considered, the derivative terms of the angles will not affect the system performance. The experimental angles are compared with the estimated values. Figure 9b shows the pitch estimation performance, and the RMSE is 0.027 degrees. Figure 9c shows the roll estimation performance, and the RMSE is 0.013 degrees. The effectiveness of the nonlinear model is confirmed.

Table 3. Coefficient values of the model.

Constant	Name and Units	Pitch (θ_p)	Roll (θ_r)
A_T	Gain [deg/A^2]	79,743	33,871
τ_{th}	Time response [s]	0.0014	0.001479
ω_n	Resonant Frequency [rad/s]	4643	4643
ζ	Damping ratio	0.00363	0.00447
J	Moment of Inertia [Kg·m^2]	6.10×10^{-15}	3.23×10^{-15}
G	Rotational Damping coefficient [N·m·s/rad]	205.6×10^{-15}	134×10^{-15}
k	Rotational Spring coefficient [N·m/rad]	132×10^{-9}	69.7×10^{-9}
a	Position of the force with respect to the axis [μm]	600	115
c_0	Constant bias of Preisach model [deg/μm]	0.99	0.38
k_0	Thermal expansion-induced force term [N/°C]	1.4×10^4	3.8×10^3

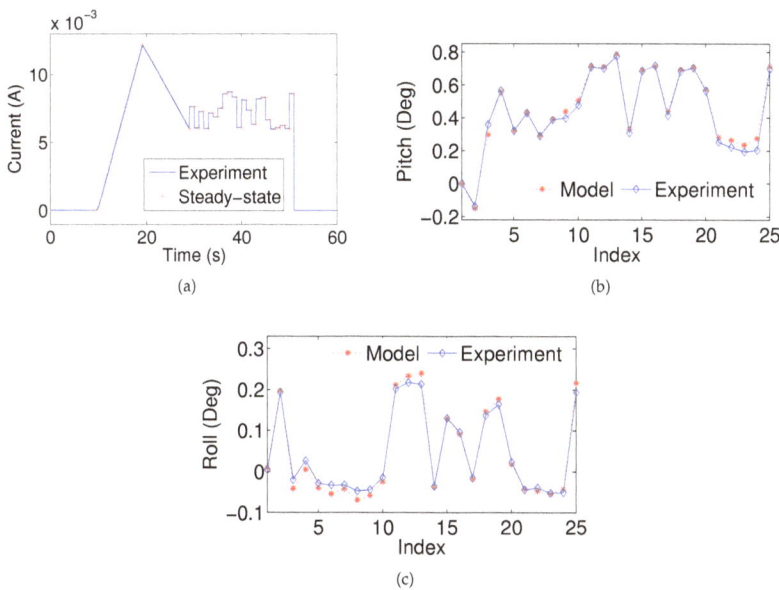

Figure 9. (**a**) A current step input for model verification; the measured and estimated steady-state (**b**) pitch angle; and (**c**) roll angle.

4.3.3. Frequency Verification

In order to verify that the model can effectively predict the performance of the mirror under dynamic inputs, sinusoidal current inputs with different frequencies are applied to the mirror, and the corresponding pitch angle and roll angle are recorded (Figure 10). As can be seen, the hysteresis relationships between the pitch angle and the current, and between the roll angle and the current, change under different frequencies. On average, the RMSE pitch angle estimation error is 0.074 degrees and the RMSE roll angle estimation error is 0.031 degrees. The model can capture the dynamic mirror motions reasonably well. It is noted that the estimation error becomes larger under higher frequencies, which is likely due to the mild discrepancies between the actual and calculated time response values.

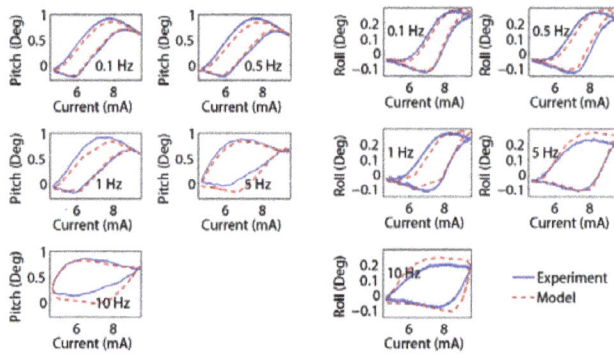

Figure 10. The pitch and roll angle verification performances for current inputs with different frequencies.

4.3.4. Multi-Frequency Verification

Furthermore, the model verification for multi-frequency inputs has been conducted. The current input $(1.35\sin(2\pi t) + 1.35\sin(10\pi t) + 1.35\sin(20\pi t) + 0.00705$ mA), as shown in Figure 11a, is applied to the system. The model estimation performances for the pitch and roll angles are shown in Figure 11b,c, respectively. The RMSE pitch angle estimation error is 0.097 degrees and the RMSE roll angle estimation error is 0.033 degrees. The effectiveness of the proposed model for the MEMS mirror is thus further validated.

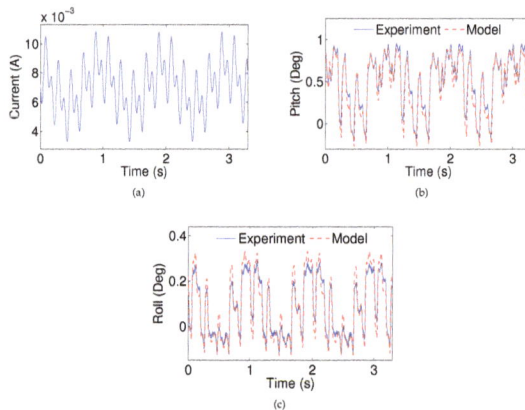

Figure 11. (a) A multi-frequency current input for model verification; the measured and estimated (b) pitch angle; and (c) roll angle.

5. Conclusions

In this article, we have derived and verified the model for MEMS mirrors actuated by phase-change materials. The model included mechanical and thermal processes, and accounted for nonlinear behavior typically found in most phase-change materials. The approach presented here involves a combination of theoretical and experimental results, resulting in a comprehensive hybrid analysis. Although the emphasis of the present work is on MEMS mirrors actuated by phase-change materials, particularly VO_2, the work can be extended to simpler electrothermal designs based on typical TEC difference or phase-change materials. Therefore, the present work presents a platform that can be adapted for the design of a broad scope of MEMS mirrors. Future work will focus on generalizing the presented model to other actuation modes and observing the effect of the other legs in the actuated leg and studies at higher frequencies,introducing a closed-loop control design, based on the present model, to accurately manipulate the different tilting angles of the mirror. Furthermore, future work will focus on incorporating the control system on each actuator of the VO_2-based MEMS mirror for different purposes such as creating a 2D image or laser tracking. Additionally, a comprehensive model will be studied to incorporate the pseudo-creep behavior of the device, which is relevant for quasi-static positioning applications.

Acknowledgments: This work was supported in part by the National Science Foundation under Grant ECCS 1306311 and Grant CMMI 1301243. The work of David Torres was supported by the Science, Mathematics, and Research for Transformation (SMART) scholarship by the Department of Defense of USA. Device development was made possible by a cooperative research and development agreement (CRADA No. 15-075-RY-01) between Air Force Research Laboratory Sensors Directorate (AFRL/RY) and Michigan State University. The fabrication of the MEMS Mirror was partially done at the Lurie Nanofabrication Facility at the University of Michigan. The SEM images were taken at the Center for Advanced Microscopy at Michigan State University.

Author Contributions: D.T. was in charge of the overall design of the devices, experiments, data gathering and analytical parts of the model. J.Z. was in charge of the nonlinear model analysis. S.D. contributed to the device design. X.T. and N.S. assisted in the conception of the initial ideas for the demonstrations, supervision and general guidance for the experiments, analysis of results, and preparation of the manuscript. X.T. focused on theoretical parts, while N.S. was focused more on the experimental parts. All the authors were included during the discussion of results and participated during the review of the manuscript.

Conflicts of Interest: The authors declare no conflict of interest.

References

1. Xie, H.; Pan, Y.; Fedder, G.K. A CMOS-MEMS mirror with curled-hinge comb drives. *J. Microelectromech. Syst.* **2003**, *12*, 450–457.
2. Koh, K.H.; Kobayashi, T.; Lee, C. A 2-D MEMS scanning mirror based on dynamic mixed mode excitation of a piezoelectric PZT thin film S-shaped actuator. *Opt. Express* **2011**, *19*, 13812–13824.
3. Hung, A.C.L.; Lai, H.Y.H.; Lin, T.W.; Fu, S.G.; Lu, M.S.C. An electrostatically driven 2D micro-scanning mirror with capacitive sensing for projection display. *Sens. Actuators A Phys.* **2015**, *222*, 122–129.
4. Naono, T.; Fujii, T.; Esashi, M.; Tanaka, S. Non-resonant 2-D piezoelectric MEMS optical scanner actuated by Nb doped PZT thin film. *Sens. Actuators A Phys.* **2015**, *233*, 147–157.
5. Yalcinkaya, A.; Urey, H.; Brown, D.; Montague, T.; Sprague, R. Two-axis electromagnetic microscanner for high resolution displays. *J. Microelectromech. Syst.* **2006**, *15*, 786–794.
6. Cho, A.R.; Han, A.; Ju, S.; Jeong, H.; Park, J.H.; Kim, I.; Bu, J.U.; Ji, C.H. Electromagnetic biaxial microscanner with mechanical amplification at resonance. *Opt. Express* **2015**, *23*, 16792–16802.
7. Wu, L.; Dooley, S.; Watson, E.; McManamon, P.F.; Xie, H. A Tip-Tilt-Piston Micromirror Array for Optical Phased Array Applications. *J. Microelectromech. Syst.* **2010**, *19*, 1450–1461.
8. Jain, A.; Qu, H.; Todd, S.; Xie, H. A thermal bimorph micromirror with large bi-directional and vertical actuation. *Sens. Actuators A Phys.* **2005**, *122*, 9–15.
9. Samuelson, S.R.; Xie, H. A Large Piston Displacement MEMS Mirror With Electrothermal Ladder Actuator Arrays for Ultra-Low Tilt Applications. *J. Microelectromech. Syst.* **2014**, *23*, 39–49.
10. Torres, D.; Wang, T.; Zhang, J.; Zhang, X.; Dooley, S.; Tan, X.; Xie, H.; Sepúlveda, N. VO_2-Based MEMS Mirrors. *J. Microelectromech. Syst.* **2016**, *25*, 780–787.

11. Sepulveda, N.; Rua, A.; Cabrera, R.; Fernández, F. Young's modulus of VO$_2$ thin films as a function of temperature including insulator-to-metal transition regime. *Appl. Phys. Lett.* **2008**, *92*, 1913.
12. Zylbersztejn, A.; Mott, N. Metal-insulator transition in vanadium dioxide. *Phys. Rev. B* **1975**, *11*, 4383.
13. Barker, A., Jr.; Verleur, H.; Guggenheim, H. Infrared optical properties of vanadium dioxide above and below the transition temperature. *Phys. Rev. Lett.* **1966**, *17*, 1286.
14. Mlyuka, N.R.; Niklasson, G.A.; Granqvist, C.G. Mg doping of thermochromic VO$_2$ films enhances the optical transmittance and decreases the metal-insulator transition temperature. *Appl. Phys. Lett.* **2009**, *95*, 171909.
15. Cao, J.; Gu, Y.; Fan, W.; Chen, L.; Ogletree, D.; Chen, K.; Tamura, N.; Kunz, M.; Barrett, C.; Seidel, J.; et al. Extended mapping and exploration of the vanadium dioxide stress-temperature phase diagram. *Nano Lett.* **2010**, *10*, 2667–2673.
16. Breckenfeld, E.; Kim, H.; Burgess, K.; Charipar, N.; Cheng, S.F.; Stroud, R.; Piqué, A. Strain Effects in Epitaxial VO$_2$ Thin Films on Columnar Buffer-Layer TiO$_2$/Al$_2$O$_3$ Virtual Substrates. *ACS Appl. Mater. Interfaces* **2017**, *9*, 1577–1584.
17. Merced, E.; Tan, X.; Sepúlveda, N. Strain energy density of VO$_2$-based microactuators. *Sens. Actuators A Phys.* **2013**, *196*, 30–37.
18. Rúa, A.; Fernández, F.l.E.; Sepúlveda, N. Bending in VO$_2$-coated microcantilevers suitable for thermally activated actuators. *J. Appl. Phys.* **2010**, *107*, 074506.
19. Cabrera, R.; Merced, E.; Sepúlveda, N. Performance of Electro-Thermally Driven VO$_2$-Based MEMS Actuators. *J. Microelectromech. Syst.* **2014**, *23*, 243–251.
20. Cabrera, R.; Merced, E.; Sepúlveda, N. A micro-electro-mechanical memory based on the structural phase transition of VO$_2$. *Phys. Status Solidi* **2013**, *210*, 1704–1711.
21. Merced, E.; Cabrera, R.; Dávila, N.; Fernández, F.E.; Sepúlveda, N. A micro-mechanical resonator with programmable frequency capability. *Smart Mater. Struct.* **2012**, *21*, 035007.
22. Bai, Y.; Yeow, J.T.W.; Wilson, B.C. A Characteristic Study of Micromirror with Sidewall Electrodes. *Int. J. Optomech.* **2007**, *1*, 231–258.
23. Isikman, S.O.; Urey, H. Dynamic Modeling of Soft Magnetic Film Actuated Scanners. *IEEE Trans. Magn.* **2009**, *45*, 2912–2919.
24. Han, F.; Wang, W.; Zhang, X.; Xie, H. Modeling and Control of a Large-Stroke Electrothermal MEMS Mirror for Fourier Transform Microspectrometers. *J. Microelectromech. Syst.* **2016**, *25*, 750–760.
25. Zhang, J.; Merced, E.; Sepúlveda, N.; Tan, X. Optimal compression of generalized Prandtl–Ishlinskii hysteresis models. *Automatica* **2015**, *57*, 170–179.
26. Zhang, J.; Merced, E.; Sepúlveda, N.; Tan, X. Modeling and Inverse Compensation of Nonmonotonic Hysteresis in VO$_2$-Coated Microactuators. *IEEE/ASME Trans. Mech.* **2014**, *19*, 579–588.
27. Zhang, J.; Torres, D.; Ebel, J.L.; Sepúlveda, N.; Tan, X. A Composite Hysteresis Model in Self-Sensing Feedback Control of Fully Integrated VO$_2$ Microactuators. *IEEE/ASME Trans. Mech.* **2016**, *21*, 2405–2417.
28. Merced, E.; Torres, D.; Tan, X.; Sepúlveda, N. An Electrothermally Actuated VO$_2$-Based MEMS Using Self-Sensing Feedback Control. *J. Microelectromech. Syst.* **2015**, *24*, 100–107.
29. Xie, H. Vertical Displacement Device. US Patent 6,940,630, 6 September 2005.
30. Wu, L.; Xie, H. A large vertical displacement electrothermal bimorph microactuator with very small lateral shift. *Sens. Actuators A Phys.* **2008**, *145*, 371–379.
31. Zhang, L.; Tsaur, J.; Maeda, R. Residual Stress Study of SiO$_2$/Pt/Pb(Zr,Ti)O$_3$/Pt Multilayer Structure for Micro Electro Mechanical System Applications. *Jpn. J. Appl. Phys.* **2003**, *42*, 1386.
32. Matsui, Y.; Hiratani, M.; Kumagai, Y.; Miura, H.; Fujisaki, Y. Thermal Stability of Pt Bottom Electrodes for Ferroelectric Capacitors. *Jpn. J. Appl. Phys.* **1998**, *37*, L465.
33. Kinbara, A.; Haraki, H. Internal Stress of Evaporated Thin Gold Films. *Jpn. J. Appl. Phys.* **1965**, *4*, 243.
34. Kebabi, B.; Malek, C.; Ladan, F. Stress and microstructure relationships in gold thin films. *Vacuum* **1990**, *41*, 1353–1355.
35. Leo, D.J. *Engineering Analysis of Smart Material Systems*; John Wiley & Sons: Hoboken, NJ, USA, 2007.
36. Mayergoyz, I. *Mathematical Models of Hysteresis and Their Applications*; Springer: New York, NY, USA, 1991.
37. Tan, X.; Baras, J. Modeling and control of hysteresis in magnetostrictive actuators. *Automatica* **2004**, *40*, 1469–1480.
38. Nye, J. *Physical Properties of Crystals: Their Representation by Tensors and Matrices*; Oxford University Press: Oxford, UK, 1985.

39. Wortman, J.J.; Evans, R.A. Young's Modulus, Shear Modulus, and Poisson's Ratio in Silicon and Germanium. *J. Appl. Phys.* **1965**, *36*, 153–156.

40. Gall, K.; Dunn, M.L.; Zhang, Y.; Corff, B.A. Thermal cycling response of layered gold/polysilicon MEMS structures. *Mech. Mater.* **2004**, *36*, 45–55.

41. Gall, K.; West, N.; Spark, K.; Dunn, M.L.; Finch, D.S. Creep of thin film Au on bimaterial Au/Si microcantilevers. *Acta Mater.* **2004**, *52*, 2133–2146.

42. Tsai, K.Y.; Chin, T.S.; Shieh, H.P.D.; Ma, C.H. Effect of as-deposited residual stress on transition temperatures of VO$_2$ thin films. *J. Mater. Res.* **2004**, *19*, 2306–2314.

43. Case, F.C. Modifications in the phase transition properties of predeposited VO$_2$ films. *J. Vac. Sci. Technol. A Vac. Surf. Films* **1984**, *2*, 1509–1512.

micromachines

MDPI

Article

Multi-Response Optimization of Electrothermal Micromirror Using Desirability Function-Based Response Surface Methodology

Muhammad Mubasher Saleem [1,*], Umar Farooq [1], Umer Izhar [2] and Umar Shahbaz Khan [1]

[1] Department of Mechatronics Engineering, National University of Sciences and Technology, Islamabad 44000, Pakistan; umerfarooq.nust@gmail.com (U.F.); u.shahbaz@ceme.nust.edu.pk (U.S.K.)
[2] School of Engineering and Technology, Central Queensland University, Mackay Ooralea 4740, Australia; u.izhar@cqu.edu.au
* Correspondence: mubasher.saleem@ceme.nust.edu.pk; Tel.: +92-51-9247544

Academic Editor: Huikai Xie
Received: 20 December 2016; Accepted: 23 March 2017; Published: 1 April 2017

Abstract: The design of a micromirror for biomedical applications requires multiple output responses to be optimized, given a set of performance parameters and constraints. This paper presents the parametric design optimization of an electrothermally actuated micromirror for the deflection angle, input power, and micromirror temperature rise from the ambient for Optical Coherence Tomography (OCT) system. Initially, a screening design matrix based on the Design of Experiments (DOE) technique is developed and the corresponding output responses are obtained using coupled structural-thermal-electric Finite Element Modeling (FEM). The interaction between the significant design factors is analyzed by developing Response Surface Models (RSM) for the output responses. The output responses are optimized by combining the individual responses into a composite function using desirability function approach. A downhill simplex method, based on the heuristic search algorithm, is implemented on the RSM models to find the optimal levels of the design factors. The predicted values of output responses obtained using multi-response optimization are verified by the FEM simulations.

Keywords: micromirror; Micro-Electro-Mechanical Systems (MEMS); bimorph; optimization; biomedical; desirability function; response surface models

1. Introduction

Micro-Electro-Mechanical Systems (MEMS) technology-based micromirrors are microscale devices used in optical systems to project light over a wide range of reflection angles. Micromirrors are generally used in various applications depending upon their geometric configuration, actuation mechanism, and output performance characteristics. The major application areas of micromirrors include optical switches [1], optical communications [2], optical displays [3,4], microscopic topometry [5], barcode scanning [6], biomedical imaging [7,8], and optical interconnects [9]. The deflection angle of a micromirror can be adjusted statically or dynamically by an actuation mechanism that allows the rotation of the mirror surface. The actuation mechanisms for the deflection of micromirror plates for optical scanning are mainly divided into four categories: electrostatic, piezoelectric, electromagnetic, and electrothermal. The choice of an actuation mechanism is generally dependent on the maximum angular displacement, device size, input power, input voltage, and microfabrication process [10].

Electrothermal actuation for micromirrors allows us to achieve relatively large angular deflection in the mirror plate at low actuation voltages. Moreover, the electrothermal micromirrors have almost linear response between the deflection angle and actuation voltage, high fill factor, simple design, and easy fabrication process. These characteristics make electrothermal micromirrors a suitable choice for biomedical imaging applications [11–14]. The electrothermal actuators may be designed using either a single thin-film metal structural layer to achieve an in-plane or out-of-plane deflection corresponding to an applied voltage [15,16] or a combination of two material layers (typically a metal and dielectric) bonded at an interface with a significant difference in their coefficients of thermal expansion (CTE) [17–19]. When an increase in temperature is applied, the thermal bi-layer actuator bends towards the side of the material that has a lower CTE value. The displacement caused by bending is used in micromirror designs to rotate the mirror surface. A mirror plate is usually attached to the end of the actuator and deflects at an angle equal to the tangential angle of the bimorph end. In optical imaging applications, like an Optical Coherence Tomography (OCT) system, the optical scanning angle has twice the mechanical deflection angle of the mirror plate. Earlier work on the application of the bimorph thermal actuators for the micromirror was presented by Bulher et al. [20]. Ataka et al. [21] reported a bimorph actuator based on a dual-layer polyimide material for the distributed micromotion systems. Yang et al. [22] reported a precise position tracking based on SiO_2/doped silicon bimorph actuator, where the proposed micromirror can be vertically actuated by 1 μm at an input power of 3 mW. Jain et al. [23] demonstrated an electrothermally actuated micromirror design with optical scan angles larger than $\pm 30°$ in two dimensions with driving voltages of less than 12 V. The bimorph layers used in this design were aluminum and silicon dioxide. Singh et al. [24] demonstrated an electrothermal micromirror based on an aluminum/silicon bimorph actuator with a reflecting metal-coated silicon mirror plate. The device size was 2.5 mm × 2.5 mm and achieved 17° angular mechanical deflection at an actuation voltage of 1.6 V. Xie et al. [25] presented a micromirror design using Al/SiO_2 bimorph thermal actuators for laser beam scanning in an OCT system. Izhar et al. [26] presented an electrothermally actuated multi-axis micromirror design for OCT systems and reported an optical scanning angle of 32° with an applied voltage of 6 V and input power of 12 mW. Liu et al. [27] presented a micromirror with aluminum/tungsten bimorphs for fast thermal response with an optical scanning angle of $\pm 11°$ at 0.6 V. A large optical scanning angle of $\pm 60.4°$ is reported in [28], at an actuation voltage of 9.8 V, corresponding to a mechanical deflection angle of 18.1° only. The large optical scanning angle is achieved by submerging the mirror into a mineral oil with a refractive index of 1.47 and utilizing the "Snell's window effect". Samuelson et al. [29] reported a micromirror actuated by ladder actuators showing 0.25° lateral mechanical deflection angle at 90 μm piston mode displacement, with an actuation voltage of 1.2 V. Jang et al. [30] reported a MEMS-based parallel plate-rotation (PPR) device for a single-imager-based stereoscopic endoscope. The fabricated MEMS PPR device rotates an optical plate with a rotation angle up to 37°. Recently, Duan et al. [31] presented a microendoscopic OCT probe with a tilted electrothermal micromirror, directly integrated on a silicon optical bench. The micromirror consists of a two axis scanning single-crystal-silicon (SCS) mirror tilted using an Al/SiO_2 bimorph thermal actuator. The maximum scan angle of the mirror plate is 40° at an actuation voltage of 5.5 V for both axes. The main performance characteristics of a micromirror design discussed in the literature for biomedical applications in general, and for an OCT system specifically, include micromirror plate deflection angle and input power. For an OCT system, a higher micromirror deflection angle allows us to scan a large area from a certain distance. The power dissipated in the micromirror due to the electrothermal actuation results in a temperature rise in the device, which adversely affects the output power of the laser integrated in the OCT system [26].

The main challenge in the design of a MEMS device is to obtain the optimal geometric configuration of the device while considering multiple performance constraints. Conventionally, optimization of MEMS devices is carried out by developing analytical models, FEM models, topology optimization, artificial neural networks, and genetic algorithms. These techniques for multiple output responses become impractical due to the complex geometry and high computational costs involved, especially for electrothermal micromirrors, which involve complex structural-thermal-electric interactions. A multi-response optimization using Design of Experiments (DOE) allows for investigating the design space of a MEMS device at different sample points using FEM models with less time, effort, and computational costs and facilitates the analysis of the effect of different parameters on output responses in detail. Previously, authors have discussed the application of the DOE technique for single-response optimization of a RF-MEMS switch to achieve a reliable and optimized design considering the microfabrication process uncertainties and residual stresses [32]. In this paper, a DOE-technique-based multi-response optimization for the scanning electrothermal micromirror design, to be integrated in the sample arm of an OCT system, is presented, considering mirror plate optical deflection angle, input power, and temperature rise from the ambient in the mirror plate.

2. Design and Working Principle of the Proposed Micromirror

The proposed micromirror design consists of a mirror plate and four bimorph electrothermal actuators, which are symmetrically connected to the mirror plate on four sides through flexural connectors, as shown in Figure 1. The bimorphs consist of two structural layers of aluminum and silicon with an embedded platinum heater. An oxide layer is used for electrical insulation between the structural layers and the heater. The heater pads are exposed to apply the input voltage. To achieve a high out-of-plane displacement, the elecrothermal actuators are optimized with a rectangular notch at the end. Once a voltage difference is applied to the exposed heater pads of a bimorph, a current passes through it and heats it up due to joule heating. The bimorph tends to deflect out-of-plane because of the significant difference in values of CTEs of both constituent structural layers. As a result, the micromirror rotates with a certain angular deflection. For piston mode motion, all four actuators are excited to make a vertical out-of-plane displacement of the micromirror. The micromirror design presented in this paper is optimized considering the microfabrication process presented in [26].

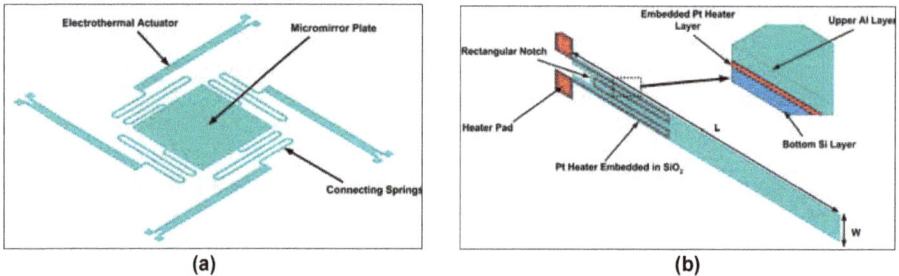

(a)   (b)

Figure 1. (a) Schematic of the micromirror design consisting of flexural springs with a reflecting plate in the center; (b) schematic of the electrothermal actuator with bottom Si layer, top Al layer, and embedded Pt heater. The presence of the rectangular notch at the heater end allows for achieving higher vertical deflection in the actuator.

Figure 2 shows a basic layout of a time domain OCT system consisting of a monolithically integrated broadband light source, silicon germanium photodiode (used as a photodector), waveguides, and micromirrors in the reference and sample arms. In its simplest form, an OCT system operates by splitting a single beam of light into two with a beam splitter. One of the beams travels towards the target sample through the sample arm and the other beam travels to the reference mirror through the reference arm, and reflects back towards the beam splitter from a movable reference mirror. This reflected light from a reference mirror then interacts with the light reflected from the target and produces interference fringes. These signals are then read and electronically processed to determine the reflectivity values of the target as a function of the depth into the tissue. The scanning depth can be swept by changing the path length of the reference mirror.

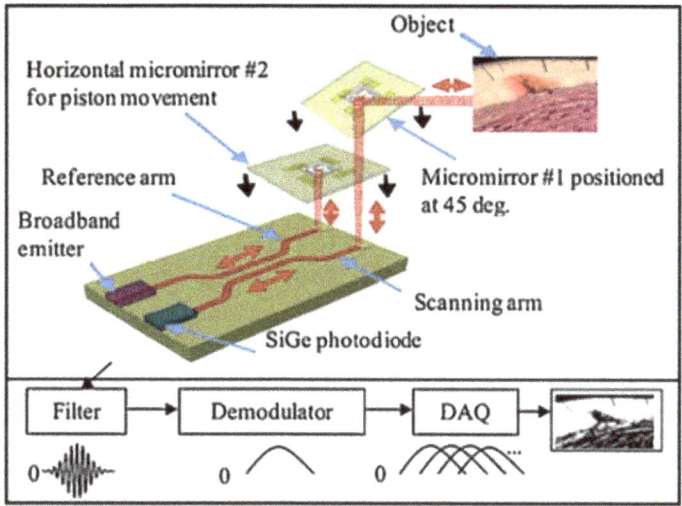

Figure 2. A basic layout of an Optical Coherence Tomography (OCT) system with axial and transverse scanning micromirrors [26].

3. Micromirror Design Optimization Using Design of Experiments (DOE)

A DOE-based design matrix consists of a sequence of FEM simulations to be carried out in terms of design factors set at pre-defined levels. The rows and columns of the design matrix represent the simulation runs and initial design factors settings, respectively. Initially, the identification of the important design factors, affecting a particular output response, is carried out using a screening design matrix. The screening of the significant design factors allows for analyzing and optimizing an output response with respect to design factors in detail using response surface methodology. Figure 3 shows a complete layout for the multi-response optimization using DOE based FEM simulations. The output responses considered in the optimization of the micromirror design, presented in this paper, are the optical deflection angle (twice the mechanical deflection angle), input power, and the temperature rise in micromirror plate from the ambient. Initially, nine design factors that may affect these output responses are considered on two levels, as shown in Table 1. The levels of the design factors are decided based on the previous designs presented in the literature for electrothermally actuated micromirrors. The size of the micromirror plate is considered to be 500 μm × 500 μm, large enough to allow the easy focus of the laser beam spot in optical imaging applications. The low and high levels of the electrothermal actuator length (L) and width (W), shown in Table 1, depict a minimum and maximum L/W ratio of 10 and 16, respectively. The out-of-plane deflection of the actuator can be increased by further increasing this L/W ratio. However, the maximum length of the actuator is limited by the chip size and fill factor, while the width of actuator is dependent on the width of the embedded Pt heater. Moreover, since the thermal response time of the electrothermal actuator is proportional to the square of the actuator length [33], a larger value of actuator length results in lower switching rates for electrothermal micromirrors.

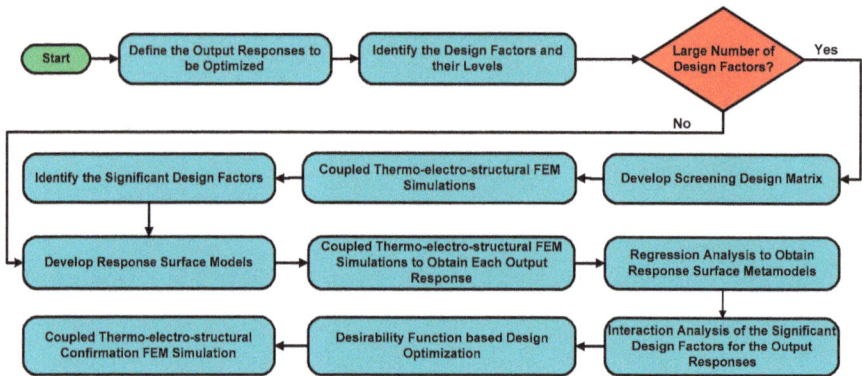

Figure 3. Flowchart showing the schematic layout of the steps implemented for the optimization of the micromirror using Response Surface Models (RSM)-based Design of Experiments (DOE).

Table 1. Design factors with their respective codes, selected at two levels, for the optimization of the micromirror.

Code	Design Factor (µm)	Low Level (−1)	High Level (+1)
X_1	Actuator Length (L)	500	800
X_2	Actuator Width (W)	50	100
X_3	Silicon Thickness (SiT)	1	1.5
X_4	Heater Thickness (HT)	0.1	0.5
X_5	Heater Length (HL)	200	300
X_6	Metal Thickness (MT)	0.5	1.5
X_7	Spring Length (SpL)	400	500
X_8	Spring Width (SpW)	8	10
X_9	Mirror Thickness (MIRT)	5	10

3.1. Screening Design Matrix for Significant Design Factors

Screening designs are the most important DOE design matrices that determine the most significant design factors in the optimization process. The Placket–Burman design matrix is the most common screening design matrix used to identify the significant factors in a minimal number of simulation runs with a good degree of accuracy [34]. The Placket–Burman design matrix is based on the first-order model given by:

$$Y = \beta_0 + \sum_{i=1}^{i=n} \beta_i X_i, \tag{1}$$

where Y is the output response, β_0 is the model intercept, β_i is the linear coefficient, and X_i is the level of the design factor. The Placket–Burman design matrix, with 20 simulations and the corresponding three output responses, is shown in Supplementary Table S1. The output responses, for the different combinations of the design factors, are obtained using FEM-based structural-thermal-electric coupled analysis in ANSYS. The structural parts are modeled using SOLID98, which is a coupled field tetrahedral solid element. The micromirror is constrained at the electrothermal actuator ends for both structural and thermal boundary conditions. The material properties used in the FEM simulations are summarized in Table 2. The variation in the material properties with a change in the temperature was previously discussed by the authors in [26] and the temperature coefficient of resistance for the embedded Pt heater in the bimorph actuator was observed to be significantly affected by the temperature. However, in the present work, it is assumed that all the material properties exhibit a linear elastic behavior and remain constant despite changes in the temperature. In general, the heat transfer modes for the electrothermal actuators include conduction, natural convection, and radiation.

For a similar micromirror design, presented in [26], the heat transfer due to conduction, convection, and radiation was simulated to be 85%, 14%, and 1%, respectively. These results show a negligible effect of convection and radiation as compared to conduction, and a similar effect has also been presented for electrothermal actuators in [22,35–38]. Therefore, to reduce computational time during the FEM simulations, only heat transfer due to the conduction is considered, with the assumption that most of the heat transfer occurs along the bimorph actuator and connecting springs as compared to the heat loss from the air. A fixed input voltage of 0.8 V is applied across the electrothermal actuator pads and the corresponding deflection angle, input power, and temperature rise from the ambient in the micromirror plate are obtained. Based on the desired performance of the micromirror, an overall figure of merit (FOM) is obtained considering all three output responses:

$$\text{FOM} = \frac{\text{Deflection angle}}{\text{Input power (mW)} \times \text{Temperature rise from the ambient (°C)}}. \tag{2}$$

Table 2. Material properties used in the FEM simulations [22,24,39,40].

Material Properties	Aluminum	Platinum	Silicon	Silicon Dioxide
Young's modulus (GPa)	70	170	162	70
Poisson ratio	0.33	0.38	0.22	0.17
Density (kg/μm³)	2.3×10^{-15}	21.4×10^{-15}	2.32×10^{-15}	2.66×10^{-15}
Specific heat (pJ/kg K)	9.02×10^{14}	1.3×10^{14}	7.53×10^{14}	10×10^{14}
Resistivity (TΩ·μm)	2.83×10^{-14}	10.9×10^{-14}	1.32×10^{-14}	1.0×10^{10}
CTE (1/K)	23.1×10^{-6}	8.8×10^{-6}	2.66×10^{-6}	0.5×10^{-6}
Thermal conductivity (pW/μm K)	23.7×10^{7}	7.1×10^{7}	1.5×10^{8}	0.1×10^{7}

3.2. Mean Effect Model and Analysis of Variance for the Screening Design

The output responses obtained using the FEM simulations for the screening design matrix can be described by a linear statistical model given as:

$$y_{ij} = \mu_i + \varepsilon_{ij} \begin{cases} i = 1,2,,,a \\ j = 1,2,,,n \end{cases}, \tag{3}$$

where y_{ij} is the *ij*th response value, *n* is the number of times the design factor level appears in the design matrix, μ_i is the *ij*th design factor level, and ε_{ij} are the random errors. The means for each design factor at low and high level are obtained for each output response, as shown in Figure 4. The horizontal axis for each design factor is the low and high level value, while the vertical axis is the mean value of the output responses for each design factor level. A large difference in the means of two design factor levels shows that the design factor has a significant effect on the output response. Figure 4a shows a steeper slope for the two levels of the design factors L, W, SiT, MT, HL, and HT. Figure 4b shows that the only heater length (HL), heater thickness (HT), and actuator width (W) have a significant effect on the temperature rise of the micromirror plate. For the input power, the slope of the heater length (HL) and heater thickness (HT) is high compared to the other design factors, as shown in Figure 4c. Since the mean effect of the design factors at low and high levels is not the same for all three output responses, the mean effect plot for the figure of merit is obtained using Equation (2). Figure 4d shows the mean effect plots of the design factors for the overall FOM.

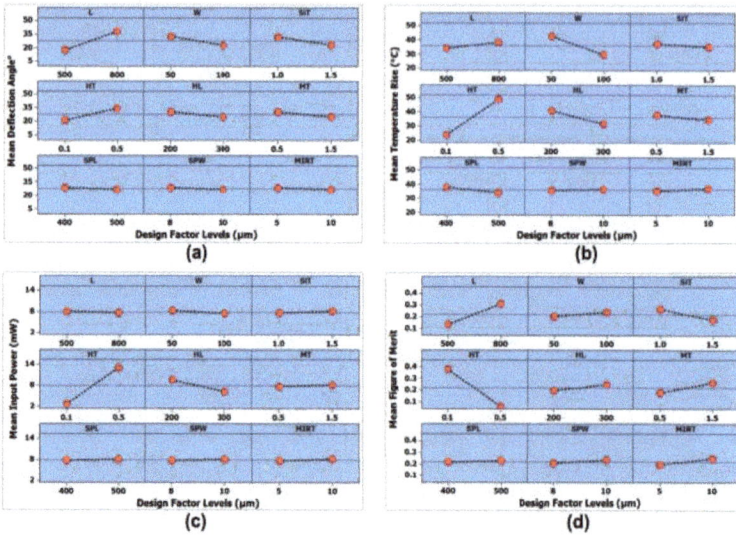

Figure 4. Mean effect plots for the output responses: (**a**) mean effect plot for the deflection angle. Design factor actuator length (L) has the highest change in the mean value with change from low level to high level. (**b**) Mean effect plot for the micromirror central plate temperature rise from the ambient; The design factor heater thickness (HT) has a highest deviation from the mean; (**c**) Mean effect plot of the input power. The design factor HT and heater length (HL) have a visible change in the mean values while all other factors have negligible effect on mean at two different levels; (**d**) Mean effect plot of the figure of merit showing the design factors HT and L to have the highest change in the mean value going from the low to high factor level.

Analysis of variance (ANOVA) is a collection of statistical design models that analyze the effect of considered design factors on a specific output response. This technique is based on the assumption that the sources of variability in the output response variables can be attributed to the design factors as well as to the random noise in the experiments. The total variation in the output response for each design factor is calculated in the form of the total sources of variance SS_T, which is a combination of variable sum of squares (variance due to design factors effects) SS_A and error sum of squares (random error) SS_E. These total source variance can be represented as [41]:

$$SS_T = SS_A + SS_E \tag{4}$$

$$SS_A = \sum_{i=1}^{a} \sum_{j=1}^{n_i} (\bar{y}_i - \bar{y})^2 \tag{5}$$

$$SS_E = \sum_{i=1}^{a} \sum_{j=1}^{n_i} (y_{ij} - \bar{y}_i)^2, \tag{6}$$

where $\bar{y}_i$ is the design factor level group mean, $\bar{y}$ is the overall mean, a is the number of levels of the design factor, y_{ij} is the the ijth response in the ith variable level, and n_i is the number for which the variable is at i level. The assumptions for ANOVA (that the random errors are normally distributed with mean zero and constant variance) are initially verified using Anderson–Darling [42] and Levene tests [41]. A detailed description of these tests is provided by the authors in [43]. p-values > 0.05 are obtained for both these tests, thus verifying the basic ANOVA assumptions. ANOVA results are generally described in terms of p-value. A p-value ≤ 0.05 for a design factor means that it can be concluded with 95% confidence level that the considered design factor has a significant effect on the output response. For the angular deflection, p-value ≤ 0.05 is obtained for the design factors L, W, SiT,

HT, HL, and MT. Similarly, the analysis showed p-value ≤ 0.05 for the design factors W, HT, and HL in the case of micromirror temperature rise from ambient. For the input power, the design factors HT and HL showed p-value ≤ 0.05. The results obtained using ANOVA are further verified using half-normal probability plots [44]. The half-normal probability plots are used to find out whether and to what extent the distribution of the design factors follow the normal distribution. The estimates for the significant design factors do not follow the normal distribution. Figure 5 shows the half-normal probability plots for the three output responses of the screening design matrix.

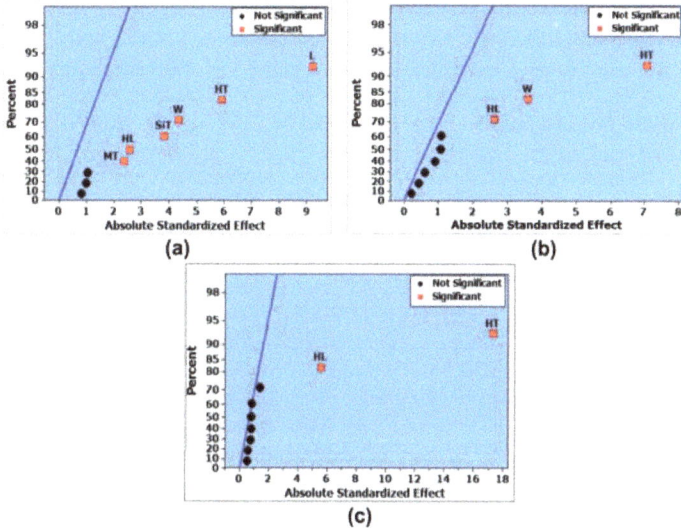

Figure 5. (a) Half-normal probability plot for deflection angle. The effect of the actuator length (L) is highest among the significant factors, while that of the mirror thickness (MT) is lowest; (b) Half-normal probability plot for the micromirror temperature rise from the ambient. Among the three significant factors, heater thickness (HT) has the highest effect on temperature rise and silicon thickness (SiT) has the lowest; (c) Half-normal probability plot of the input power. Heater thickness (HT) and heater length (HL) are the two significant design factors.

3.3. Design Matrix for Multi-Response Optimization

The ANOVA and half-normal probability plots show that the significant design factors for the angular deflection also include the design factors that were proven to be significant for the micromirror temperature rise and input power. So, the significant design factors L, W, HL, HT, SiT, and MT, obtained using a Plackett–Burman based screening design matrix for the angular deflection, are further investigated using response surface metamodels. The response surface method is based on a statistical approach to develop an appropriate relationship between an output response and the design factors using the following second-order model:

$$y = \beta_0 + \sum_{i=1}^{k} \beta_i x_i + \sum \sum_{i<j} \beta_{ij} x_i x_j + \sum_{i=1}^{k} \beta_{ii} x_i^2 + \epsilon \qquad (d = 2), \qquad (7)$$

where ϵ is the random error and the β coefficients are obtained by the method of least squares regression, such that the sum of the squares of the predicted values and the actual values are minimized. In matrix form, Equation (7) can be written as:

$$Y = bX + E, \qquad (8)$$

where Y is the matrix of the measured output response values and X is the matrix of the design factors. The matrix b of the β coefficients can be obtained as:

$$b = \left(X^T X\right)^{-1} X^T Y. \tag{9}$$

The selection of a proper design matrix for the response surface-based optimization is very important. In this work, we have selected Central Composite Design (CCD) design matrix for multi-response optimization. The CCD requires only a fraction of all the possible combinations of the design factors. The number of simulation runs required for the CCD design matrix are $N = 2^k + 2k + C_0$ where k is the number of the design factors and C_0 is the number of central points. Table 3 shows the significant design factors at three levels used for the response surface metamodels.

Table 3. Design factors and their three levels for the Central Composite Design (CCD) design matrix.

Code	Design Factor (um)	Low Level (−1)	Medium Level (0)	High Level (+1)
X_1	Actuator Length (L)	500	650	800
X_2	Heater Thickness (HT)	0.1	0.3	0.5
X_3	Actuator Width (W)	50	75	100
X_4	Silicon Thickness (SiT)	1.0	1.25	1.5
X_5	Heater Length (HL)	200	250	300
X_6	Metal Thickness (MT)	0.5	1.0	1.5

The CCD design matrix with 53 simulation runs and the corresponding output responses (deflection angle, input power, and temperature rise in the mirror) obtained through FEM simulations are shown in Supplementary Table S2. The non-significant design factors SpL, SpW, and MIRT are kept at the levels that gave a maximum value of FOM in the Plackett–Burman screening design. Polynomial equations for the responses Y_1 = deflection angle, Y_2 = Input power, Y_3 = micromirror temperature rise obtained, using the second-order model and calculating the value of the β coefficients (using Equations (7)–(9)), are given as:

$$\begin{aligned}
Y_1 &= 23.59 + 9.86X_1 + 7.39X_2 - 4.66X_3 - 5.23X_4 - 4.62X_5 - 1.02X_6 + 2.84X_1X_2 - \\
&\quad 1.24X_1X_3 - 1.81X_1X_4 - 1.27X_1X_5 + 0.048X_1X_6 - 1.59X_2X_3 - 1.03X_2X_4 - \\
&\quad 0.88X_2X_5 - 0.24X_2X_6 + 0.97X_3X_4 + 0.67X_3X_5 + 0.29X_3X_6 + 0.89X_4X_5 + 2.10X_4X_5 + \\
&\quad 0.088X_5X_6 + 0.37X_1^2 - 1.52X_2^2 + 5.71X_3^2 + 0.28X_4^2 + 0.60X_5^2 - 1.94X_6^2.
\end{aligned} \tag{10}$$

$$\begin{aligned}
Y_2 &= 7.48 + 5.18X_2 - 0.20X_3 - 1.65X_5 - 0.13X_2X_3 - 1.10X_2X_5 + 0.080X_3X_5 \\
&\quad -2.953e^{-4}X_1^2 + 2.047e^{-4}X_2^2 - 0.075X_3^2 - 2.953e^{-4}X_4^2 + 0.38X_5^2 - 2.953e^{-4}X_6^2
\end{aligned} \tag{11}$$

$$\begin{aligned}
Y_3 &= 27.06 - 0.17X_2 + 13.65X_2 - 7.50X_3 - 1.33X_4 - 5.64X_5 - 3.70X_6 + 0.1X_1X_2 - \\
&\quad 0.015X_1X_3 + 9.906e - 003X_1X_4 + 0.026X_1X_5 + 0.16X_1X_6 - 3.35X_2X_3 - 0.70X_2X_4 - \\
&\quad 1.64X_2X_5 - 2.15X_2X_6 + 0.53X_3X_4 + 1.21X_3X_5 + 1.32X_3X_6 + 0.24X_4X_5 + 0.77X_4X_6 + \\
&\quad 0.38X_5X_6 - 0.33X_1^2 - 0.030X_2^2 + 7.68X_3^2 - 0.23X_4^2 + 0.62X_5^2 + 0.74X_6^2.
\end{aligned} \tag{12}$$

3.4. Regression Analysis for the CCD Design Matrix

To verify that the developed response surface models for the three output responses, given in Equations (10)–(12), provide an adequate approximation of the true behavior of the micromirror, a regression analysis is carried out. The first step in the regression analysis is to ensure that none of the least squares assumptions, i.e., errors in the model, are normally distributed with mean zero and variance σ^2, are violated [41]. These assumptions are verified by analyzing the residuals from the least squares fit defined by $e_i = y_i - \hat{y}_i$, $i = 1, 2, ..., n$, where y_i is the vector of actual observed values and $\hat{y}_i$

is the vector of the fitted values. The relationship of the vector of fitted values to the vector of actual observed values is given as:

$$\hat{y} = Xb = X\left(X^T X\right)^{-1} X^T y. \tag{13}$$

To verify the assumptions for the response surface models obtained from the CCD design matrix, standardized residuals-based normal probability plots and fitted values versus standardized residual plots are obtained, which verified the basic assumptions for the regression analysis. The test for the regression analysis is based on the following null hypothesis [41]:

$$\left. \begin{array}{l} H_0 : \beta_1 = \beta_2 = \ldots = \beta_k = 0 \\ H_1 : \beta_j \neq 0 \text{ for at least one } j \end{array} \right\}. \tag{14}$$

The regression analysis test is used to verify if a statistical relationship exists between the output response and at least one of the design factors. If the null hypothesis is rejected, then it means that at least one of the design factors significantly affects the output response surface model. The test for the hypothesis is carried out using the following F-test ratio:

$$F_0 = \frac{MS_R}{MS_E}, \tag{15}$$

where MS_R and MS_E are the regression and residual mean square, respectively. The null hypothesis is rejected if the calculated $F_0 > F_{\alpha,k,n-k-1}$ (or p-value $< \alpha$), where α is the level of significance, k is the number of design factors, and n is the number of observations [41]. A regression analysis for the three output responses is performed and F-test ratios and corresponding p-values for each output response model are obtained. Supplementary Table S3 shows the regression analysis results.

3.5. Interaction Analysis of the Design Factors for Angular Deflection

The regression analysis results, given in Supplementary Table S3, for the output response angular deflection (Y_1) show that the design factor interactions $X_1 X_2$, $X_1 X_3$, $X_1 X_4$, $X_1 X_5$, $X_2 X_3$, $X_2 X_4$, $X_2 X_5$, $X_3 X_4$, $X_3 X_5$, $X_4 X_5$, and $X_4 X_6$ are significant with p-value < 0.05. These interactions for the design factors can be further analyzed with respect to the output response using 3D surface and contour plots. In this paper, the design factor interaction $X_1 X_2$ with the highest F-value of 83.2 is further investigated using 3D surface and contour plots as an example. Figure 6 shows that the deflection angle increases with an increase in both the electrothermal actuator length L (X_1) and heater thickness HT (X_2). The deflection angle is more sensitive to the change in L as compared to HT. When the actuator length is 500 μm, the change in the deflection angle is less affected by the change in the heater thickness as compared to when the actuator length is at 800 μm.

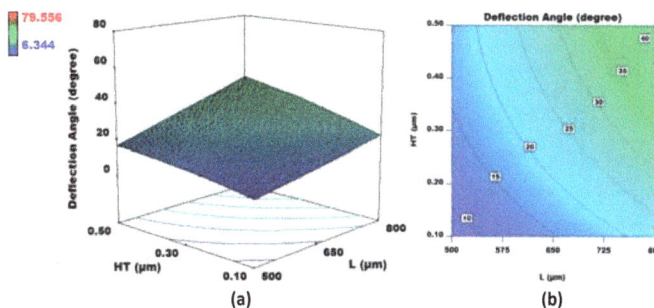

Figure 6. (**a**) 3D surface plot and (**b**) contour plot for L and HT for fixed values of W, HL, SiT, and MT. The output response considered is angular deflection.

3.6. Interaction Analysis of the Design Factors for Input Power

In Supplementary Table S3, the *p*-values < 0.05 for X_2X_3, X_2X_5, and X_3X_5 show that there is a significant relationship between the W and HT, HT and HL, and HL and W for the output response input power (Y_2). The interaction between HT and HL has the highest *F*-value among the three significant design factor interactions. Figure 7 shows the 3D surface and contour plots for HT and HL, with all other design factors set at their medium levels. The plots show that input power decreases with the increase in the heater length and a decrease in the heater thickness. The change in the output power is more sensitive to the change in the heater thickness as compared to the heater length. Moreover, the change in the input power value is less than the change in the HL when HT = 0.1 μm, as compared to when HT = 0.5 μm.

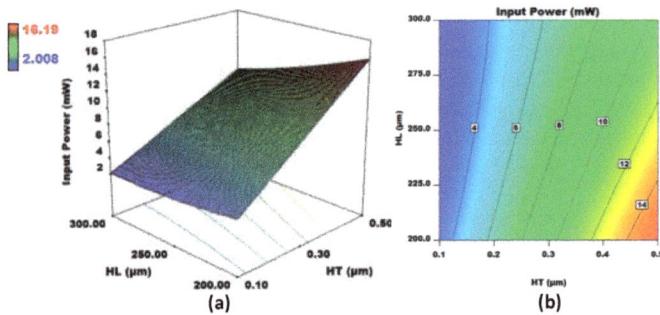

Figure 7. (a) 3D surface plot and (b) contour plot for HL and HT for fixed values of W, L, SiT, and MT. The output response considered is input power.

3.7. Interaction Analysis of the Design Factors for Temperature Rise

The design factor interactions X_2X_3, X_2X_5, X_2X_6, X_3X_5, X_3X_6, and X_4X_6 are observed to be the significant interactions for the temperature rise in the micromirror plate from the ambient. Figure 8 shows the 3D surface and contour plots for the interaction between the electrothermal actuator width W (X_3) and heater thickness HT (X_2). The interaction between W and HT has the highest *F*-value of 117.4 for the temperature rise as compared to all other significant design factor interactions. The interaction plots between W and HT are highly non-linear, with a noticeable curvature. The plots show that the temperature rise in the micromirror plate is less sensitive to the change in the electrothermal actuator width W as compared to the heater thickness HT for a fixed value of all other design factors. However, for W ≤ 75 μm, the temperature rise in the micromirror is more influenced by the change in the HT as compared to when 75 μm ≤ W ≤ 100 μm.

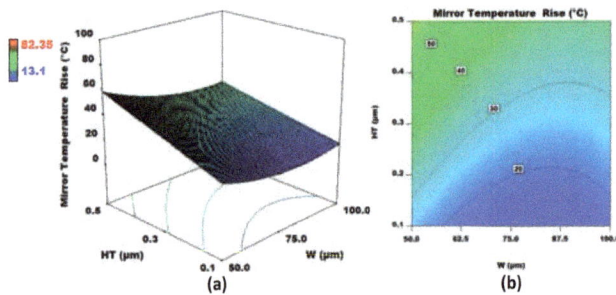

Figure 8. (a) 3D surface plot and (b) contour plot for W and HT for fixed values of L, HL, SiT, and MT. The output response considered is micromirror central plate temperature rise from the ambient.

3.8. Multi-Response Optimization

The design optimization of electrothermally actuated micromirror, considered in this paper, involves optimization of three output responses simultaneously for a given set of design factors. For the multi-response optimization of the micromirror, an optimization objective function is initially defined, which is given as:

Maximize deflection angle
Minimize input power :
Temperature rise in the micromirror plate $\leq 30\,^{\circ}\text{C}$
such that :

$$
\begin{aligned}
500\ \mu\text{m} &\leq L \leq 800\ \mu\text{m} \\
0.1\ \mu\text{m} &\leq HT \leq 0.5\ \mu\text{m} \\
50\ \mu\text{m} &\leq W \leq 100\ \mu\text{m} \\
1\ \mu\text{m} &\leq SiT \leq 1.5\ \mu\text{m} \\
200\ \mu\text{m} &\leq HL \leq 300\ \mu\text{m} \\
0.5\ \mu\text{m} &\leq MT \leq 1.5\ \mu\text{m}.
\end{aligned}
\tag{16}
$$

One of the traditional methods for multi-response optimization is overlaid contour plots. This method is mainly useful when there are two or three design factors, since in higher dimensions it loses its efficiency [45]. The most practical method to optimize multiple output responses was proposed by Derringer and Suich and is based on the desirability function approach [46]. The desirability function allows us to find suitable values for the design factors to simultaneously reach an optimal solution for all the output responses considering the desired objective function. Initially, an individual desirability function $d_i(y_i)$ for each response y_i is calculated using the developed response surface models and defined objective function. If the output response y_i is at the goal defined in the objective function then $d_i = 1$, and if it is outside an acceptable region then $d_i = 0$. If the objective function is to maximize the output response then $d_i(y_i)$ is given as [45]:

$$
d_i(y_i(x)) = \begin{cases} 0 & \text{if } y_i(x) < L_i \\ \left(\frac{y_i(x)-L_i}{U_i-L_i}\right)^{r_1} & \text{if } L_i \leq y_i(x) \leq U_i \\ 1 & \text{if } y_i(x) > U_i \end{cases}.
\tag{17}
$$

If the objective function is to minimize the output response then $d_i(y_i)$ is given as:

$$
d_i(y_i(x)) = \begin{cases} 1 & \text{if } y_i(x) < L_i \\ \left(\frac{U_i-y_i(x)}{U_i-L_i}\right)^{r_2} & \text{if } L_i \leq y_i(x) \leq U_i \\ 0 & \text{if } y_i(x) > U_i \end{cases}.
\tag{18}
$$

When the output response is to be optimized with respect to some target T then $d_i(y_i)$ is given as:

$$
d_i(y_i(x)) = \begin{cases} 0 & \text{if } y_i(x) < L_i \\ \left(\frac{y_i(x)-L_i}{T_i-L_i}\right)^{r_1} & \text{if } L_i \leq y_i(x) \leq T_i \\ 1 & \text{if } y_i(x) = T_i \\ \left(\frac{y_i(x)-U_i}{T_i-U_i}\right)^{r_2} & \text{if } T_i \leq y_i(x) \leq U_i \\ 0 & \text{if } y_i(x) > U_i \end{cases},
\tag{19}
$$

where U_i is the upper value of the desired output response range, L_i is the lower value of the output response range, and T_i is the target value for the output response. The parameters r_1 and r_2 define the importance of the output response to be close to the desired value. The optimum solution can be obtained by combining the individual desirability functions, given as:

$$D(d_1[y_1(x)], d_2[y_2(x)], \cdots, d_n[y_n(x)]) = \left(\prod_{i=1}^{n} d_i[y_n(x)] \right)^{\frac{1}{n}}. \tag{20}$$

The desirability values for the multiple output responses can be maximized by using the well-known Nelder–Mead downhill simplex algorithm-based heuristic search algorithm [47]. This search algorithm finds a local optimum solution to a problem with multiple variables and iteratively narrows down to a design factor value that maximizes the desirability of the objective function. Figure 9 shows the optimal solution for three output responses and the corresponding values of the design factors with respect to the optimization objective function defined in Equation (14). The values of the simultaneously optimized deflection angle, input power, and micromirror temperature rise from the ambient are 43.9°, 2.85 mW, and 29.3 °C, respectively. The value of the combined overall desirability function is 0.72.

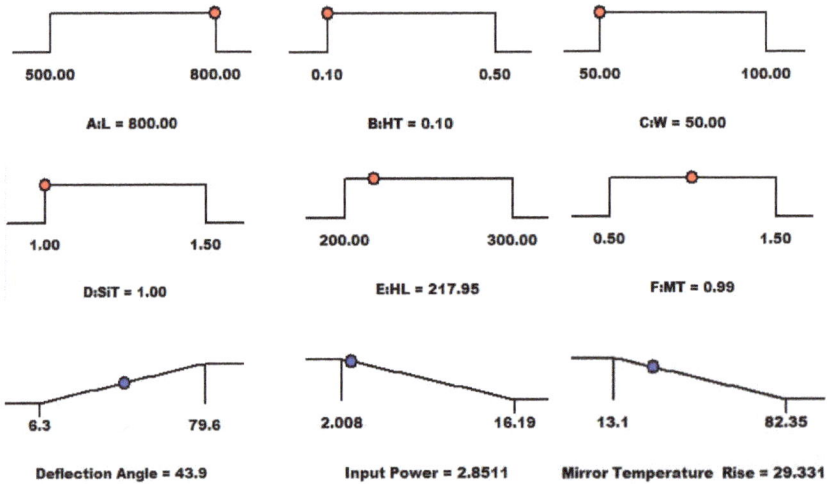

Figure 9. The optimal values of the design factors obtained using the Nelder–Mead downhill simplex-based heuristic search algorithm.

The regression analysis results for all three output responses, given in Supplementary Table S3, show that the interaction between the design factors W and HT is a significant interaction with p-value < 0.05. This gives an opportunity to further explore the effect of these two factors on the individual output responses and overall desirability by keeping all other design factors at the optimized values predicted by the desirability function approach. Figure 10 shows the contour plots for the effect of W and HT on the deflection angle, input power, and micromirror temperature rise from the ambient. For the deflection angle, the interaction between the electrothermal actuator width W and heater thickness HT is highly non-linear, with large contours. The deflection angle increases considerably with the increase in the HT up to 0.5 µm. The change in the deflection angle is more sensitive to the change in HT than W. For the input power, the contour plot shows linear behavior. The input power changes sharply with the increase in the HT, while the effect of the change in W is negligible. The contour plots for the micromirror temperature rise from the ambient show that the output response is very sensitive to the increase in HT as compared to W. However, for $W \geq 62.5$ µm, change in W has an almost negligible effect on the temperature rise.

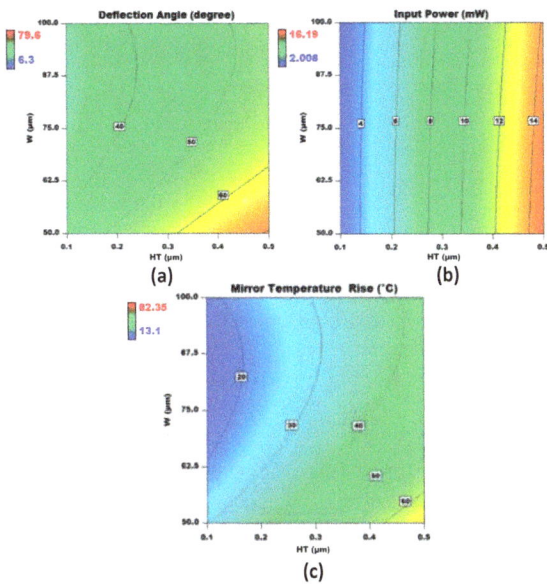

Figure 10. Contour plots for (**a**) deflection angle; (**b**) input power; and (**c**) micromirror plate temperature rise from the ambient. The contour plots show interaction between W and HT for the final optimized design, while all other design factors are kept at the optimized values predicted by the direct search algorithm.

Figure 11 shows the overlay contour plots for the three output responses with respect to HT and W. The individual contour plots are obtained using response surface models by using the predicted design factor values. The yellow region shows all the feasible solutions that lie within the defined objective function of Equation (16).

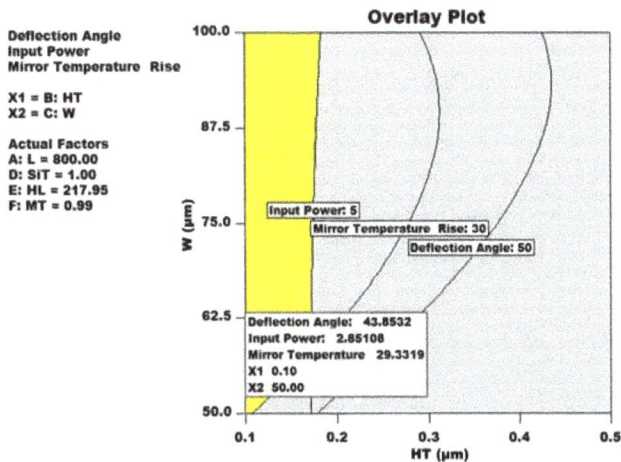

Figure 11. Overlay contour plots for all three output responses. The yellow region shows the acceptable solutions according to the objective function defined in Equation (16). However, the best solution with highest desirability is highlighted in the text box.

3.9. Verification of the Multi-Response Optimization

The results obtained using desirability function-based multi-response optimization are verified using FEM simulations. The micromirror model is developed with the optimal values of the design factors, shown in Figure 9. Figure 12a shows the vertical deflection in the micromirror plate with a maximum upward and downward deflection of 18.4 µm and 263.4 µm respectively, in opposite corners of the micromirror plate. An absolute deflection angle of 43.4° is calculated for the micromirror plate deflection using trigonometric functions. Figure 12b shows the temperature distribution in the micromirror. The temperature rise in the micromirror plate from the ambient is 27.4 °C at an actuation voltage of 0.8 V. The calculated input power for the optimized design for an actuation voltage of 0.8 V is 2.94 mW. These actual values of the deflection angle, micromirror plate temperature rise from ambient, and input power, obtained using FEM simulations, lie within the 95% confidence interval of the predicted output responses, thus verifying the accuracy of the developed response surface models-based multi-response optimization.

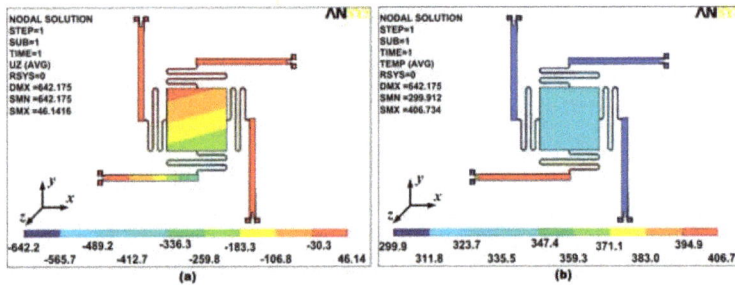

Figure 12. (a) Vertical deflection plot for the final optimized design; (b) temperature distribution within the final optimized design. The temperature rise in the micromirror plate is much less than the bimorph actuator temperature.

The DOE-based optimization technique discussed in this paper is implemented only for the scanning micromirror used in the sample arm of an OCT system. However, for the use of a micromirror in the reference arm of an OCT, or for in-depth tissue scanning, an out-of-plane displacement (piston mode operation) is desired. To analyze the possibility of using the optimized tilting micromirror in the reference arm of an OCT system, all four bimorph actuators are simultaneously actuated in the FEM simulations. A vertical displacement of nearly 500 µm is obtained in the micromirror plate, as shown in Figure 13. The temperature rise in the micromirror plate from the ambient is observed to be 110 °C for the piston mode, which is much higher than the case of angular deflection.

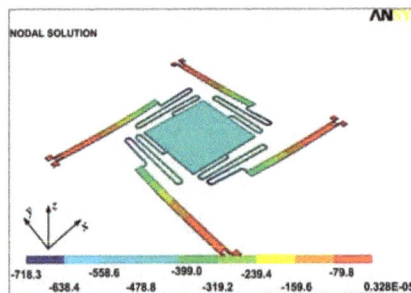

Figure 13. Vertical deflection in the final optimized design for piston mode operation. All four electrothermal actuators are simultaneously actuated.

4. Discussion

The output responses considered in the optimization of the electrothermal micromirror using Response Surface Models (RSM) are deflection angle, input power, and micromirror temperature rise from the ambient. However, for an OCT system, the scanning speed and overall device area of the electrothermal micromirror are also important output responses. The scanning speed is dependent on the thermal time response of the bimorph actuator, while the overall device is decided by the electrothermal actuator and micromirror reflecting plate dimensions. These output responses may also be considered in the multi-response optimization of the electrothermal micromirror, following the optimization steps discussed earlier. For the final optimized design presented in this paper, the overall micromirror size is 1.65 mm × 1.65 mm, with a mirror plate size of 500 μm × 500 μm. For a fixed mirror plate size, the device area may be minimized by decreasing the electrothermal actuator length. For example, with an actuator length of 400 μm, the overall device area is 0.825 mm × 0.825 mm and the corresponding RSM-based deflection angle, input power, and micromirror temperature rise from the ambient are 18.3°, 3.3 mW and 34.8 °C, respectively. These values are obtained by modifying the objective function given in Equation (16) with L = 400 μm and repeating all the optimization steps.

The performance of metal thin-films-based MEMS devices is significantly affected by the time-dependent accumulation of plastic strain under the influence of applied stress and temperature (the creep effect). This leads to a change in both the static and dynamic response of the device [48]. In MEMS, the creep effect was initially reported in electrostatically actuated digital micromirror devices (DMD), fabricated using a series of aluminum metal depositions [49]. The micromirrors were tested at a temperature of 65 °C and a change in the static response was observed. In the electrothermally actuated micromirrors, the temperature in the actuators is generally higher than room temperature. Mu et al. [50] have reported a temperature of 90 °C in a Al/Si bimorph thermal actuator with a relatively low temperature of 30 °C in the mirror plate. Bauer et al. [51] have implemented Au/Si bimorph actuators in a scanning micromirror design for an initial offset of the angular vertical comb-drives. The temperature in the bimorph actuators is simulated to be 380 °C, while a very high temperature of 770 °C in the comb-drive supporting beam is reported. For an electrothermally actuated micromirror, temperature values of nearly 160 °C and 65 °C in the Al/W bimorph actuator and mirror plate, respectively, are reported in [14]. For the optimized micromirror design, presented in this paper, the temperature distribution in the micromirror (Figure 12) shows a temperature increase of nearly 106 °C and 27.4 °C from the ambient in the bimorph actuator and micromirror plate, respectively. These high temperature values in the metal thin-film-based electrothermal micromirrors may initiate the creep phenomenon and affect long-term reliability. The other mechanical reliability issue related to micromirrors is the formation of residual stress during the microfabrication process, which may result in curling in both the thermal actuators and central plate, like all other MEMS devices [52]. Similarly, the application areas of the micromirrors involve cyclic loading, which may deteriorate the flexural stiffness if the device is operated for a large number of cycles (the fatigue phenomenon) [53]. Generally, during the design and optimization phase of the MEMS in general, and electrothermal micromirrors in particular, these reliability issues are not considered. A reliable design of an electrothermally actuated micromirror requires a robust design optimization considering both thermal and mechanical reliability issues. A DOE-based robust multi-response-based design optimization using the dual response surface method [54] or mixed array design [36] can be a good alternative to the conventional design optimization methodologies for electrothermal micromirrors.

5. Conclusions

A DOE-based multi-response design optimization methodology for MEMS devices is presented. The device considered for optimization is an electrothermally actuated micromirror for OCT system applications. Three output responses, deflection angle, input power, and micromirror temperature rise from the ambient, are considered for simultaneous optimization and the respective response surface models are developed through regression analysis. A desired objective function is defined

Micromachines **2017**, *8*, 107

and the optimal values of the design factors and corresponding output responses, satisfying the objective function, are obtained using combined desirability functions and a Nelder–Mead downhill simplex-based heuristic search algorithm. A deflection angle of 44° with an input power of 2.85 mW and a temperature rise of 29.3 °C from ambient, with an overall device size of 1.65 mm × 1.65 mm, is predicted by the developed RSM model at the optimal level of the design factors. These predicted values are verified using FEM-based confirmation simulation. The proposed multi-response design optimization methodology can be implemented for the optimization and detailed interaction analysis of different design factors of MEMS devices at the design level.

Acknowledgments: The authors would like to thank National University of Sciences and Technology for providing a high-performance computing facility for the micromirror simulations.

Author Contributions: Muhammad Mubasher Saleem and Umar Farooq carried out the design and FEM simulations of the micromirror. Umer Izhar worked on the selection of the important design factors, considering microfabrication process constraints. Umar Shahbaz Khan worked on the DOE-based design optimization along with Muhammad Mubasher Saleem. Muhammad Mubasher Saleem supervised the work and wrote the paper. All authors read and approved the final manuscript.

Conflicts of Interest: The authors declare no conflict of interest.

References

1. Tsai, C.H.; Tsai, J.C. MEMS optical switches and interconnects. *Displays* **2015**, *37*, 33–40. [CrossRef]
2. Solgaard, O.; Godil, A.A.; Howe, R.T.; Lee, L.P.; Peter, Y.A.; Zappe, H. Optical MEMS: From micromirrors to complex systems. *J. Microelectromech. Syst.* **2014**, *23*, 517–538. [CrossRef]
3. Chong, J.; He, S.; Mrad, R.B. Development of a vector display system based on a surface-micromachined micromirror. *IEEE Trans. Ind. Electron.* **2012**, *59*, 4863–4870. [CrossRef]
4. Silva, G.; Carpignano, F.; Guerinoni, F.; Costantini, S.; De Fazio, M.; Merlo, S. Optical detection of the electromechanical response of MEMS micromirrors designed for scanning picoprojectors. *IEEE J. Sel. Top. Quantum Electron.* **2015**, *21*, 147–156. [CrossRef]
5. Proll, K.P.; Nivet, J.M.; Körner, K.; Tiziani, H.J. Microscopic three-dimensional topometry with ferroelectric liquid-crystal-on-silicon displays. *Appl. Opt.* **2003**, *42*, 1773–1778. [CrossRef] [PubMed]
6. Yalcinkaya, A.D.; Ergeneman, O.; Urey, H. Polymer magnetic scanners for bar code applications. *Sens. Actuators A Phys.* **2007**, *135*, 236–243. [CrossRef]
7. Jung, W.; Zhang, J.; Wang, L.; Wilder-Smith, P.; Chen, Z.; McCormick, D.T.; Tien, N.C. Three-dimensional optical coherence tomography employing a 2-axis microelectromechanical scanning mirror. *IEEE J. Sel. Top. Quantum Electron.* **2005**, *11*, 806–810. [CrossRef]
8. Pengwang, E.; Rabenorosoa, K.; Rakotondrabe, M.; Andreff, N. Scanning micromirror platform based on MEMS technology for medical application. *Micromachines* **2016**, *7*, 24. [CrossRef]
9. Wang, K.; Nirmalathas, A.; Lim, C.; Skafidas, E.; Alameh, K. High-speed reconfigurable free-space card-to-card optical interconnects. *IEEE Photonics J.* **2012**, *4*, 1407–1419. [CrossRef]
10. Lin, L.Y.; Keeler, E.G. Progress of MEMS scanning micromirrors for optical bio-imaging. *Micromachines* **2015**, *11*, 1675–1689. [CrossRef]
11. Jain, A.; Qu, H.; Todd, S.; Xie, H. A thermal bimorph micromirror with large bi-directional and vertical actuation. *Sens. Actuators A Phys.* **2005**, *122*, 9–15. [CrossRef]
12. Jia, K.; Pal, S.; Xie, H. An electrothermal tip-tilt-piston micromirror based on folded dual S-shaped bimorphs. *J. Microelectromech. Syst.* **2009**, *18*, 1004–1015.
13. Adams, D.C.; Wang, Y.; Hariri, L.P.; Suter, M.J. Advances in endoscopic optical coherence tomography catheter designs. *IEEE J. Sel. Top. Quantum Electron.* **2016**, *22*, 210–221. [CrossRef]
14. Pal, S.; Xie, H. A curved multimorph based electrothermal micromirror with large scan range and low drive voltage. *Sens. Actuators A Phys.* **2011**, *170*, 156–163. [CrossRef]
15. Ogando, K.; La Forgia, N.; Zarate, J.J.; Pastoriza, H. Design and characterization of a fully compliant out-of-plane thermal actuator. *Sens. Actuators A Phys.* **2012**, *183*, 95–100. [CrossRef]
16. Kim, Y.S.; Dagalakis, N.G.; Gupta, S.K. Creating large out-of-plane displacement electrothermal motion stage by incorporating beams with step features. *J. Micromech. Microeng.* **2013**, *23*, 055008. [CrossRef]

17. Kim, D.H.; Park, Y.C.; Park, S. Design and fabrication of twisting-type thermal actuation mechanism for micromirrors. *Sens. Actuators A Phys.* **2010**, *159*, 79–87. [CrossRef]
18. Zhang, X.; Zhou, L.; Xie, H. A fast, large-stroke electrothermal MEMS mirror based on Cu/W bimorph. *Micromachines* **2015**, *12*, 1876–1889. [CrossRef]
19. Tsai, C.H.; Tsai, C.W.; Chang, H.T.; Liu, S.H.; Tsai, J.C. Electrothermally-actuated micromirrors with bimorph actuators—Bending-type and torsion-type. *Sensors* **2015**, *15*, 14745–14756. [CrossRef] [PubMed]
20. Bühler, J.; Funk, J.; Paul, O.; Steiner, F.P.; Baltes, H. Thermally actuated CMOS micromirrors. *Sens. Actuators A Phys.* **1995**, *47*, 572–575. [CrossRef]
21. Ataka, M.; Omodaka, A.; Takeshima, N.; Fujita, H. Fabrication and operation of polyimide bimorph actuators for a ciliary motion system. *J. Microelectromech. Syst.* **1993**, *4*, 146–150. [CrossRef]
22. Yang, J.P.; Deng, X.C.; Chong, T.C. An electro-thermal bimorph-based microactuator for precise track-positioning of optical disk drives. *J. Micromech. Microeng.* **2005**, *15*, 958–965. [CrossRef]
23. Jain, A.; Xie, H. A single-crystal silicon micromirror for large bi-directional 2D scanning applications. *Sens. Actuators A Phys.* **2006**, *130*, 454–460. [CrossRef]
24. Singh, J.; Teo, J.H.S.; Xu, Y.; Premachandran, C.S.; Chen, N.; Kotlanka, R.; Olivo, M.; Sheppard, C.J.R. A two axes scanning SOI MEMS micromirror for endoscopic bioimaging. *J. Micromech. Microeng.* **2007**, *18*, 025001. [CrossRef]
25. Xie, H.; Pan, Y.; Fedder, G.K. Endoscopic optical coherence tomographic imaging with a CMOS-MEMS micromirror. *Sens. Actuators A Phys.* **2003**, *103*, 237–241. [CrossRef]
26. Izhar, U.; Izhar, A.B.; Tatic-Lucic, S. A multi-axis electrothermal micromirror for a miniaturized OCT system. *Sens. Actuators A Phys.* **2011**, *167*, 152–161. [CrossRef]
27. Liu, L.; Pal, S.; Xie, H. MEMS mirrors based on a curved concentric electrothermal actuator. *Sens. Actuators A Phys.* **2012**, *188*, 349–358. [CrossRef]
28. Zhang, X.; Zhang, R.; Koppal, S.; Butler, L.; Cheng, X.; Xie, H. MEMS mirrors submerged in liquid for wide-angle scanning. In Proceedings of the 18th IEEE International Conference on Solid-State Sensors, Actuators and Microsystems (TRANSDUCERS), Anchorage, AK, USA, 21–25 June 2015; pp. 847–850.
29. Samuelson, S.R.; Xie, H. A large piston displacement MEMS mirror with electrothermal ladder actuator arrays for ultra-low tilt applications. *J. Microelectromech. Syst.* **2014**, *23*, 39–49. [CrossRef]
30. Jang, K.W.; Yang, S.P.; Baek, S.H.; Lee, M.S.; Park, H.C.; Seo, Y.H.; Kim, M.H.; Jeong, K.H. Electrothermal MEMS parallel plate rotation for single-imager stereoscopic endoscopes. *Opt. Express* **2016**, *24*, 9667–9672. [CrossRef] [PubMed]
31. Duan, C.; Tanguy, Q.; Pozzi, A.; Xie, H. Optical coherence tomography endoscopic probe based on a tilted MEMS mirror. *Biomed. Opt. Express* **2016**, *7*, 3345–3354. [CrossRef] [PubMed]
32. Saleem, M.M.; Somà, A. Design optimization of RF-MEMS switch considering thermally induced residual stress and process uncertainties. *Microelectron. Reliab.* **2015**, *55*, 2284–2298. [CrossRef]
33. Gimzewski, J.K.; Gerber, C.; Meyer, E.; Schlittler, R.R. Observation of a chemical reaction using a micromechanical sensor. *Chem. Phys. Lett.* **1994**, *217*, 589–594. [CrossRef]
34. Plackett, R.L.; Burman, J.P. The design of optimum multifactorial experiments. *Biometrika* **1946**, *33*, 305–325. [CrossRef]
35. Liew, L.A.; Tuantranont, A.; Bright, V.M. Modeling of thermal actuation in a bulk-micromachined CMOS micromirror. *Microelectron. J.* **2000**, *31*, 791–801. [CrossRef]
36. Song, R.C.; Wang, G.L.; Wang, J.B.; Shen, Y.Y.; Lv, C. Design and Simulation of High Efficient MEMS Electrothermal Actuator. *Key Eng. Mater.* **2013**, *562*, 504–508. [CrossRef]
37. Pan, C.S.; Hsu, W. An electro-thermally and laterally driven polysilicon microactuator. *J. Micromech. Microeng.* **1997**, *7*, 7–13. [CrossRef]
38. Lerch, P.; Slimane, C.K.; Romanowicz, B.; Renaud, P. Modelization and characterization of asymmetrical thermal micro-actuators. *J. Micromech. Microeng.* **1996**, *7*, 7–13. [CrossRef]
39. Todd, S.T.; Jain, A.; Qu, H.; Xie, H. A multi-degree-of-freedom micromirror utilizing inverted-series-connected bimorph actuators. *J. Opt. A Pure Appl. Opt.* **2006**, *8*, 352–359. [CrossRef]
40. Ali, A.; Azim, R.A.; Khan, U.S.; Syed, A.A.; Izhar, U. Design, simulation and optimization of electrothermal micro actuator. *Appl. Mech. Mater.* **2012**, *229*, 1939–1943. [CrossRef]
41. Montgomery, D.C. *Design and Analysis of Experiments*, 5th ed.; Wiley: New York, NY, USA, 1987.
42. Anderson, T.W.; Darling, D.A. A test of goodness of fit. *J. Am. Stat. Assoc.* **1954**, *49*, 765–769. [CrossRef]

43. Saleem, M.M.; Somá, A. Design of experiments based factorial design and response surface methodology for MEMS optimization. *Microsyst. Technol.* **2015**, *21*, 263–276. [CrossRef]

44. Daniel, C. Use of Half-Normal Plots in Interpreting Factorial Two-Level Experiments. *Technometrics* **1959**, *1*, 311. [CrossRef]

45. Myers, R.H.; Montgomery, D.C.; Anderson-Cook, C.M. *Response Surface Methodology: Process and Product Optimization Using Designed Experiments*; John Wiley & Sons: Hoboken, NJ, USA, 2016.

46. Derringer, G.; Suich, R. Simultaneous optimization of several response variables. *J. Q. Technol.* **1980**, *12*, 214–219.

47. Nelder, J.A.; Mead, R. A simplex method for function minimization. *Comput. J.* **1965**, *7*, 308–313. [CrossRef]

48. Somà, A.; Saleem, M.M.; De Pasquale, G. Effect of creep in RF MEMS static and dynamic behavior. *Microsyst. Technol.* **2016**, *22*, 1067–1078. [CrossRef]

49. Douglass, M.R. Lifetime estimates and unique failure mechanisms of the digital micromirror device (DMD). In Proceedings of the IEEE 36th Annual International Reliability Physics Symposium, Reno, NV, USA, 31 March–2 April 1998; pp. 9–16.

50. Mu, X.; Sun, W.; Feng, H.; Yu, A.; Chen, K.W.S.; Fu, C.Y.; Olivo, M. MEMS micromirror integrated endoscopic probe for optical coherence tomography bioimaging. *Sens. Actuators A Phys.* **2011**, *168*, 202–212. [CrossRef]

51. Bauer, R.; Li, L.; Uttamchandani, D. Dynamic properties of angular vertical comb-drive scanning micromirrors with electrothermally controlled variable offset. *J. Microelectromech. Syst.* **2014**, *23*, 999–1008. [CrossRef]

52. Somà, A.; Saleem, M.M. Modeling and experimental verification of thermally induced residual stress in RF-MEMS. *J. Micromech. Microeng.* **2015**, *25*, 055007. [CrossRef]

53. Somà, A.; De Pasquale, G. MEMS mechanical fatigue: Experimental results on gold microbeams. *J. Microelectromech. Syst.* **2009**, *18*, 828–835. [CrossRef]

54. Myers, R.H.; Carter, W.H. Response surface techniques for dual response systems. *Technometrics* **1973**, *15*, 301–317. [CrossRef]

![micromachines logo] *micromachines*

MDPI

Article

In-Plane Optical Beam Collimation Using a Three-Dimensional Curved MEMS Mirror [†]

Yasser M. Sabry [1,2,]*, Diaa Khalil [1,2], Bassam Saadany [2] and Tarik Bourouina [2,3]

[1]   Department of Electronics and Communication Engineering, Faculty of Engineering, Ain-Shams University, 1 Elsarayat St., Abbassia 11517, Egypt; diaa_khalil@eng.asu.edu.eg
[2]   Si-Ware Systems, 3 Khaled Ibn El-Waleed Street, Heliopolis, Cairo 11361, Egypt; bassam.saadany@si-ware.com (B.S.); tarik.bourouina@esiee.fr (T.B.)
[3]   Paris-Est, Laboratoire ESYCOM, ESIEE Paris, Cité Descartes, F-93162 Noisy-le-Grand CEDEX, France
*   Correspondence: yasser.sabry@eng.asu.edu.eg; Tel.: +20-100-183-4833
[†]   This paper is an extended version of our paper published in the MOEMS and Miniaturized Systems XIII conference, 3–6 February 2014, San Francisco, CA, USA.

Academic Editor: Huikai Xie
Received: 25 March 2017; Accepted: 18 April 2017; Published: 25 April 2017

Abstract: The collimation of free-space light propagating in-plane with respect to the substrate is an important performance factor in optical microelectromechanical systems (MEMS). This is usually carried out by integrating micro lenses into the system, which increases the cost of fabrication/assembly in addition to limiting the wavelength working range of the system imposed by the dispersion characteristic of the lenses. In this work we demonstrate optical fiber light collimation using a silicon micromachined three-dimensional curved mirror. Sensitivity to micromachining and fiber alignment tolerance is shown to be low enough by restricting the ratio between the mirror focal length and the optical beam Rayleigh range below 5. The three-dimensional curvature of the mirror is designed to be astigmatic and controlled by a process combining deep, reactive ion etching and isotropic etching of silicon. The effect of the micromachining surface roughness on the collimated beam profile is investigated using a Fourier optics approach for different values of root-mean-squared (RMS) roughness and correlation length. The isotropic etching step of the structure is characterized and optimized for the optical-grade surface requirement. The experimental optical results show a beam-waist ratio of about 4.25 and a corresponding 12-dB improvement in diffraction loss, in good agreement with theory. This type of micromirror can be monolithically integrated into lensless microoptoelectromechanical systems (MOEMS), improving their performance in many different applications.

Keywords: curved micromirrors; three-dimensional fabrication; Gaussian beams; surface roughness

---

1. Introduction

Optical microelectromechanical systems (MEMS) technology has attracted great attention over the past couple of decades because of its reduced size, light weight and low cost [1]. There are two main architectures in the optical MEMS, namely in-plane architecture [2], where the light propagates from one component to another parallel to the substrate, and out-of-plane architecture [3], where the light hits the optical component either perpendicular to or with inclination on the substrate. For many applications, such as in optical telecommunication [1], optical coherence tomography [4] and on-chip sensing [5], the light source is connected to the optical MEMS device through a single-mode optical fiber, where the optical beam output from the fiber behaves as a Gaussian beam [2]. In this case, the propagation can be associated with beam size expansion before detection, leading to optical losses. This is even more serious in optical MEMS due to the size limit of the optical components [6,7]. Several solutions were introduced as shown in Figure 1 to overcome this challenge, such as the use of a lensed fiber [4] or an external lens integrated into the system in the form of a graded-index (GRIN) lens or a ball lens [6–11]. The lensed fiber solution is costly due to the piece-by-piece process of lens formation on the fibers, in addition to the reliability issue to possible fiber tip breakage. The external lens solution suffers from the cost and complexity of the assembly. In addition, refractive lenses have chromatic aberration and require anti-reflective coating to eliminate the reflection. The aberration and the coating both lead to limited working wavelength range.

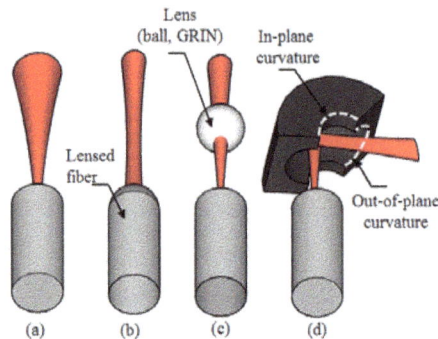

Figure 1. Optical beam propagation for the different architectures of (**a**) a cleaved fiber; (**b**) an integrated lens fiber; (**c**) an external lens; and (**d**) the proposed solution in this work.

Reflecting curved micromirrors are achromatic and can provide much a wider spectral response, but they need special attention during fabrication to obtain the curved surface. The common non-planar micro surfaces fabrication techniques are gray-tone mask [12], excimer laser [13], Reactive Ion Etching (RIE) lag effect [14] and photo resist (PR) reflow [15,16]. On one hand, non-silicon curved micromirrors were reported using a polymer dispensing and sucking technique [17], residual internal material stress resulting from deposition of gold on polysilicon for the purpose of light focusing [18], trapping of gas bubbles during melting a stack of small borosilicate glass tubes under a nitrogen atmosphere and further grinding and polishing for atomic studies [19] and deep silicon etching and PR reflow targeting optical interconnects [20]. On the other hand, silicon curved micromirrors fabricated on the wafer top surface were reported using isotropic chemical etching for the sake of optical detection of single atom [21], selective polishing method on the top of MEMS tunable vertical-cavity surface-emitting laser [22] and ion beam irradiation and electrochemical etching for atomic studies as well as optical interconnects [23]. The principal axis of the aforementioned micromirrors is oriented out-of-plane with the respect to the wafer substrate. This rendered the micromirror incompatible with silicon micro-optical bench systems where the light is propagating in-plane with respect to the substrate. Three-dimensional (3-D) micro optical bench systems requiring further assembly or mounting steps after fabrication were introduced in the literature. The most common is to use rotational assembly to create micro-optical subsystems that process free-space beams travelling above the surface of the chip [24]. Non-monolithically integrated mechanical mounting systems for connecting and aligning optical components on a micro optical bench (OB) were also reported [25,26]. This is, however, not compatible with the monolithic integration efforts for the microoptoelectromechanical systems (MOEMS) [27–30].

In this work, we demonstrate optical beam collimation and propagation loss reduction using a monolithic micromachined curved mirror with an in-plane principal axis, which is compatible with silicon micro-optical bench technology [31]. The paper is organized in the following manner. In Section 2, a theoretical study is carried out for the possibility of Gaussian beam collimation using curved surfaces exhibiting microscale focal lengths, i.e., not so large compared with the incident

Gaussian beam Rayleigh range. The design of astigmatic micromirror curvatures is related to incidence angle of the incident Gaussian beam in order to generate a stigmatic collimated beam. The effect of the surface roughness of the micromirror is analyzed in Section 3. Then, the fabrication steps of the micromirror and the resulting structure are presented in Section 4. Finally, optical measurements are presented and discussed in Section 5 using the introduced curved micromirror for single-mode fiber output collimation and propagation loss reduction where the fiber axis lies in-plane with the substrate.

2. Theoretical Analysis of Optical Beam Collimation

Consider the incidence of a Gaussian beam on a curved micromirror as shown in Figure 2. The parameters of the reflected beam are related to the incident beam by:

$$G_c = \frac{w_{out}}{w_{in}} = \frac{1}{\sqrt{(1 - d_{in}/f)^2 + z_0^2/f^2}} \tag{1}$$

$$\frac{d_{out}}{f} = \frac{z_0^2/f^2 - d_{in}/f(1 - d_{in}/f)}{(1 - d_{in}/f)^2 + z_0^2/f^2} \tag{2}$$

where w_{in} and w_{out} are the min waist radii for the incident and reflected beams, respectively, d_{in} and d_{out} are the distances between the beam waist location and the mirror surface at the point of incidence for the incident and reflected beams, respectively, f is the focal length of the mirror and z_0 is the Rayleigh range of the incident beam. The beam-waist ratio w_{out}/w_{in} is denoted by G_c and represents the collimation gain. The dependences of the beam-waist ratio and the ratio d_{out}/f on the ratio d_{in}/f for different ratios of f/z_0 are shown in Figure 3. The beam-waist ratio has a maximum value occurring when the input distance and the focal length are equal. The maximum beam-waist ratio is given by:

$$G_c = \frac{f}{z_0} \tag{3}$$

The variation of the beam-waist ratio around $d_{in}/f = 1$ is symmetric. The variation of the ratio d_{out}/f possess odd symmetry around the point $(d_{in}/f = 1, d_{out}/f = 1)$. The output beam waist location doesn't change with the input beam Rayleigh range when the input beam waist is located at the focus of the mirror. Negative values of d_{out}/f are obtained when $d_{in}/f < 1$, which means the output beam waist is located virtually behind the mirror and the beam is diverging after reflection. The opposite case occurs when $d_{in}/f > 1$ and the beam is reflected in a converging state. The output beam waist may have its waist located just at the mirror surface for a single value of d_{in}/f when $z_0/f = 2$ and for two value of d_{in}/f when $z_0/f = 0.5$; one time for a very small value of d_{in}/f and the second time for a d_{in}/f that is slightly smaller than unity.

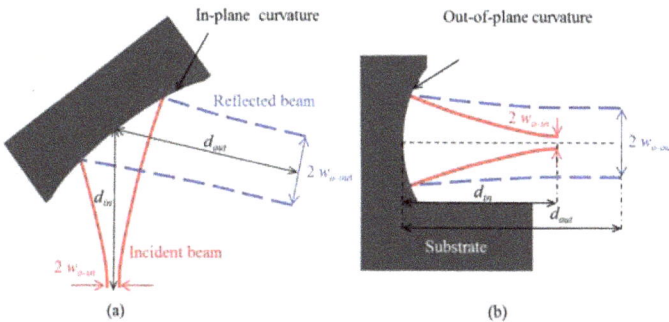

Figure 2. Three-dimensional curved micromirror used in beam collimation. (a) In-plane cross section; (b) out-of-plane cross section.

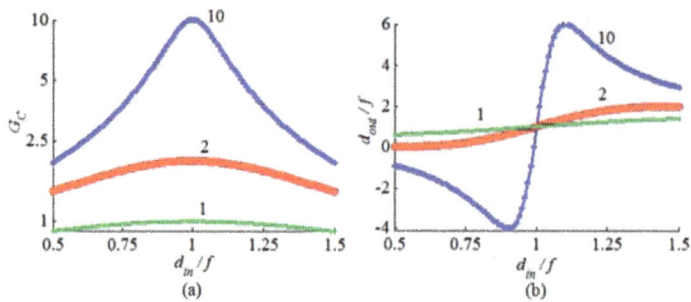

Figure 3. Dependence of the beam-waist ratio G_c and the ratio d_{out}/f on the ratio d_{in}/f in (**a**) and (**b**), respectively, for different f/z_0 ratios.

The microfabrication process tolerance may result in a variation of the curved micromirror radius of curvature, which affects the obtainable beam's beam-waist ratio. The impact depends on the gain sensitivity to the curved surface focal length. The corresponding change is determined by:

$$\Delta G_c = \frac{\Delta f}{f} \frac{d_{in}/f(1-d_{in}/f)+(z_0/f)^2}{[(1-d_{in}/f)^2+(z_0/f)^2]^{3/2}}$$
$$= \frac{\Delta f}{f}\left(\frac{z_0}{f}\right)^{-1}, d_{in}/f \approx 1 \tag{4}$$

For a given percentage change in the focal length, the gain sensitivity becomes very high when the ratio z_0/f is very small. As depicted in Figure 4a, the beam-waist ratio is less sensitive to the focal length variation when z_0/f is larger than 0.2. The output beam waist location is, however, very sensitive to the variations as shown in Figure 4b. In the case of $z_0/f > 0.2$, the fabrication tolerance impact on the output beam waist location can be compensated by active axial alignment.

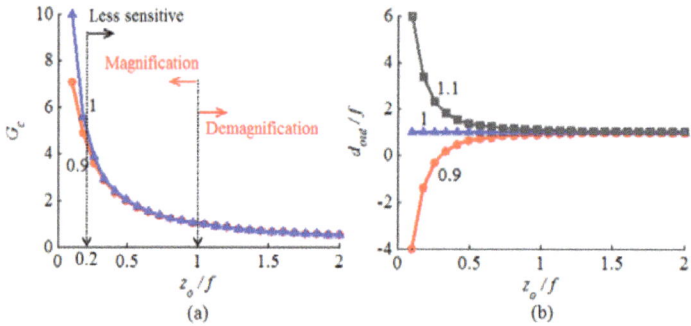

Figure 4. Dependence of the beam-waist ratio Gc and the ratio d_{in}/f on the ratio z_0/f in (**a**) and (**b**), respectively, for different d_{in}/f ratios.

The inclined incidence of the beam on the mirror in a tangential plane, while being normal to the sagittal plane, has the effect of splitting the focal length as well as the input ratio d_{in}/f of the mirror each into two different values:

$$f_{ip} = 0.5R_{ip}\cos(\theta_{inc}) \tag{5}$$

$$f_{op} = 0.5R_{op}/\cos(\theta_{inc}) \tag{6}$$

$$\left(\frac{d_{in}}{f}\right)_{ip} = \frac{2d_{in}}{R_{ip}\cos(\theta_{inc})} \tag{7}$$

$$\left(\frac{d_{in}}{f}\right)_{op} = \frac{2d_{in}\cos(\theta_{inc})}{R_{op}} \tag{8}$$

where the subscripts "*ip*" and "*op*" are used for the in-plane and out-of-plane directions, respectively, and *R* is the radius of curvature of the mirror in the indicated plane. The inclined incidence has the effect of effectively increasing the out-of-plane focal length of the curved surface while at the same time decreasing its in-plane focal length, and therefore, a stigmatic inclined curved surface should have non-equal radii of curvature in the two orthogonal planes. As will be shown in the fabrication section, the out-of-plane plane radius of curvature can be limited to 100 μm. Fortunately, increasing angle of incidence compensates for this limit. For instance, focal length matching occurs at incidence angles $\theta_{inc} = 0°$, 45° and 60° for $R_{op}/R_{ip} = 1$, 0.5 and 0.25 respectively. Away from the stigmatic beam generation angle, the reflected beam exhibits an elliptical cross section as well different beam waist location in the two orthogonal planes. This can be of particular interest in beam shaping/matching applications.

3. Effect of Surface Roughness

The effect of the surface roughness expected from the micromachining of the 3-D curved surface on the collimated optical beam profile is investigated in this section. For this purpose, the overall phase transformation of the 3-D mirror is divided into the phase curvature responsible for the collimation of the beam, which is already considered in Section 2, and a random phase due to the surface roughness. The phase curvature corresponding to the curvature of the mirror surface is given by:

$$\phi = \frac{2\pi}{\lambda}\frac{x^2 + y^2}{2f} \tag{9}$$

where *f* is the equivalent focal length of the mirror. The random phase is given by:

$$\phi_n = \frac{2\pi}{\lambda}z_n \tag{10}$$

where $z_n = f(x,y)$ is the random height variation of the surface due to the surface roughness. In our analysis, $f(x,y)$ is assumed a random rough surface that has a Gaussian height distribution function and Gaussian autocovariance functions (in both *x*- and *y*-direction). The surface is assumed to have an RMS height σ_{rms} and assumed to be isotropic in the sense that the correlation length L_c in the *x*- and *y*-direction are assumed equal.

The simulation procedure is carried out using the Fourier optics approach as follows [32]. The field at the mirror surface, denoted by $E_{in}(x,y,d_{in})$, is multiplied by the phase transformation function and the new output field is denoted by $E_o(x,y,d_{in})$:

$$E_o(x,y,d_{in}) = E_i(x,y,d_{in})\exp(-j\phi_n - j\phi) \tag{11}$$

A fast Fourier transform (FFT) is applied to get this output field in the spatial frequency domain:

$$G_o(f_x,f_y,d_{in}) = FFT\{E_o(x,y,d_{in})\} \tag{12}$$

The field is propagated a distance d_{out} by phase multiplication in the spatial frequency domain:

$$G_o(f_x,f_y,d_{out}) = G_o(f_x,f_y,d_{in})\exp(-jk_z d_{out}) \tag{13}$$

where k_z is the axial components of the wave vector. Finally, the output field profile after propagating the distance d_{out} is obtained by inverse Fourier transform:

$$E_o(x,y,d_{out}) = IFFT\{G_o(f_x,f_y,d_{out})\} \tag{14}$$

A simulation study was carried out to analyze the effect of the surface roughness of the etched mirror on the collimated beam. The effect is evaluated by calculating the coupling efficiency (overlap integral) between the resulting and the ideal beam. The radius of curvature of the mirror in the in-plane direction is assumed 300 µm, while the out-of-plane radius of curvature is 150 µm, similar to the value obtained practically as will be shown in the next section. The incident beam has a minimum waist radius of 5 µm, a wavelength of 1550 nm and located at the focal plane of the mirror in a 45-degree incidence orientation. The RMS roughness σ_{rms} is assumed in the range of 0 to $\lambda/10$. Three values of the correlation were assumed: 5λ, 10λ and 20λ.

The resulting coupling efficiency is depicted in Figure 5a. Since the roughness generation is a stochastic process, the simulation was repeated 20 times for each point and the average was taken. The coupling efficiency decreases with the increase of the RMS value of the roughness, as expected. It reaches about 75% for the case of $L_c = 10\lambda$ and $\sigma_{rms} = \lambda/10$. If we would like to maintain at least 95% of the coupling efficiency, then σ_{rms} should be less than 0.04λ, 0.06λ and 0.1λ for $L_c = 5\lambda$, 10λ and $L_c = 20\lambda$, respectively. Example resulting beam profiles for the case of $\sigma_{rms} = 0.1\lambda$ are shown in Figure 5b. The x-axis is normalized to the waist of the resulting beam profile in case of The loss in efficiency is resulting from the asymmetry in the beam profile in addition to the widening of the profiles out of the $\pm 4w$ limit due to the surface roughness.

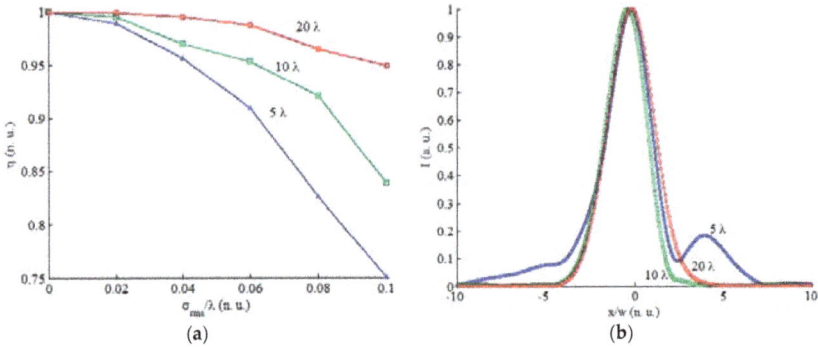

Figure 5. Effect of surface roughness on coupling efficiency and collimated beam profile. (a) Coupling efficiency versus RMS roughness normalized to the wavelength at different roughness correlation lengths; (b) collimated beam profile versus the transverse dimension normalized to the ideal beam waist radius.

4. Silicon Micromirror Fabrication

The optical axis of the target 3-D curved micromirror lies in-plane with respect to the wafer substrate to collimate the optical beam generated from single-mode optical fibers located horizontally on the wafer substrate or any other light source integrated in the system. It enables the use of the fiber-mirror configuration to replace the lensed fiber as previously shown in Figure 1d. The fabrication of the micromirror was carried out into six main steps [33]. First the definition of the in-plane profile of the micromirror with a 300-µm radius of curvature was performed using standard photolithography (see top view in Figure 6a). The lithographic process ends with a patterned SiO$_2$ mask layer for the following etching. Second, anisotropic deep reactive ion etching of the silicon was carried out, ending with a deeply etched cylindrical surface as shown in Figure 6b [34]. By this anisotropic etching step, the central line of the out-of-plane curvature (principal axis) is defined. The axis depth with respect to the wafer top surface was chosen to be large enough that optical fiber can be inserted and aligned with micromirror. Then, side wall protection was carried out using a Teflon-like layer to prevent sidewall etching from top and ensure the following isotropic etching starts at the mirror principal axis as shown in Figure 6c. The protection step was followed by a long isotropic etching step using SF$_6$ plasma to achieve the desired out-of-plane profile of the micromirror as shown in Figure 6d, in a similar way to that used to fabricate micro fluidic channels reported in [35]. The out-of-plane radius of curvature of the micromirror surface is about 150 µm. Achieving larger radii of curvatures requires deeper etching, which may result in a fragile wafer. The protective layer was removed in the fifth step as shown in Figure 6e using a high-temperature oxygen plasma ashing process. As will be shown below, the resulting surface roughness was about 22 nm RMS. Therefore, the surface was post-processed for optical quality requirement by smoothing and Aluminum metallization as shown in Figure 6f. Top and tilted views of the fabricated micromirror after step 5 are shown in Figure 7a,b, recorded using a scanning electron microscope (SEM).

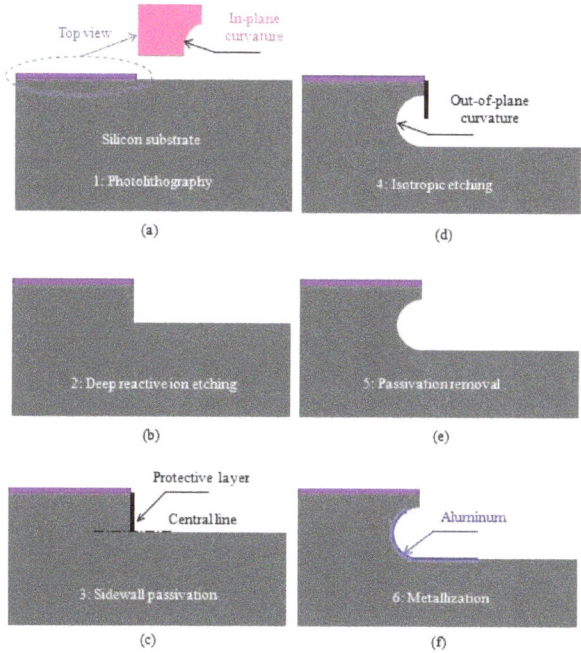

Figure 6. The fabrication steps of the collimating 3-D curved micromirror. (**a**) Photolithography, (**b**) deep reactive ion etching, (**c**) sidewall passivation, (**d**) isotropic etching, (**e**) passivation removal, and (**f**) metallization.

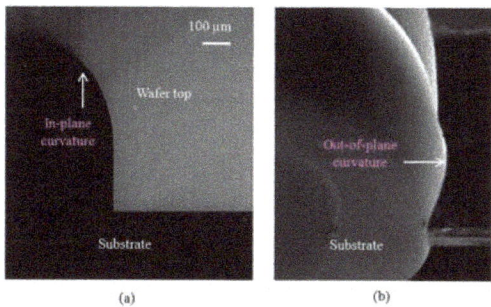

Figure 7. Scanning electron microscope (SEM) images of the fabricated micromirror. (**a**) Top view where the in-plane curvature is emphasized; (**b**) tilted view where the out-of-plane curvature is emphasized.

More than one effect was encountered regarding the isotropic etching of silicon using SF_6. First, a significant dependence of the etch rate on the trench width was observed, as shown in Figure 8. The etch rate is normalized with respect to the etch rate of the largest trench width. The data markers represent the measured normalized data while the solid line is a logarithmic fitting. This kind of logarithmic behavior is well-known for a diffusion-limited etching process [14]. The etch rate for a 10 μm trench width is about one fifth the rate for a 500 μm trench width. The second observation is the correlation between the mask opening width and the isotropic etching roughness as shown in Figure 9. The smaller the mask opening is, the higher the roughness. Considerable roughness can be observed in the smallest opening by inspecting the SEM images with the naked eye, while the roughness in the largest opening is much less, but still observable. The atomic force microscope (AFM) was used in order to get a quantitative measurement for the roughness of the largest opening. The top and 3-D tilted views of the surface topology, obtained using the AFM on an area of 10 μm by 10 μm, are shown in Figure 10a,b respectively. The measured roughness has a peak of 319 nm, an average of 16 nm and an RMS 22 nm. The lag effect as well as the surface roughness of the isotropic etching roughness can be interpreted knowing that a diffusion process governs the transport of the etching radicals from the plasma, where it is created, to the substrate, where chemical etching occurs. Due to this diffusion process, a lower amount of etchants is received in thinner trenches. This directly relates to the lag effect. At the same time, when the amount of etchants is not enough, a rough surface results from the etching process because the surface is not overwhelmed by the etchants.

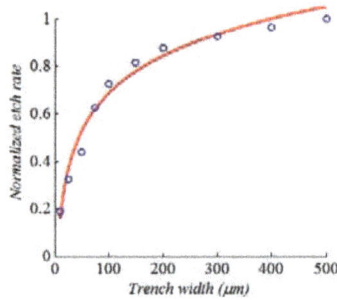

Figure 8. Normalized isotropic etching rate versus the etched trench opening width while. The trench length is 300 μm. The measured data (in markers) is fitted to a logarithmic function (in line).

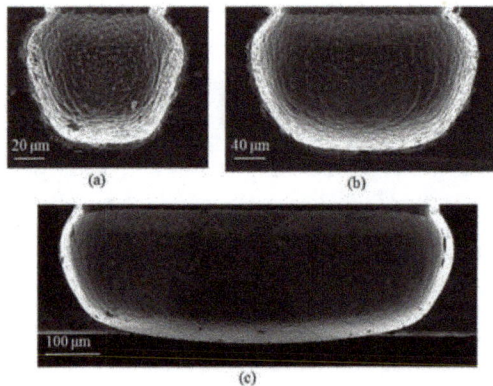

Figure 9. SEM images showing the roughness of the isotropically-etched trenches. The opening widths are 75 μm in (**a**); 150 μm in (**b**) and 500 μm in (**c**).

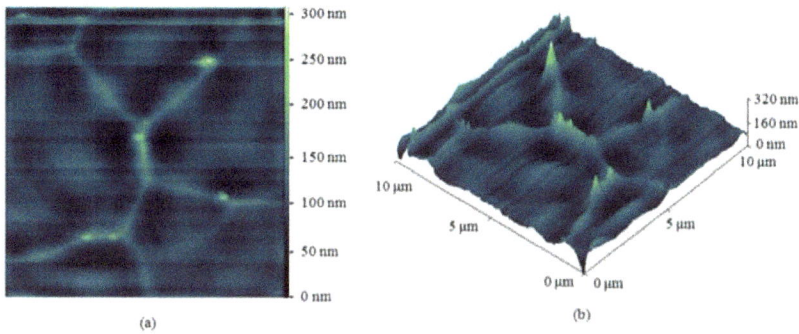

Figure 10. The isotropic etching roughness measured in a 500 μm trench using the atomic force microscope (AFM). A top view of the measured surface is shown in (**a**) while a tilted 3-D view is shown in (**b**).

5. Measurement Results and Discussion

In this section, the manufactured 3-D curved micromirror is utilized for collimating the output beam of single-mode fibers and propagation loss reduction thereof. Consider the arrangement shown in Figure 11.

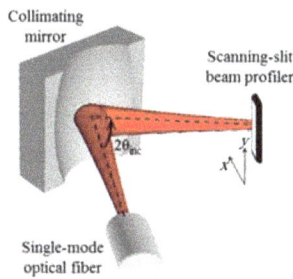

Figure 11. Measurement setup of the reflected beam from the fabricated mirror.

A single-mode optical fiber is inserted on the silicon substrate such that its optical axis is parallel to the silicon substrate and tilted with respect to the mirror principal axis. For the sake of optical spot characterization, the reflected beam is captured in the far field on a scanning-slit beam profiler. The observed beam ellipticity, defined by the ratio of the spot size in the in-plane direction to the out-of-plane direction, is adjusted to be close to unity (about 1.05) by letting the incidence angle of the beam on the mirror be about 45°. The axial distance between the optical fiber and the mirror was adjusted such that the fiber tip is located at the micromirror focal plane by minimizing the observed output beam diameter at the far field. The collimated output beam spot diameter was measured at different locations away from the micromirror and compared to the measurements of the optical fiber output beam without using the micromirror.

In the case of using a standard single-mode fiber with a core radius of 4.5 μm fed from 1550 nm laser source, a reduction in the divergence angle of the beam by a factor of 2 was achieved by the micromirror. The output beam has a minimum waist radius of about 10 μm, which is a typical value for many optical MEMS applications. A typical captured beam profile at one location d is shown in Figure 12a. The profile was fitted to a Gaussian profile with average root mean square errors smaller than 1% and 1.5% in the x- and y- directions respectively as shown in Figure 12b,c. This is an indication of the good performance offered by the fabricated micromirror, using the presented method, in terms of its phase front transformation function. This experiment was repeated with a special single-mode fiber with a core radius of 2 μm working at a 675 nm wavelength. The special fiber is positioned at the same location used for the standard one because of the constant focal length of the mirror independent of the wavelength value. A reduction in the divergence angle of the beam by a factor of 4.25 was achieved. The resulting output beam has a minimum waist radius of about 10 μm as well. This visible beam will be used hereinafter for evaluating the propagation loss reduction offered by the micromirror.

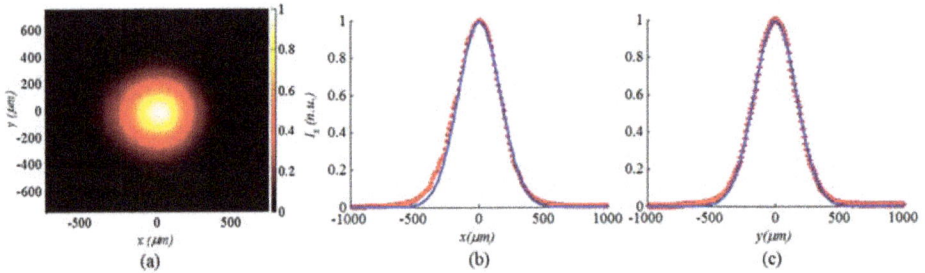

Figure 12. Measured spot profile: (**a**) contour plot; (**b**) in-plane beam profile (markers) fitted to a Gaussian profile (line), and (**c**) out-of-plane beam profile (markers) fitted to a Gaussian profile (line).

The collimation of the beam by the micromirror was also evaluated by measuring the detected power in free space with a limited-aperture detector as shown in Figure 13.

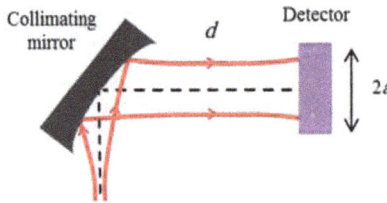

Figure 13. Measurement setup of the power on a detector with aperture radius *a*.

Theoretically, the transmitted power in terms of the system aperture radius *a* and the beam spot radius at the detector is given by [36]:

$$P = 1 - \exp\left(-2\frac{a^2}{w^2}\right) \tag{15}$$

The power collected by a detector with 3.5 mm aperture radius is shown in Figure 14a. The power was measured at different distance *d* in the far field away from the beam waist. The measurements were carried out one time for the collimated beam by the micromirror, denoted by P_c, and another time for beam originally emitted by the single-mode fiber, denoted by P_0. The experimental data are depicted using markers while the theoretical data are depicted using lines. The power is normalized with respect to the initially maximum power. The measured power clearly starts to fall when the beam diameter starts to exceed the detector aperture as given by Equation (9). The micromirror significantly reduces the propagation losses with respect to the original fiber output. The detected power from the micromirror has a slower roll-off and drops to half its maximum value 25 cm far from the micromirror compared to less than 8 cm without using the micromirror. The ratio between the two detected powers is depicted in Figure 14b, where the improvement reaches about 11–12 dB. Indeed, in the far field, the ratio between the detected powers is given by:

$$G_p = \frac{P_c}{P_o} = \frac{1 - \exp\left(-2\frac{a^2}{\theta_{div-c}^2 d^2}\right)}{1 - \exp\left(-2\frac{a^2}{\theta_{div-o}^2 d^2}\right)} \tag{16}$$

where the beam spot radius in the far field was replaced by $wd/z_o = d/\theta_{div}$. The maximum improvement is achieved when the spot radius becomes much larger than the detector aperture. In this case, Taylor expansion of the exponential terms can be applied to second order and Equation (16) becomes:

$$G_{p-\max} = \frac{\theta_{div-c}^2}{\theta_{div-o}^2} = G_c^2 \tag{17}$$

The maximum power gain due to the usage of the collimating mirror is given by the beam-waist ratio squared. For the fabricated micromirror and using the single-mode fiber at 675 nm, the power gain is $G_p = (4.25)^2 = 18$ that is about 12.5 dB, in good agreement with the measured data. This value is independent of the specific sizes of the beam spot and the detector aperture, as long as significant truncation loss is encountered.

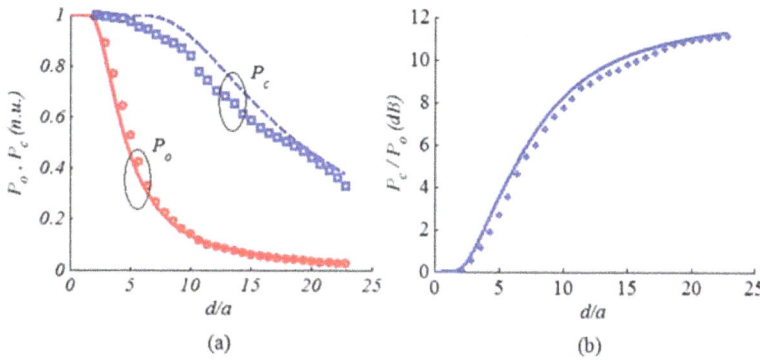

Figure 14. (**a**) The normalized power collected by the detector; (**b**) diffraction loss reduction in dB using the collimating micromirror. The measured data is given in markers when the theoretical one is given in lines.

6. Conclusions

Optical beam collimation was analyzed and successfully carried out using a micro-reflector with a three-dimensional curved surface. The surface was etched in silicon by a technique combining deep reactive ion etching and isotropic etching technologies. The produced surface is astigmatic with an out-of-plane radius of curvature that is about half the in-plane radius of curvature. Having the incident beam in-plane and inclined by 45° with respect to the principal axis, the reflected beam is kept stigmatic with about a 4.25-fold reduction in the beam expansion angle in free space and about 12-dB reduction in propagation losses. The fibre–mirror configuration may serve as a potential replacement for the lensed fibers widely used in the MOEMS system. This replacement has the advantage of producing monolithically integrated systems with a wider-band spectral response.

Author Contributions: Y.M.S. carried out the theoretical analysis, studied the roughness effect using optical simulations, fabricated the structure, performed the experiments and wrote the manuscript. D.A. participated in the design of the optical setup and the method of roughness simulation. B.S. participated in the idea and design of the fabrication steps. T.B. participated in the idea, revised the paper and supervised the overall work.

Conflicts of Interest: The authors declare no conflict of interest.

Abbreviations

The following abbreviations are used in this manuscript:

RMS	Root mean square
MOEMS	Micro-opto-electro-mechanical systems
RIE	Reactive ion etching
PR	Photo resist
3-D	Three-dimensional
OB	Optical bench
SEM	Scanning electron microscope
AFM	Atomic force microscope

References

1. Wu, M.; Solgaard, O.; Ford, J. Optical MEMS for Lightwave Communication. *J. Lightwave Technol.* **2006**, *24*, 4433–4454. [CrossRef]
2. Sabry, Y.M.; Khalil, D.; Bourouina, T. Monolithic silicon micromachined free-space optical interferometers on chip. *Laser Photonics Rev.* **2015**, *9*, 1–24. [CrossRef]
3. Holmstrom, S.T.; Baran, U.; Urey, H. MEMS laser scanners: A review. *J. Microelectromech. Syst.* **2014**, *23*, 259–275. [CrossRef]
4. Elhady, A.; Sabry, Y.M.; Yehia, M.; Khalil, D. Dual-fiber OCT measurements. In Proceedings of the SPIE 8934, Optical Coherence Tomography and Coherence Domain Optical Methods in Biomedicine XVIII, San Francisco, CA, USA, 1 February 2014; p. 89343J.
5. Gaber, N.; Sabry, Y.M.; Marty, F.; Bourouina, T. Optofluidic Fabry-Pérot micro-cavities comprising curved surfaces for homogeneous liquid refractometry—Design, simulation, and experimental performance assessment. *Micromachines* **2016**, *7*, 62. [CrossRef]
6. Sabry, Y.M.; Omran, H.; Khalil, D. Intrinsic improvement of diffraction-limited resolution in optical MEMS fourier-transform spectrometers. In Proceedings of the 31st Radio Science Conference (NRSC), Cairo, Egypt, 28–30 April 2014; pp. 326–333.
7. Erfan, M.; Sabry, Y.M; Sakr, M.; Mortada, B.; Medhat, M.; Khalil, D. On-Chip Micro–Electro–Mechanical System Fourier Transform Infrared (MEMS FT-IR) Spectrometer-Based Gas Sensing. *Appl. Spectrosc.* **2016**, *70*, 897–904. [CrossRef] [PubMed]
8. Syms, R. Principles of free-space optical microelectromechanical systems. *Proc. Inst. Mech. Eng. Part C J. Mech. Eng. Sci.* **2008**, *222*, 1–18. [CrossRef]
9. Nussbaum, P.; Völkel, R.; Herzig, H.P.; Eisner, M.; Haselback, S. Design, fabrication and testing of microlens arrays for sensors and microsystem. *Pure Appl. Opt.* **1997**, *6*, 617–636. [CrossRef]
10. Zickar, M.; Noell, W.; Marxer, C.; de Rooij, N. MEMS compatible micro-GRIN lenses for fiber to chip coupling of light. *Opt. Express* **2006**, *14*, 4237–4249. [CrossRef] [PubMed]
11. Lee, S.; Huang, L.; Kim, C.; Wu, M.C. Free-space fiber-optic switches based on MEMS vertical torsion mirrors. *J. Lightwave Technol.* **1999**, *17*, 7–13.
12. Oppliger, Y.; Sixt, P.; Stauffer, J.M.; Mayor, J.M.; Regnault, P.; Voirin, G. One-step 3D shaping using a gray-tone mask for optical and microelectronic applications. *Microelectron. Eng.* **1994**, *23*, 449–454. [CrossRef]
13. Mihailov, S.; Lazare, S. Fabrication of refractive microlens arrays by excimer laser ablation of amorphous Teflon. *Appl. Opt.* **1993**, *32*, 6211–6218. [CrossRef] [PubMed]
14. Bourouina, T.; Masuzawa, T.; Fujita, H. The MEMSNAS Process: Microloading Effect for Micromachining 3D structures of Nearly All Shapes. *IEEE/ASME J. Microelectromech. Syst.* **2004**, *13*, 190–199. [CrossRef]
15. Kogel, B.; Debernardi, P.; Westbergh, P.; Gustavsson, J.S.; Haglund, A.; Haglund, E.; Bengtsson, J.; Larsson, A. Integrated MEMS-Tunable VCSELs Using a Self-Aligned Reflow Process. *IEEE J. Quantum Electron.* **2012**, *48*, 144–152. [CrossRef]
16. Park, S.; Jeon, H.; Sung, Y.; Yeom, G. Refractive Sapphire Microlenses Fabricated by Chlorine-Based Inductively Coupled Plasma Etching. *Appl. Opt.* **2001**, *40*, 3698–3702. [CrossRef] [PubMed]
17. Hsiao, S.; Lee, C.; Fang, W. The implementation of concave micro optical devices using a polymer dispensing technique. *J. Micromech. Microeng.* **2008**, *18*, 085009. [CrossRef]

18. Burns, D.M.; Bright, V.M. Micro-electro-mechanical focusing mirrors. In Proceedings of the Eleventh Annual International Workshop on Micro Electro Mechanical Systems, Heidelberg, Germany, 25–29 January 1998; pp. 460–465.

19. Cui, G.; Hannigan, J.M.; Loeckenhoff, R.; Matinaga, F.M.; Raymer, M.G. A hemispherical, high-solid-angle optical micro-cavity for cavity-QED studies. *Opt. Express* **2006**, *14*, 2289–2299. [CrossRef] [PubMed]

20. Lee, M.; Choi, C.; Lim, K.; Beom-Hoan, O.; Lee, S.G.; Park, S.; Lee, E. Novel fabrication of a curved micro-mirror for optical interconnection. *Microelectron. Eng.* **2006**, *83*, 1343–1346. [CrossRef]

21. Moktadir, Z.; Koukharenka, E.; Kraft, M.; Bagnall, D.M.; Powell, H.; Jones, M.; Hinds, E.A. Etching techniques for realizing optical micro-cavity atom traps on silicon. *J. Micromech. Microeng.* **2004**, *14*, S82–S85. [CrossRef]

22. Kanbara, N.; Tezuka, S.-I.; Watanabe, T. MEMS tunable VCSEL with concave mirror using the selective polishing method. In Proceedings of the IEEE/LEOS International Conference on Optical MEMS and Their Applications Conference, Big Sky, MT,USA, 21–24 August 2006; pp. 9–10.

23. Ow, Y.; Breese, M.B.H.; Azimi, S. Fabrication of concave silicon micro-mirrors. *Opt. Express* **2010**, *14*, 14511–14518. [CrossRef] [PubMed]

24. Wu, M.C.; Lin, L.-Y.; Lee, S.-S.; Pister, K.S.J. Micromachined freespace integrated micro-optics. *Sens. Actuators A Phys.* **1995**, *50*, 27–134. [CrossRef]

25. Flanders, D.; Whitney, P.; Masghati, M.; Racz, L. Mounting and Alignment Structures for Optical Components. U.S. Patent US006625372B1, 23 September 2003.

26. Do, K.; Sell, J.; Kono, R.; Jones, D.; Torro, R.; Kozlovsky, W.; Gupta, B.; Pace, D.; Chapman, W.; Sawyer, K. Micro Optical Bench for Mounting Precision Alighned Optics, Optical Assembly and Method of Mounting Optics. U.S. Patent US006775076B2, 10 August 2004.

27. Sabry, Y.M.; Khalil, D.; Saadany, B.; Bourouina, T. Integrated Wide-Angle Scanner Based on Translating a Curved Mirror of Acylindrical Shape. *Opt. Express* **2013**, *21*, 13906. [CrossRef] [PubMed]

28. Sabry, Y.M.; Bourouina, T.E.; Saadany, B.A.; Khalil, D.A.M. Integrated Monolithic Optical Bench Containing 3-D Curved Optical Elements and Methods of its Fabrication. U.S. Patent Application 20130100424, 25 April 2013.

29. Omran, H.; Medhat, M.; Mortada, B.; Saadany, B.; Khalil, D. Fully integrated Mach-Zhender MEMS interferometer with two complementary outputs. *IEEE J. Quantum Electron.* **2012**, *48*, 244–251. [CrossRef]

30. Omran, H.; Sabry, Y.M.; Sadek, M.; Hassan, K.; Shalaby, Y.M.; Khalil, D. Deeply-etched optical MEMS tunable filter for swept laser source applications. *Photonic Technol. Lett.* **2013**, *26*, 37–39. [CrossRef]

31. Sabry, Y.M.; Khalil, D.; Saadany, B.; Bourouina, T. Three-dimensional collimation of in-plane-propagating light using silicon micromachined mirror. In Proceedings of the SPIE 8977, MOEMS and Miniaturized Systems XIII, San Francisco, CA, USA, 1 February 2014; p. 89770J.

32. Goodman, J.W. *Introduction to Fourier Optics*; Roberts and Company Publishers: Englewood, CO, USA, 2005.

33. Sabry, Y.M.; Saadany, B.; Khalil, D.; Bourouina, T. Silicon micromirrors with three-dimensional curvature enabling lens-less efficient coupling of free-space light. *Light Sci. Appl.* **2013**, *2*, e94. [CrossRef]

34. Marty, F.; Rousseau, L.; Saadany, B.; Mercier, B.; Français, O.; Mita, Y.; Bourouina, T. Advanced etching of silicon based on deep reactive ion etching for silicon high aspect ratio microstructures and three-dimensional micro- and nanostructures. *Microelectron. J.* **2005**, *36*, 673–677. [CrossRef]

35. Boer, M.; Tjerkstra, R.; Berenschot, J.; Jansen, H.; Burger, G.; Gardeniers, J.; Elwenspoek, M.; Berg, A. Micromachining of Buried Micro Channels in Silicon. *IEEE/ASME J. Microelectromech. Syst.* **2000**, *9*, 94–103. [CrossRef]

36. Goldsmith, P. *Quasioptical Systems: Gaussian Beam Quasioptical Propagation and Applications*; Wiley-IEEE Press: Hoboken, NJ, USA, 1997.

MDPI

St. Alban-Anlage 66

4052 Basel, Switzerland

Tel. +41 61 683 77 34

Fax +41 61 302 89 18

http://www.mdpi.com

Micromachines Editorial Office

E-mail: micromachines@mdpi.com

http://www.mdpi.com/journal/micromachines

www.ingramcontent.com/pod-product-compliance
Lightning Source LLC
Chambersburg PA
CBHW051845210326

41597CB00033B/5780